MATERNAL EFFECTS IN DEVELOPMENT

THE FOURTH SYMPOSIUM OF
THE BRITISH SOCIETY
FOR DEVELOPMENTAL BIOLOGY

Maternal effects in development

EDITED BY

D. R. NEWTH & M. BALLS

Regius Professor of Zoology
University of Glasgow

Senior Lecturer in
Human Morphology
University of
Nottingham

CAMBRIDGE UNIVERSITY PRESS

CAMBRIDGE

LONDON   NEW YORK   NEW ROCHELLE

MELBOURNE   SYDNEY

Published by the Press Syndicate of the University of Cambridge
The Pitt Building, Trumpington Street, Cambridge CB2 1RP
32 East 57th Street, New York, NY 10022, USA
296 Beaconsfield Parade, Middle Park, Melbourne 3206, Australia

First published 1979

Printed in Great Britain at the
University Press, Cambridge

Library of Congress Cataloguing in Publication Data

British Society for Developmental Biology.
Maternal effects in development.

Papers from the symposium held in Exeter, Sept. 19–21, 1978.
Includes index.
1. Maternal–fetal exchange – Congresses.
2. Developmental biology – Congresses. I. Newth,
D. R. II. Balls, Michael. III. Title.
[DNLM: 1. Fetus – Congresses. 2. Maternal–
fetal exchange – Congresses. 3. Embryology –
Congresses. WQ210.5 B862m 1978]
QL971.B77 1979   591.3   78-73812
IBSN 0 521 22685 6

Contents

Contents

Preface

D. R. NEWTH and MICHAEL BALLS

This volume contains papers which were read and discussed at the fourth symposium of the British Society for Developmental Biology which was held in Exeter between 19 and 21 September 1978.

We would like to express our thanks, and those of the British Society for Developmental Biology, to the contributors; to our hosts the University of Exeter, and more particularly to Dr Elizabeth Deuchar, who was both contributor and local organizer; and to the staff of the Cambridge University Press. The financial support of the Royal Society of London, the British Council, and the International Society of Developmental Biologists, made the symposium possible.

February 1979
While this volume was in press we were saddened to hear of the death of Dr E. M. Deuchar from cancer long bravely endured. Her scientific colleagues were also her personal friends, and the sense of loss will be felt in all the laboratories at home and abroad in which she worked. In addition to her distinguished scientific work, she was one of the founders, and the first secretary, of the British Society for Developmental Biology.

Maternal constraints on development

JACK COHEN

Dept of Zoology and Comparative Physiology,
University of Birmingham, B15 2TT, UK

The restrained increase of complexity during development has presented a puzzle to philosophers at least since Aristotle; explanation has ranged from the frankly preformationist to the metaphysical entelechist (see Needham (1959) for a review, Bertalanffy & Woodger (1933) for an exposition). For the preformationist, complexity is not *really* increasing; it is becoming manifest, extensive, where previously (e.g. in the anlage or in the genome) it was intensive. The medieval 'Chinese boxes' image, which even managed to include the contribution of both parents by a logical trick, foundered on the rocks of atomism: there was a finite divisibility to matter, so an infinite number of generations could not be nested within the first creation. But modern preformationism, with replication, is still very much with us, as in such phrasing as 'to the extent that all the structures and performances of organisms result from the structures and activities of the proteins composing them, one must regard the total organism as the ultimate epigenetic expression of the genetic message itself' (Monod, 1974). In this phrase 'epigenetic' is used wrongly, to refer only to this extension of previously intensive message. The implication is clear: the zygote genetic message produces the embryo, and there is no room for, for example, maternal effects.

Classically, it has been supposed that preformationist philosophies have been superseded by real epigenetic explanations, in which limited complexity is generated determinately by interaction of egg components with each other and with 'the environment', including cytoplasmic and other 'environment' provided by mother. But thinking which claims to be epigenetic frequently has other bases. We must analyse this further. For example,

I

phrases like 'the organism as it grows from its bare genetic
beginnings represented by the initial egg' (Monod, 1974), which
purport to illustrate epigenetic thinking, suggest the moulding of
a plastic embryo by an almost dialectical 'evolution'. Indeed the
word 'evolution' used to mean 'development' (as in evolution of
a gas from a liquid), and many embryologists still use phrases, and
retain prejudices, from this pre-Darwinian 'evolutionism'.
According to this view, the simple egg *proceeds*, and each onward
step brings the embryo into the realm of new, more complex,
geometrical possibilities and constraints: it cleaves because that
is its nature, and makes a hollow sphere, the blastula which,
perhaps constrained by its enveloping membrane, buckles (gast-
rulates) becoming 2-layered, and so on; each stage, or *horizon*, is
the entry to a new set of external (indeed metaphysical) rules.
Parallels between the evolution of the embryo into more complex
geometries and the phyletic evolution of animals, or the *scala
naturae* before phyletics, have bedevilled biological thought since
Aristotle. Bonnet (1786) and Haeckel (1874) had phyletic history,
again an *external* constraint, determining development: the embryo climbed its family tree, up the trunk, boughs and branches
of its own lineage – yet of course for each embryo the tree is but
one railway route, with all the points set by external historical
factors. Again, maternal effects are reduced to a minimum –
mother is but the last ancestor! De Beer (1958) exposed all these
problems lucidly in modern biological terms, and Gould (1977)
has examined the relations between ontogeny and phylogeny,
beguiled by the apparent control of ontogeny by phylogeny in
some crucial evolutionary examples, bringing both simplification
and wisdom to bear on them. We should be aware, as Gould
emphasizes too, that some modern explanations of ontogeny
conceal preformationist and evolutionist ideas while claiming to
be epigenetic, and so exclude the kind of maternal effects described
in the following pages.

  Genetics has been, on a simplistic level, completely successful
in explaining differences between organisms of the same species;
they resulted from differences in the genetic message ('genetic')
or from differences in the external biasing of the developmental
route of the organisms ('environment'). Inheritance was considered to provide not characters, but a spectrum of potential
developmental responses to a range of environmental possibilities.

Bounding the viable was a whole penumbra, a *terra incognita*, whose routes led to early death (see Waddington (1957) for a sophisticated version and Jacob (1974) for a naive version). Many environments could warp a normal genetic programme into viable abnormality (lithium, thalidomide, the uterus of a starving woman or one who smokes heavily), and conversely many mutant or unbalanced genetic programmes would result in anomalous development in a range of 'normal' environments. Epigenetic interactions were mostly seen as such dialectically 'genetic'/ 'environmental' examples. When abnormal environment or genetics resulted in a more normal embryo than expected, the anomaly was explained by the metaphysics either of 'regulation' (e.g. Waddington, 1957) or of 'penetrance' (e.g. Levine, 1962). Usually, the mistaken view has been promulgated that when allelic differences at a known genetic locus cause differences in phenotype, the gene concerned controls (indeed, even 'causes') the character concerned; see Monod (1974) for innumerable examples and Waddington (1957) for the argument; see also Malacinski & Spieth in this volume. The morphogenetic action of each gene is 'confessed' or 'discovered' by the failure of its mutant alleles to generate normality. The assumption follows that the 'wild-type' allele produces normal development of the character. So overall development came to be seen as the effect of the sum of the normal genes. Maternal-effect genes were included, and I will refer to this as the 'jigsaw' model of development.

After the era of the organizer, which saw the development of constrained complexity by interaction of different parts of the egg, the epigenetic activity came to be seen as a hierarchy whose 'lowest level' was this action of individual genes. The primary organizer (dorsal lip of blastopore, or Hensen's node) produced notochord, which induced neural plate, which caused the development of the axial system; the somites appeared, and then within each region more inductions caused more predictable heterogeneity down to tissue or even cellular scale. Where the organizer got its power was not an important question until the 1950s; attention was directed almost entirely to how the organizer exerted its effects. The post-Spemann experimental embryology was very well suited to the 'physiological' bias of biology in the 1930s, and to the genetic models then current in which non-mutant genes caused normal development (the 'jigsaw' model, for example,

Blizard & Randt, 1974). But its discoveries and problems eclipsed an older view, more germane to our present cause (see Bertalanffy & Woodger 1933; Waddington 1957; Cohen 1977).

At the turn of the century, Weissman's germ plasm, Driesch's entelechies ('principles' directing development within each embryo), and especially Conklin's and Wilson's elegant studies of spirally cleaving and other mosaic eggs had produced a reproductive philosophy in which the egg seemed more and more complex. The embryologists were describing the eggs, even of 'simple' creatures, as vastly complex vessels carrying the specific organization across the generations. On the other hand, the early geneticists like Janssen and even Morgan were focusing attention upon simple linear chromosomes as the hereditary agents. With the re-discovery of Mendel, and the vastly impressive coincidence of behaviour of segregating hereditary Mendelian determinants and the meiotic chromosomes, most biologists considered the parental contribution to the next generation to have been elucidated (Bertalanffy & Woodger, 1933); so the nucleus and its chromosomes became the whole matter of heredity, and the embryological view fell out of fashion and has remained out, especially in student textbooks. For example: 'they inherit the ability to produce these phenotypes. This ability resides in the genotype and it is the material of the genotype that is transmitted from one generation to the next' (Levine, 1962).

Conklin (1918) wrote of the egg having an organization which 'roughed out' the major organization of the embryo, leaving the details to be filled in by the Mendelian genes, and this duality in the provision of embryonic structure has been a recurrent, but always a minor, theme in both embryological and evolutionary theory (Bertalanaffy & Woodger, 1933): Raven (e.g. 1959), emphasizing the extra-nuclear organization of eggs, especially the cortex; Curtis (1965) with a near-demonstration of separable heredity of the amphibian grey crescent; the maternal inheritance of shell coiling sense in *Lymnaea peregra* (Boycott, Diver, Garstang & Turner, 1930); Tartar and especially Paul Weisz (e.g. 1951) with their work on the inheritance of cortical structure in *Stentor* and its analogy with metazoan embryology; Moore's hybrid frogs (e.g. 1955), whose development only resembled that of the nuclear species after the blastula stage – the egg 'cytoplasm' controlled development till then; my own revival (Cohen, 1963) of the eccentric embryological notion that development in general was separable into a maternally (egg cytoplasm) controlled early stage, linked to a later control by zygote genes via a recognizable 'phyletic stage'; these all were minor characters whose small contributions served only to point up the dramatic unfolding of the nuclear inheritance story. The Jacob–Monod operon 'explanation' of cellular differentiation held centre stage during the 1960s and early 1970s (e.g. Waddington, 1966; Cohen, Hambley, Newth & Thomas, 1973) while embryological 'detail' like ooplasmic segregation, and animalization/vegetalization effects, became part of the historical literature. Students found the molecular biology of development far more relevant to the rest of their biology than were the classical embryological series of examples like determinate/indeterminate cleavage or variations in the amount or siting of yolk. Ontogeny had become the working-out of the sequential chemistry of the dogmatic DNA story; Brenner (e.g. Brenner, 1974; Ward, Thomson, White & Brenner, 1975) was in perfect accord with the times in choosing nematode to illumine all of development because it was both simple and perfectly regular. It had good chromosomes, a suitable reproductive cycle for 'bucket biochemistry' of successive stages, and simple repeatable anatomy. If its genes could only be made to 'confess' their actions by their mutant alleles, we could for the first time correlate genes, development, nervous circuitry and behaviour. That the organism concerned was so anomalous as to be, for example, determinate to the point of eutely, was irrelevant; for those accustomed to argue from *E. coli* to *H. sapiens* and even to *X. laevis*, the fact that nematodes had the same 'biochemical machinery' made it ideal. The highly determinate egg cytoplasm, carrying maternal organization sufficient to carry the embryo right through gastrulation, was unimportant.

The complexity of eggs

Indeed, the birth of a new consensus about maternal constraints on development was preceded by a conception in 'molecular' biology: 'informosomes', or stable oocyte messenger RNA, were the plot device which brought the minor characters back into the limelight. Two diametrically opposed views of egg function were current in the 1950s and 1960s. One saw the egg only as the passive 'haploid gamete', which functioned by remaining relatively immobile while awaiting discovery and penetration by the sperm, after which the restored diploid genotype of the zygote could go on to make the new organism. The other viewed the egg as a complex, balanced machinery awaiting activation by the sperm, a kind of 'Sleeping Beauty' model. By and large geneticists and DNA-is-God-and-RNA-is-his-prophet molecular biologists took the first approach, and embryologists saw the egg as a beautiful but wilful maiden awaiting the Prince's kiss. This lyricism is not untrue to history; for Dalq (1957), how 'The deep and perilous inertia of the egg, and the somewhat fussy mobility of the sperm, contrive between them to animate a new individual' was the central question of embryology. They were not equal gametes, as they effectively were for geneticists.

While all mitochondria, and one centriole, had long been supposed to be transmitted through egg cytoplasm, the ribosomal complement of the oocyte, together with the concept of 'bound' or 'masked' messenger RNA, drew the attention of many biologists in the 1960s to transmission via oogenesis. The theoretical implications of this carry-over of maternal-genome-produced machinery were not new or serious. Raven's books (e.g. 1966), Monroy's articles (e.g. Monroy & Tyler, 1967) and nucleolusless *Xenopus* (Elsdale, Fischberg & Smith, 1958) had prepared the ground; and in general it was believed that only 'housekeeping' machinery was involved, not really 'informational molecules', even by Spirin at first (1966). The spirality of *Lymnaea* had been relegated to nurse-cell asymmetry, and *grandchildless* in *Drosophila* to change of polar granule region in eggs, and both variations were known to be controlled by 'ordinary' mutant alleles. The nucleus, therefore, maintained its central place in the dogma of heredity, and maternal-effect genes, which exerted their action via the ovary and oogenesis, were not seen to be remarkable. Indeed, because of their existence it was generally assumed that all of the egg's cytoplasmic structures were equally the result of comparable genes – the egg was simply a highly differentiated cell, the result of activities of maternal genes as were the mother's erythrocytes or liver cells.

Eventually evidence accumulated, even at the molecular level, which showed the egg to be much more complicated, more heterogeneous in time and space, than any somatic cell: gene amplification of nucleolar genes, lampbrush chromosomes, meiotic and metabolic arrest, yolk synthesis and organization, and the relationship with nurse/follicle cells were all unique, and none were explicable in DNA–RNA–ribosome–protein dogmatic terms only (e.g. Cohen, 1971). The realization is beautifully illustrated by the differences between the first and second editions of Davidson's review book: in the first edition (1968) the nucleus is primary; in the second (1976) nuclei are controlled by cytoplasmic factors during oogenesis, and are reactive passengers during cleavage and into early gastrulation. The oocyte is shown to be loaded, not only with 'housekeeping utensils' but also with the explicit spatial and temporal instructions for the first phase of development. Unlike cells, it has a melange of components and organization from many sources, with a built-in programme of cleavage divisions to apportion this complexity, independently of its own nuclei. Indeed, the heterogeneity of cytoplasms in which later nuclei find themselves determines *their* developmental paths.

The egg as a device for the transfer of information

Raven gave a paper with (nearly) that title in 1961 and emphasized the information-carrying *capacity* of cytoplasm and, especially, cortex in comparison to the nucleus (he called eggs 'egg-cells', a usage which I find unnecessary and probably misleading). His point was well taken, despite Wolpert's reminder (in the discussion following the Raven paper) that very little information is needed to prescribe the initial development of eggs: Raven quoted 10^{17} bits for cytoplasm, 10^{21} for cortex (10^{12} for nuclear DNA), but Wolpert only needed about 10^1 to prescribe his early echinoderm morphogenesis! Perhaps more important than the simple amount of this information is its provenance, especially in evolutionary (or at least *scala naturae*) context. What is the source of this information transmitted by eggs, or indeed by all agents whose organization results in progeny? Bonner (1974) has emphasized the asynchrony of reduction of differing informational systems in all organisms to their smallest compass for transmission across the generations, but I would go further: most of what is transmitted is not zygotic but parental organization.

Even in viruses the various components need not be of the same generation: coat can be prescribed by *parental* genome, yet contain filial nucleic acid. Parental genome, not contained genome, determines infectivity of some virus particles; so also for most bacteria. But this is also true for eggs, nearly all of whose structure and material is maternally prescribed and indeed transcribed during oogenesis from the diploid (but probably crossed-over) maternal chromosome set, or from diploid nurse or follicle cells. To digress somewhat, sperms may be a remarkable exception to the parental-genome control, for post-meiotic transcription has been clearly shown (Kirszenbaum & Tres, 1975) as has segregation of products (Yanagisawa *et al.*, 1974; Festenstein, Halim & Arnaiz-Villena, 1978); haploid gene effects, both minor and demonstrated (Bennett, 1975; Festenstein *et al.*, 1978) and major but hypothetical (Cohen, 1974) appear in recent literature. And many plants, of course, show some haploid effects in the gametophyte; but even here most of the effectiveness or viability of both the pollen grain and the embryo-sac seems to be parentally determined by genes or plastids (see Tilney-Bassett & Abdel-Wahab, this volume). Eggs, of course, are very rarely haploid, so there is no chance of the actual gametic chromosomes affecting egg structure differentially (but T^{Hp} may disprove this assertion: see Johnson (1974) and McLaren, this volume).

Detailed studies of the control of production and transmission of this maternally derived egg machinery, especially cytoplasmic localization and germ-cell determinants, are to be found elsewhere in this volume; some classical embryological problems will be considered here, in an attempt at a new view. The existence, or rather the evolutionary development, of a system of non-genetic inheritance will be argued from this new version of an old philosophy of development.

The egg as a maternal product

What then does the egg carry across the generations and to what extent is development determined, constrained or permitted by the maternal structures in the egg?

The nuclear, genetic, DNA-based part of the embryonic inheritance is unarguable; that development is affected by environmental accidents as diverse as temperature change, thalidomide, parasite or predator is not in doubt. But two questions remain.

Does the additional material, information and influence inherited from the parents, especially from the mother via the oocyte, provide the ambience within which genome and environment impinge on each other constraining both, or is it no more than a trivial 'fine tuning' of a determinate genetic programme? And if it does provide this developmental ambience, is all of this maternal heredity coded directly from nuclear genes of a previous generation? If it is not, then 'genetics' and 'environment' do not exhaust the kinds of influences on development; the universe in which their interaction occurs is itself specifically constraining, and *its* properties too are inherited. I propose to show that the egg contains three extra kinds of inheritance:

1. A subtle, egg-line-specific, set of developmental instructions: biochemical stores and functions and, perhaps especially, limiting provenances which *prevent* the continuing increase of complexity and permit true epigenesis; for example the 'ooplasms' in *Styela*, the 'polar lobe' organization in *Dentalium*, those instructions which result in the 'phyletic stage' of complexity without the direct morphogenetic involvement of zygote genes.

2. A very general, entirely obvious and thus generally unmentioned (overlooked?) set of biological 'rules for that species, defined by the choice made through 4000 million years of that individual's phyletic history; for example the (probable) use of HLA/H2/T/F9 membrane determinants by mammals for early embryonic organization, use of 'ordinary' mitosis for cleavage of nuclei (but *not* by *Ascaris* or cecidomyids), reliance on the mitochondrial remnant of DNA for provision of amphibian blastular cells with these organelles but nuclear genes (with amplification or repetition) for provision of ribosomes.

3. Yolk and other secondary effects of maternal phenotype donated to all her embryos independently of their zygote genotypes, and referred to collectively as 'privilege'; contrast with primary maternal inheritance, from her genotype, for example shell sense in *Lymnaea* and the axolotl's array of maternal gene-effects.

Whether these may usefully be related to a previous genome will be considered; in general my prejudices will suggest that each zygote genome expresses itself within an egg organization embedded within the maternal phenotype physiology, most of which is so causally remote from any genome that it is best considered as a structured ambience inherited by each embryo in addition to its own genome.

The organization of eggs

Except in mammals, specific egg properties are now thought to account for the acquisition of basic blastula and, to a large extent, gastrula organization in metazoans (e.g. Malacinski & Spieth, this volume). They are usually 'simple' properties, which interact with the general metazoan cytoplasmic properties to generate the characteristic 'phyletic stage' (Cohen, 1963, 1977). The zygotic chromosomal genome has not been involved in this earliest morphogenesis (e.g. Gurdon, 1974).

However, it is certain that many gene products are involved, both as building material and as control elements. During cleavage, not only must chromosomes be replicated, often amazingly quickly (e.g. Laskey, Gurdon & Trendelenburg, this volume), but also centrioles too; whether these have nucleic acid genomes is not known (but a possible RNA basis for their inheritance correlates with reverse transcriptase activity in 'uninfected' metazoan cells). The mitochondria, too, are commonly divided among the cleavage products of the dividing egg (Billett, this volume), as are ribosomes and other 'household utensils' accumulated during oogenesis (Ford, this volume). Some mitochondrial proteins are prescribed by their own DNA rings, but many (differing among organisms) are produced in the pre-vitellogenic oocyte from nuclear DNA transcripts. Ribosomes are entirely built of RNA and protein produced during oogenesis from nuclear transcripts, either of the oocyte (or even oogonium) or of nurse cells. Provision of these essential elements for a cleaving embryo from maternal genome is a necessary, but not a sufficient, prelude to their involvement in morphogenesis.

In ciliate protozoa, much of the protein complex used to construct membrane/cortex/ciliary apparatus is doubtless nuclear in provenance (via transcribed messenger RNA); nevertheless, the cortical *pattern* is replicated without involvement of nuclear morphogenesis: experimentally or naturally peculiar patterns are repeated in subsequent multiplication of the organisms (e.g. Weisz, 1954; Curtis, 1965). There are ciliate nuclear genes known which modify cortical pattern, just as a nuclear gene in *Lymnaea* modifies cleavage sense. But only the most naive biologist would argue that this maternal-effect spirality gene is actually that locus, that length of DNA, which 'makes the egg cleave'.

This is a central issue when discussing maternal-effect inheritance. Are all the oocyte gradients (Graham, 1976), ooplasms, cortical specializations (Dohmen & Verdonk, this volume) produced anew in each ovary as part of the differentiation of 'egg-cells'? Or are they, like the ciliate cortex, a separate hereditary system? That mitochondria have separate multiplication is now certain, and much extra-cellular inheritance is by such organelles (Jinks, 1964). Is the organization of mitochondria in the frog's egg determined by their own responses, nuclear dominance of cytoplasm, or by a separate oocyte architecture of characteristically germ cell kind and separate heredity? Although no definitive answer can yet be given for any oocyte morphogenetic system (Brachet, 1977), the 'sequential hierarchy' models of control molecules packed by oocyte nucleus (e.g. Davidson, 1976) could probably work in a polarized oocyte to produce the increasing heterogeneity we observe (see the beautiful description by Graham, 1976). It may well be, then, that all the oocyte's 'deep and perilous inertia' is only one cell cycle old, and *all* of the oocyte's organization is referable back to the maternal nuclear (and perhaps mitochondrial) genomes.

The egg is still, however, two systems at least – one extensive and one intensive. The current tape player is last year's model made from instructions on last year's tape, and the initial processes of development consist of preparation for playing the current tape. To continue the analogy perhaps too far, the tape player must be connected, switched on, laced, and the right buttons set and levels adjusted. Only then can the zygote's own tape be played by the maternally provided machinery. There is no way by which the message on a tape can be deciphered, even it it includes instructions for building the tape player, without knowing the construction of a tape player. So, to understand the first steps of morphogenesis we must understand not only the code but also the decoding machine.

Such understanding can, of course, be at several levels. It may be profitable, if naive, to make distinction here between 'specific egg properties' and 'givens'. An ooplasm in the *Tubifex* egg is a specific property; the use of tubulin/centromeres/mitosis to

cleave the *Tubifex* cytoplasm and deploy the chromosomes is a
'given'. In an evolutionary sense, 'givens' are ancient, 'specifics'
are modern, but the distinction is a relative one. 'Spiral cleavage'
is a 'specific' of molluscs or annelids, but a 'given' of *Ostrea* or
Pomatoceros. Many 'givens' are molecular in nature, and some are
'simple' chemistry. Just as specification for a sodium chloride
crystal need not also specify that it be cubical, so many of the
'given' molecules like (particularly) histones, actins, myosins,
tubulins have structural properties which could serve as a basis
for morphogenesis. But *which* morphogenesis, using these same
structures, presumably depends on specifics: tilted or animal–
vegetal mitoses, formation of polar lobes, use of germinal vesicle
contents to define part of an egg.

Even with the same genome and the same 'givens', very
different organisms can be built, depending on phase relations
between 'specifics' and morphogenesis, between tapes and the
tape-player parts they prescribe: caterpillar and butterfly, the
successive larvae of digeneans or crustaceans, tadpole and frog,
even baby and adult human share all the 'givens' and the same
genome. The egg organization makes a miracidium with the same
genome that the redial organization uses to make a cercaria.

It is surprisingly difficult to list the 'givens'; under scrutiny
many turn into 'specifics'. We should, perhaps, expect this from
Moore's results with hybrid and andromerogone amphibia
(1955); nuclei may be incompatible with cytoplasm even of closely
related species, and this may not be reciprocal. For each species,
as suggested in the introduction to this paper, there has been but
one unbranched phyletic history of some 4000 million years. Each
decision is enshrined in co-adapted mechanisms. So every
ontogeny is unique. Nevertheless closely related forms share most
givens, and many specifics. Harris's cell fusion experiments (e.g.
1970) show that many intracellular communication systems are
common even to mammals and birds, especially cytoplasmic
control of nuclear function which is like the most basic maternal
effect. Von Baer's law works on the ultrastructural and biochemical
levels as well as the gross morphological. In fact, it works better,
because the earliest, pre-phyletic morphology is often (as in the
various vertebrate classes) biased by yolkiness and egg adaptations
in general. Vertebrate eggs converge in morphology until the
phyletic stage (early neurula, say) and then diverge, according to

Von Baer. However, the early biochemical mechanisms might be much more similar, producing the same neurulae. This argument results from application of Ockham's razor to what we know of pre-phyletic development (with a due consideration of Gurdon and Harris); it *could* be prescribed afresh from maternal genome in each generation, using the 'givens' of that group of organisms as a basis. Curtis's work (e.g. 1965) stands almost alone in suggesting a parallel heredity for the 'reference points' (Graham, 1976; Brachet, 1977), as in the ciliates. I am reluctant to relinquish my prejudice that there is usually such a parallel heredity (compare Nieuwkoop, 1977), until Curtis's experiments have been repeated, preferably with other organisms too, and different results have been obtained.

Yolk and other maternal donations

Very nearly all animals pass a variety of food materials to their offspring before the offspring are developed enough to obtain it for themselves. An alternative way of putting this is to show parents making profit on their metabolic transactions with the environment and investing this profit in their offspring as a variable 'privilege' (Cohen, 1977). Embryos with much yolk are privileged to develop much further before they are thrown upon their own resources, by then greater in comparison with less well-endowed embryos. But there is usually a developmental price to pay for this privilege. Except in forms like gastropods and eutherian mammals, whose nutrient is absorbed from without, yolk is included within the egg-membrane and modifies that egg-organization which can be achieved in the ovary. So differing yolkiness of eggs could be informative about the extent to which embryo-structure derives from egg-structure without intervention of zygote nucleus. The classical comparisons of development of many different vertebrates, including dipnoans and Gymnophiona (e.g. Kellicott, 1914) or a few very different forms (e.g. Wadding- ton, 1952; Pasteels, 1940) was supposed to do just this.

Although many puzzles and gaps in our knowledge remain, I will assume that (except in mammals) achievement of the very early neurula stage of vertebrates usually occurs without significant operation of the zygote genome; at the very least, the 'germ layers' are segregated and possibly the closed neural tube can be achieved (Moore, 1955; Cohen, 1977; Neyfakh, 1971). The various classes of vertebrates have different blastulae, and gastrulate differently, yet achieve very similar neurulae. Classically, this has been attributed to different topological relationships of yolk to embryo (e.g. Pasteels, 1940; Waddington, 1952). But experimental change of the yolk pattern of frogs by centrifugation of the eggs makes their development resemble teleosts in many respects (Ballard, 1964; Ballard & Dodes, 1968; Cohen, 1977). This enables us to discount the necessity for precision in transmission of oocyte yolk/cytoplasm structure. What, then, causes vertebrate eggs to gastrulate convergently? Some structure is clearly necessary, as we know from experiments which delete or transfer the grey crescent of amphibians or the comparable structures in birds (see above). But in these indeterminate forms, like the echinoids worked on by Gustavson & Wolpert (1961), general properties of nuclear/cytoplasmic systems ('givens') are elicited by simple cues and the yolk can in most respects be regarded as a simple 'packed lunch' added to the structured egg in different topologies (Waddington, 1957).

In this section, however, I wish only to suggest a series of examples relating to the provision of yolk analogues, and not to developmental effects. Yolk is, of course, only one of the ways in which the mother's adult metabolism can re-direct materials and energy to her progeny: successive ovulations (in the shark *Lamna*); brothers and sisters (trophic embryos, as in the whelk *Buccinum*); albumen (as in birds); mother's tissues (some gall midges, most scale insects); and many other kinds of transformed maternal material are passed to embryos. Some mothers avoid such metabolic involvement and only use their adult abilities to acquire special food for their helpless (see below) offspring: a tarantula spider (left by the tarantula wasp mother); insects, fish or seeds (birds); even breakfast cereals. The penultimate block of examples in this series has mothers only using their adult abilities to put their offspring in that privileged environment in which less metabolic effort is required: parasitic wasps laying eggs in other insect's larvae; aphid eggs overwintering on a second host plant found by the mother (e.g. spindle tree/bean aphid); light oil droplets in eggs of benthic marine forms, carrying them up into the plankton to hatch (e.g. plaice); even seasonal breeders whose young hatch, or are born, into a time of plenty; the human mother laying her piteous bundle on the steps of the Foundling Home has many natural counterparts.

Maternal shielding of embryonic defect

The above tedious, and obvious, list is given to emphasize these additional maternal roles as *part of the inheritance of the zygote*, independent of its genome but entirely determined by its specific parentage (see Cohen (1977) for a more extended but elementary account). Most of the examples above are post-embryonic, when the 'helplessness' of some juveniles renders the need for maternal aid very obvious. I wish to suggest that the long embryology, and the juvenile helplessness of those forms with advanced reproductive strategies, like birds and mammals, can only be understood in terms of such maternal 'cossetting'.

First, however, a digression into evolutionary genetics allows presentation of an old paradox in modern form. When it was believed that genes, in aggregate effect, caused phenotype by their interaction with the embryo's environment (I simplify) the genome was seen as mostly homozygous with a few mutations, on the way out or on the way in, giving occasional heterozygosity (e.g. Haldane, 1941, or even 1957). This 'classical' genome picture, with other genetic naiveties like 'those which produce more progeny thereby ensure the passage of their genes' (any elementary biology text), were rendered untenable by Lack (e.g. 1954) and especially by the isozyme studies of natural breeding populations started by Lewontin & Hubby (1966). It is now abundantly clear that about a third of structural loci in most populations are occasionally heterozygotic, while the few individuals which survive to breed (often very far from the average individual) are heterozygotes at about 10% of their structural loci (Lewontin, 1974; Powell, 1975). In some organisms with simple reproductive strategies it is common to find a high level of abnormality (e.g. flatfish: Bannister, Harding & Lockwood, 1974; oysters: Galtsoff, 1964), manifested *after* the phyletic stage (see next section), and it is very tempting to suggest that this constitutes visible evidence for segregational genetic load (Wallace, 1970; Cohen, 1977). The same high levels of heterozygosity are found in 'higher' forms, and this almost certainly means that the apparent regularity of development of these advanced forms conceals a diversity of genetic programme. Waddington's epigenetic landscape (1957, and in Mintz, 1958) accurately portrayed the 'balanced genome' model more usual in modern genetics, and which can include the isozyme results. But even with 'co-adaptive complexes' and all the cytogenetic tricks used by *Drosophila* and maize to retain advantageous allele combinations (inversions, translocations, chiasma localization, etc.: see John & Lewis (1968) for a clear account) neither genetic assimilation (e.g. Waddington, 1957) nor its necessary basis of embryonic versatility can be explained satisfactorily by modern genetic models (e.g. Lewontin, 1974). Here, then, is the paradox: if embryos were variable, reflecting their segregational diversity, so that only a few survived in any specific environment and genetic assimilation was really only stringent selection, this variability should be expressed as embryonic mortality or at least diversity in any large clutch. But this is *not* observed to any extent in any of the commonly used genetic or developmental species. If on the other hand each embryo is versatile (i.e. chreodes show considerable elasticity) then, while developmental conservatisms would be explained, neither the variety of caenogenetic adaptations nor assimilation would be expected; the paradox lies in the apparent opposition of versatility and stringency of selection.

This paradox may be resolved by distinguishing two kinds of apparent developmental versatility. The first is true versatility: *Rana temporaria* eggs, in the wild, develop normally while the temperature varies from 4 °C to 18 °C daily; *Tenebrio molitor* larvae (mealworms) can pupate normally after eight weeks of plenty or two years of virtual starvation; all the classical examples of late embryonic regulation, even after accidental or experimental wounding, show many routes culminating in normality. But the second is only apparent versatility; the embryo is in a privileged state: *Ascaris* embryos are permitted to develop in a variety of toxic media by the protective shell; freshwater fish eggs can develop in virtually pure water, retaining sufficient ions until their cellular barrier is complete, largely because of the stockpile complexed in the yolk; branchiopods (e.g. *Artemia*), and even fish (e.g. some *Aphyosemion*) which inhabit temporary pools have eggs which can be desiccated several times, yet still hatch competent juveniles, because *maternal* materials protect them, not zygotic devices (Wourms, 1974). The most dramatic apparent versatility is of early, pre-phyletic development. All the classical examples of early embryonic regulation, of indeterminacy and pluripotency, puzzle us because we cannot imagine a built-in variety of genetic programmes to cope with such meddling as fission or fusion or centrifugation and still produce normality. We now realize that the maternal egg-making programme, rather than the zygotic, accounts for this normalizing; we see it as the re-balancing of a sophisticated set of gradients (Graham, 1976) or ooplasmic localizations (Nieuwkoop, 1977) or cortical fields or patches (Curtis, 1965; Raven, 1966), and not as the miraculous self-healing of a newly created zygote just setting up its developmental path. So, too, the amazingly high proportion of plaice or oyster eggs, or *Echinus* plutei, which develop despite supposed genetic anomaly (showing 'genetic' as well as 'environmental' versatility) can be explained as maternal programme, fuelled by maternal provisions, working itself out (Cohen, 1977).

The evolution of dependence

If there is any reality in this division between real and apparent developmental versatility, then both viviparity and parental care need further examination. It has been long supposed that disadvantageous gene effects occurring later than breeding can only have selective effects in special circumstances, and are therefore selectively neutral. Equally, if maternal cossetting results in a privileged first phase of the life history, then many genetic anomalies will have little or no effect: delay of effective food-gathering mechanism does not matter if yolk is available; being born blind does not matter in a dark nest with mother in attendance. The converse is more interesting: virtually all worm eggs are guaranteed to make trochophores, echinoid eggs to achieve pluteus, anurans to reach tail-bud state, because mother provides both food and guidance.

Presumably, as in the post-reproductive period, precise but unnecessary control of the privileged early phase of life history will decay by accumulation of genomic differences. Those differences whose later advantages outweigh disabilities for the protected embryo will be selected. These genomic differences will be of two kinds. Some will be early-effect genes, which should equip the zygote with its own abilities; but others, more germane to our present purpose, will be genes which control the egg organization itself, manifested in the differentiated ovary. Surely by this mechanism most mammalian embryos, and birds to a lesser extent, have lost the ability to develop in a varying temperature that is still present in the less cosseted eggs of the frogs. But further, the mammalian egg (or at least the mouse egg) is not equipped with yolk, *or* with a 'specific' architecture to guide its first steps in development; it inherits the full provision of 'givens', of course. It really is close to the 'bare genetic beginnings' supposed by Monod in his evolutionist mode. The extraordinarily efficient homeostasis of the mammalian mother has so taken over the control of embryonic milieu that mammalian development has been constrained into ever narrower paths. Even embryonic nutrition is extrinsic, controlled by maternal physiology, from early cleavage on.

Such use of maternal adult physiology to control developmental milieu is not by any means unique to mammals. Even such extended control of zygotic life as in guinea-pig or gnu is not very impressive compared with *Glossina*, the tsetse fly, where the entire larval life is spent *in utero* and the offspring achieves full adult size without any more food; *Gyrodactylus*, a monogenean parasite on fish skin, already has an embryo *in utero* when it is born; the tarantula wasp lives all its adult life on food reserves from the tarantula spider provided by its mother – the imago does not feed itself, but only provides amply for its progeny.

It is tempting to believe that such use of maternal physiology, to restrict the necessities for developmental versatility, frees genetic programming space so that a more predictably complex organism can be built. This in turn can presumably regulate embryonic milieu within closer limits, so an evolutionary 'pulling oneself up by one's own bootstraps' seems feasible (Cohen (1977) argues a better case). Such a series would culminate in a gnu or a guinea-pig, with embryo passively programmed into competence while sheltered by mother's exquisite homeostasis. Eisenberg (1977) suggests such a series for marsupials as well as for eutherians, on a comparative behaviour basis rather than an embryological.

The origins of cultural inheritance

Yet many mammals do *not* use such maternal constraints to restrict their offspring to a behavioural repertoire programmed almost entirely *in utero*. Carnivores, myomorph (but not hystrico-morph) rodents, and especially primates have adopted a repro-ductive strategy which in this context seems maladaptive. They give birth to helpless offspring which must *learn* to become competent adults via a development full of rude shocks and stringent necessities compared with the peace within. Although Trivers (1972) has argued for early birth as a device to promote flow in a series production line (and it certainly seems to be that in macropid marsupials: see Sharman, 1976) these parents rarely become pregnant while lactating (see Crook (1977) for compari-sons). The post-partum oestrus of mouse may be a secondary reversion to assist its adaptation as a 'weed' (see Berry, 1977).

Guinea-pigs learn, but rats learn better and primates better yet (e.g. McLean, 1977). A case can be made that perfectly adapted, or perfectly programmed, organisms can never learn, because their behaviour is always congruent with requirements. In a changing environment, however, such perfection can never be attained anyway, and some versatility, preferably as learned behaviour, is adaptive. The very helplessness of early-born mam-mals assists this learning because competence prevents learning, and their inabilities provide a stream of aversive stimuli. Mothers, sometimes later both parents, constrain this behaviour so that learning occurs in a restricted environment, but this should be seen as instructive, i.e. using offspring sensory-motor apparatus as a link in homeostatic chains (see, for examples and argument, Crook (1977) and Simpson (1978)). Only after the embryo has surrendered control of vital functions to maternal physiology can the fetus appear as a stage in the life cycle (contrast marsupial and eutherian). Then it becomes possible for the fetus to be born earlier, and so relate to the mother via sensory-motor events rather than metabolic/hormonal ones. Infantile volition becomes im-portant, and rewarding, so a new kind of maternal effect on development appears. The simplest examples of this are various 'imprinting' phenomena: the most complex include the passage of language, myths and manners.

To illustrate this takeover into maternal constraint of evolu-tionarily previous zygotic programme, contrast the tarantula wasp

and the goose in the provenance of their mate-recognition programme. The tarantula wasp passes its whole larval life isolated,
in a burrow with a relatively enormous paralysed spider; it never
sees another wasp before it must emerge, and find and mate with
one of its own species. The recognition circuits are clearly
'hard-wired' into its imago neural circuitry during late development, by zygotic genome interaction with egg organization in a
general hymenopteran context; so, of course, are the recognition
circuits for spiders, and the motor-responses. I expect that most
wasps get it right; I suspect that very few attempt to copulate with
spiders or paralyse mates.

Lorenz (e.g. 1965) has made bird imprinting, especially the
acquisition of mother-image by new-hatched geese, very well-
known. The first large moving object seen regularly (not the first
moving object: e.g. Bateson, 1978) is followed. It is less well-
known that this also sets the bird's future mate-selection image.
A goose imprinted on Lorenz cannot breed naturally, and a mother
goose which somehow fails to imprint her offspring has not
reproduced successfully; she is like a *gs/gs Drosophila* female, her
offspring are sterile. This imprinting of future mate character is
also vital in the early-born mammals, like rat and mouse, but
possibly not in at least some long-gestation species (Larsson,
1978) and not in the cuckoo (Mayr, 1977), which emphasises the
lability of this phenomenon. So some offspring must inherit, in
the sense of 'copy and be constrained by the image of' a picture
of mother in order to mate later. The imprinting *machinery* in the
rat or goose is obviously of zygotic provenance, but the *programming* requires mother's image to be received visually. There are
in addition many well-known effects of maternal presence
mediated via olfactory (Keverne, 1978) and tactile (Simpson, 1978)
routes which affect or determine adult sexual performance. It
would be surprising if the only behaviour whose inheritance is
determined or modulated by maternal character should be sexual
behaviour. Probably all behaviour in early-born mammals is at
least modulated by direct sensory-motor interaction with mother.

By a rather different route, but with somewhat similar effects,
maternal actions may affect offspring by mediation through the
(usually related) organisms whose behaviour is controlled or
modified by mother's phenotype. Social Hymenoptera clearly
come into this category: the queen determines worker-activity in

provisioning larvae, and the determination of any larva as queen or worker (or, of course, drone, from 'voluntarily' unfertilized egg) is related to status of the present queen and the age of the colony. In a somewhat similar way the breeding chances of a particular bird may be determined by the number of siblings in the nest (e.g. Lack, 1954); again this is mother's actions determining development of her offspring via other organisms, the competing siblings. Status of mother in rabbit communities will determine status of offspring relative to age peers, especially by control of lactation (Mykytowicz & Fullager, 1973). In group-living primates like common baboons and macaques, the mother's status usually determines that of her offspring; progeny of low-status females usually remain low-status, even though all females temporarily rise to high status at oestrus and all are mated by the dominant male or males.

Such an inheritance of status is not unknown in human societies. Here the mother is clearly the most important transmitter of the baby's cultural inheritance from birth onward. As well as direct transmission of language, manners, myths and courtesies (or failure to transmit these), she largely determines the particular segment of human society with which the toddler interacts, and so restricts the cultural stimuli affecting the phenotype. Maternal constraints on human development have very largely made us all what we are.

References

BALLARD, W. W. (1964). *Comparative Anatomy and Embryology*. New York : Ronald.

BALLARD, W. W. & DODES, L. W. (1968). The morphogenetic movements at the lower surface of the blastodisc in salmonid embryos. *Journal of Experimental Zoology*, **168**, 67–84.

BANNISTER, R. C. A., HARDING, D. & LOCKWOOD, S. J. (1974). Larval mortality and subsequent year-class strength in the plaice (*Pleuronectes platessa*). In *The Early Life History of Fish*, ed. J. Blaxter, pp. 21–37. Berlin: Springer-Verlag.

BATESON, P. P. G. (1978). Early experience and sexual preferences. In *Biological Determinants of Sexual Behaviour*, ed. J. B. Hutchinson, pp. 29–54. New York: Wiley.

BENNETT, D. (1975). The *T*-locus of the mouse. *Cell*, **6**, 441–54.

BERRY, R. J. (1977). The population genetics of the house mouse. *Science Progress, Oxford*, **64**, 341–70.

BERTALANFFY, L. VON & WOODGER, J. H. (1933). *Modern Theories of Development. An Introduction to Theoretical Biology.* London: Oxford University Press.

BLIZARD, D. A. & RANDT, C. T. (1974). Genotype interaction with undernutrition and external environment in early life. *Nature, London,* **251**, 705–6.

BONNER, J. T. (1974). *On Development : the Biology of Form.* Cambridge, Mass.: Harvard University Press.

BONNET, CHARLES (1786). Palingénèsie philosophique IX, 1. In *Oeuvres,* vol. 4, p. 350. Quoted in Jacob (1974).

BOYCOTT, A. E., DIVER, C., GARSTANG, S. L. & TURNER, F. M. (1930). The inheritance of sinistrality in *Limnaea peregra. Philosophical Transactions of the Royal Society, Series B,* **219**, 51–130.

BRACHET, J. (1977). An old enigma: the gray crescent of amphibian eggs. In *Current Topics in Developmental Biology 11,* ed. A. A. Moscona & A. Monroy, pp. 133–86. New York & London: Academic Press.

BRENNER, S. (1974). The genetics of *Caenorhabditis elegans. Genetics,* **77** (1), 71–94.

COHEN, J. (1963). *Living Embryos : an Introduction to the Study of Animal Development.* Oxford: Pergamon Press.

COHEN, J. (1971). The comparative physiology of gamete populations. *Advances in Comparative Physiology and Biochemistry,* **4**, 267–380.

COHEN, J. (1974). Gamete diversity within an ejaculate. In *The Functional Anatomy of the Spermatozoon,* ed. B. A. Afzelius, pp. 329–39. Oxford: Pergamon Press.

COHEN, J. (1977). *Reproduction.* London: Butterworth.

COHEN, N. R., HAMBLEY, J., NEWTH, D. R. & THOMAS, J. N. (1973). *Genes and Development,* units 1 and 2. Milton Keynes: Open University Press.

CONKLIN, E. G. (1918). *Heredity and Environment in the Development of Men.* Princeton: Princeton University Press.

CROOK, J. H. (1977). On the integration of gender strategies in mammalian social systems. In *Reproductive Behaviour and Evolution,* ed. J. S. Rosenblatt & B. R. Komisaruk, pp. 17–38. New York: Plenum Press.

CURTIS, A. S. G. (1965). Cortical inheritance in the amphibian *Xenopus laevis*: preliminary results. *Archives de Biologie,* **76**, 523–46.

DALCQ, A. M. (1957). *Introduction to General Embryology* (trans. J. Medawar). London: Oxford University Press.

DAVIDSON, E. H. (1968). *Gene Activity in Early Development.* New York & London: Academic Press.

DAVIDSON, E. H. (1976). *Gene Activity in Early Development,* 2nd edn. New York & London: Academic Press.

DE BEER, G. (1958). *Embryos and Ancestors.* Oxford: Clarendon Press.

EISENBERG, J. F. (1977). The evolution of the reproductive unit in the Class Mammalia. In *Reproductive Behaviour and Evolution,* ed. J. S. Rosenblatt & B. R. Komisaruk, pp. 39–72. New York: Plenum Press.

ELSDALE, T. R., FISCHBERG, M. & SMITH, S. (1958). A mutation that reduces nucleolar number in *Xenopus laevis. Experimental Cell Research,* **14**, 642–3.

FESTENSTEIN, M., HALIM, K. & ARNAIZ-VILLENA, A. (1979). HLA antigens on human spermatozoa. In *Sperms, Antibodies and Infertility,* ed. J. Cohen & W. H. Hendry, pp. 11–16. Oxford: Blackwell Scientific Publications.

GALTSOFF, P. S. (1964). The American oyster *Crassostrea virginica* Gmelin. *Fisheries Bulletin of US Fish and Wildlife Service*, **64**, 1–102.

GOULD, J. (1977). *Ontogeny and Phylogeny*. Harvard University Press.

GRAHAM, C. F. (1976). The formation of different cell types in animal embryos. In *The Developmental Biology of Plants and Animals*, ed. C. F. Graham & P. F. Wareing, pp. 14–28. Oxford: Blackwell Scientific Publications.

GURDON, J. B. (1974). *The Control of Gene Expression in Animal Development*. Oxford: Clarendon Press.

GUSTAVSON, T. & WOLPERT, L. (1961). The forces that shape the embryo. *Discovery, New Series*, **22**, 470–7.

HAECKEL, E. (1874). The Gastraea-theory, the phylogenetic classification of the animal kingdom and the homology of the germ-lamellae (trans. E. P. Wright). *Quarterly Journal of Microscopical Science*, **14**, 142–65 and 223–47.

HALDANE, J. B. S. (1941). *New Paths in Genetics*. London: Allen & Unwin.

HALDANE, J. B. S. (1957). The cost of natural selection. *Journal of Genetics*, **55**, 511–24.

HARRIS, H. (1970). *Nucleus and Cytoplasm*, 2nd edn. Oxford: Clarendon Press.

JACOB, F. (1974). *The Logic of Living Systems. A History of Heredity* (trans. B. C. E. Spillman). London: Allen Lane.

JINKS, J. L. (1964). *Extrachromosomal Inheritance*. New Jersey: Prentice-Hall.

JOHN, B. & LEWIS, K. R. (1968). *Chromosome Complement*. Berlin: Springer-Verlag.

JOHNSON, D. R. (1974). Hairpin-tail: a case of post-reductional gene action in the mouse egg? *Genetics*, **76**, 797–805.

KELLICOTT, W. E. (1914). *A Textbook of General Embryology*. London: Constable & Co.

KEVERNE, E. B. (1978). Olfactory cues in mammalian sexual behaviour. In *Biological Determinants of Sexual Behaviour*, ed. J. B. Hutchinson, pp. 727–64. New York: Wiley.

KIRSZENBAUM, A. L. & TRES, L. L. (1975). Structural and transcriptional features of the mouse spermatid genome. *Journal of Cell Biology*, **65**, 258–70.

LACK, D. (1954). The evolution of reproductive rates. In *Evolution as a Process*, ed. J. Huxley, A. C. Hardy & E. B. Ford, pp. 143–56. London: Allen & Unwin.

LARRSON, K. (1978). Experimental factors in the development of sexual behaviour. In *Biological Determinants of Sexual Behaviour*, ed. J. B. Hutchinson, pp. 55–86. New York: Wiley.

LEVINE, R. P. (1962). *Genetics*. New York: Holt, Rhinehart & Winston.

LEWONTIN, R. C. (1974). *The Genetic Basis of Evolutionary Change*. New York: Columbia University Press.

LEWONTIN, R. C. & HUBBY, J. L. (1966). A molecular approach to the study of genic heterozygosity in natural populations of *Drosophila pseudo-obscura*. *Genetics*, **54**, 595–609.

LORENZ, K. (1965). *Evolution and Modification of Behaviour*. Illinois: University of Chicago Press.

MACLEAN, P. D. (1977). An evolutionary approach to brain research on prose-matic (non-verbal) behaviour. In *Reproductive Behaviour and Evolution*, ed.

J. S. Rosenblatt & B. R. Komisaruk, pp. 137–64. New York: Plenum Press.

MAYR, E. (1977). Concepts in the study of animal behaviour. In *Reproductive Behaviour and Evolution*, ed. J. S. Rosenblatt & B. R. Komisaruk, pp. 1–16. New York: Plenum Press.

MINTZ, B. (ed.) (1958). *Environmental Influences on Prenatal Development.* Illinois: University of Chicago Press.

MONOD, J. (1974). *Chance and Necessity. An Essay on the National [sic] Philosophy of Modern Biology* (trans. A. Wainhouse). Glasgow: William Collins, Fontana Books.

MONROY, A. & TYLER, A. (1967). The activation of the egg. In *Fertilization: Comparative Morphology, Biochemistry and Immunology*, vol. 1, ed. C. B. Metz & A. Monroy, pp. 369–412. New York & London: Academic Press.

MOORE, J. A. (1955). Abnormal combinations of nuclear and cytoplasmic systems in frogs and toads. *Advances in Genetics*, **7**, 132–82.

MYKYTOWICZ, R. & FULLAGER, P. J. (1973). Effect of social environment on reproduction in the rabbit *Oryctolagus cuniculus*. *Journal of Reproduction and Fertility, Supplement*, **19**, 503–22.

NEEDHAM, J. (1959). *A History of Embryology*, 2nd edn (with A. Hughes). London: Cambridge University Press.

NEYFAKH, A. A. (1971). Steps of realisation of genetic information in early development. In *Current Topics in Developmental Biology 6*, ed. A. A. Moscona & A. Monroy, pp. 45–78. New York & London: Academic Press.

NIEUWKOOP, P. D. (1977). Origin and establishment of embryonic polar axes in amphibian development. In *Current Topics in Development Biology 11*, ed. A. A. Moscona & A. Monroy, pp. 115–32. New York & London: Academic Press.

PASTEELS, J. (1950). Un aperçu comparitif de la gastrulation chez les Chordes. *Biological Reviews*, **15**, 45–94.

POWELL, J. R. (1975). Protein variation in natural populations of animals. *Evolutionary Biology*, **8**, 79–120.

RAVEN, CHR. P. (1959). *An Outline of Developmental Physiology* (trans. L. de Ruiter). Oxford: Pergamon Press.

RAVEN, CHR. P. (1961). 'The egg-cell as a device for the transfer of information.' Paper presented at Vth International Embryological Conference, 18–21 September 1961. Sponsored by *Journal of Embryology and Experimental Morphology*.

RAVEN, CHR. P. (1966). *Morphogenesis: The Analysis of Molluscan Development*, 2nd edn. Oxford: Pergamon Press.

SHARMAN, G. B. (1976). Evolution of viviparity in mammals. In *The Evolution of Reproduction, Reproduction in Mammals 6*, ed. C. R. Austin & R. V. Short, pp. 32–70. London: Cambridge University Press.

SIMPSON, M. J. A. (1978). Tactile experience and sexual behaviour: aspects of development with special reference to primates. In *Biological Determinants of Sexual Behaviour*, ed. J. B. Hutchinson, pp. 785–807. New York: Wiley.

SPIRIN, A. S. (1966). On 'masked' forms of messenger RNA in early embryogenesis and in other differentiating systems. In *Current Topics in Developmental Biology 1*, ed. A. A. Moscona & A. Monroy, pp. 2–38. New York & London: Academic Press.

TRIVERS, R. L. (1972). Parental investment and sexual selection. In *Sexual Selection and the Descent of Man 1871–1971*, ed. B. Campbell, pp. 136–79. London: Heinemann.

WADDINGTON, C. H. (1952). *The Epigenetics of Birds*. London: Cambridge University Press.

WADDINGTON, C. H. (1957). *The Strategy of the Genes*. London: Allen & Unwin.

WADDINGTON, C. H. (1966). *Principles of Development and Differentiation*. New York: Macmillan.

WALLACE, B. (1970). *Genetic Load : its Biological and Conceptual Aspects*. New Jersey: Prentice-Hall.

WARD, S., THOMSON, N., WHITE, J. G. & BRENNER, S. (1975). Electron microscopical reconstruction of the anterior sensory anatomy of the nematode *Caenorhabditis elegans*. *Journal of Comparative Neurology*, **160** (3), 313–38.

WEISZ, P. B. (1951). A general mechanism of differentiation based on morpho-genetic studies in ciliates. *American Naturalist*, **85**, 293–306.

WEISZ, P. B. (1954). Morphogenesis in protozoa. *Quarterly Review of Biology*, **29**, 207–29.

WOURMS, J. P. (1974). The developmental biology of annual fishes. II. Naturally occurring dispersion and reaggregation of blastomeres during the development of annual fish eggs. In *Genetic Studies of Fish 2. Collected Works*. MSS Information Corporation, New York.

YANAGISAWA, K., POLLARD, D. R., BENNETT, D., DUNN, L. C. & BOYSE, E. A. (1974). Transmission ratio distortion at the *T*-locus: serological identification of two sperm populations in *t*-heterozygotes. *Immunogenetics*, **1**, 91–6.

Appendix

Previous considerations of 'maternal effects' as a category have mostly been for purposes of attribution of hereditary differences. This '*post hoc*' attribution of causes has been analysed in Mather & Jinks (*Biometrical Genetics*, Chapman & Hall, 1971, p. 293) as:

(i) Cytoplasmic inheritance
(ii) Maternal nutrition...
(iii) Transmission of pathogens and antibodies...
(iv) Imitative behaviour
(v) Interaction between sibs...

and methods of statistical analysis which begin to discriminate these are described. Such methods, and many more experimental techniques are described in Beale & Knowles (*Extranuclear Genetics*, Edward Arnold, 1978) for such problems as male sterility in angiosperms, *petite* mutants of yeasts, and SR characters of *Drosophila*. This last situation, with *Drosophila* species (and even races) having different endosymbiont spiroplasmas each with associated viruses fatal to foreign spiroplasmas, extends extra-nuclear heredity into a new dimension. For developmental biologists, however, the involvement of maternal properties in offspring development is of interest more for its mechanism than for its attribution or extent. So a rather different and more detailed classification is proposed, with basic division into maternal nuclear, extra-nuclear, phenotypic and completely extra-genetic categories of effects.

Maternal effects on development: an attempt at a comprehensive listing

Direct effects of maternal genotype (nuclear)

(*a*) Alleles affecting shell spiral sense in gastropods (= direction of first alternation of spiral cleavage): abnormal allele *d* (= sinistral) in *Lymnaea*.

(*b*) Alleles affecting presence of germ plasm in oocytes (note: the special granules may be self-replicating, but their presence in oocytes is affected by maternal nuclear genes): abnormal allele *gs* in *Drosophila*.

(*c*) Nuclear transcripts during oogenesis: ribosomal, structural and control mRNA; other HnRNA functions; 'housekeeping' and morphogenetic locus transcripts.

(*d*) Those 99 + % of nuclear loci whose 'normal' alleles complement both the cytoplasmic machinery of the oocyte and the extra-nuclear genomes of organelles (includes most of *c*). Abnormal alleles: possibly T^{Hp} in mouse; abnormal or aberrant complementation in amphibian hybrids and *Amoeba* nuclear transplants.

(*e*) Viruses, and other plasmids, associated with the nuclear genome of the oocyte, which may act later in embryonic or adult life.

(*f*) Possible alleles with 100 % penetrance (i.e. irrevocably genomic in action) which affect yolk provision or distribution, gradients or cortical patterns in oocytes, nurse or follicle cell contributions, timing or other physiology of ovulation, or choice of paternal genome. That is to say, any of the maternal effects below which can plausibly be related directly to gene or gene transcript *without* interaction effects. An example could be albino alleles in negro women, which doubtless affect the choice of fathers for their offspring, as well as nutrient levels. Even here, however, some (phenotypic) cosmetic moderation is usual.

Direct or indirect effects of maternally derived genotypes (extra-nuclear)

(*a*) Effects via mitochondrial genome, which is apparently entirely maternal in derivation in animals (compare with '*poky*' genes in fungi): e.g. antibiotic resistance variability in *Paramecium*.

(*b*) Effects via chloroplast genome: e.g. plants, and animals like *Hydra*, corals, *Tridacna* and *Convoluta* whose plant symbionts are passed in the egg or egg investments.

(*c*) Effects via other symbiont genomes: e.g. *Drosophila* 'A bodies' (*Rickettsia*), termite bacteria and protozoans, ruminant symbionts.

There is clearly a gradation from these, via commensals normally transmitted from mother (eland trypanosomes, rabbit fleas) to pathogens ('hereditary' syphilis, possible pathogenic viruses, e.g. in mouse milk).

(*d*) Effects of centriolar heredity or other possible autonomous and parallel heredities (as in ciliate cortex, or amphibian grey crescent, or germinal cytoplasmic bodies).

Indirect effects of maternal genome (nuclear), i.e. maternal phenotype effects

(*a*) Effects of maternal competence (profit on metabolic transactions) on nutritional supply, e.g. amount and quality of yolk or milk.

(*b*) Effects of maternal response to environmental stimuli: e.g. timing of ovulation or birth, and regulation of clutch size to nutrient supply (but note eagles and other cases of trophic embryos or juveniles provided to be eaten by siblings).

(*c*) Effects of female sexual desirability, competence or cooperation in determination of male contribution to zygote genotypes (e.g. some humans); and effects of female discrimination in choosing fathers: e.g. choice of male wing-beat frequency by *Drosophila* females; choice of male bower ornament by female bower-birds; choice of mate by Kob antelope, or cichlid fishes, or wild canid dominant females, or most humans. (Note: this effect can only operate where males choose females by genotype-associated variation observable phenotypically.)

(*d*) Effects of maternal antibodies in: (i) selection of gametes (e.g. *t*-locus mouse); (ii) selective abortion or runting (Rhesus, T^{Hp}); (iii) the acquisition of a maternally specific spectrum of immunities and tolerances in embryonic/juvenile life.

(*e*) Effects of maternal appearance, activity level, or pathology on embryonic or juvenile development: e.g. imprinting on variant (e.g. surrogate) mothers; rearing by sedated mothers in rats; anatomical/physiological/pathological effects of diabetes, smoking or obesity in human mothers.

(*f*) Possible effects of variation in maternal regulation of embryonic environment: e.g. in viviparous poikilotherms living in very variable environments like guppies and *Lacerta vivipara*; or even homoiotherms with variable nests, like mice in below-freezing conditions.

Effects via maternal phenotype not *causally related to her genotype*

(*a*) *Nutritional effects.* While diabetes, and probably obesity and smoking, have genetic components, chronic starvation in some African Sahel tribes does not; and there is evidence of impairment of child intellect by maternal starvation. This impairment may prolong the original effect through several generations. There are presumably comparable animal and plant examples, and the chromosomal changes of progeny resulting from differing fertilizer status of maternal plant fits closest to this category.

(*b*) *Status effects*. In rabbits and wild dogs, at least, low-status mothers rarely breed successfully. In most status-breeding species with general sexual access, highest-status males sire all zygotes, so genetic effect on status should be minimal. In human status-systems too, maternal status usually determines early post-natal choices within a culture.

(*c*) *Cultural transmission*. Here a variety of maternal effects is seen in all human cultures, and some in animals too.

(i) Imprinting. The mother goose 'picture' parallels the genotype as each gosling uses its maternal imprint as a mate model. It is less well known that many mammals, including primates, imprint similarly.

(ii) Imitative behaviour (initially, usually of mother; later of both parents and conspecifics): e.g. (*a*) the acquisition of food/fear habits in most 'higher' organisms; (*b*) the acquisition of intraspecific communication behaviour in altricial mammals; (*c*) the acquisition of language; (*d*) the acquisition of 'mothering skills' (sometimes in interaction with sibs); (*e*) acquisition of specific cultural practices (e.g. drug addiction, cultural sex images).

(iii) Cultural diversion of physiology and/or behaviour ('unnatural practices'): e.g. (*a*) peculiar practices during pregnancy (abortifacients (including IUCDs – they affect development), starvation, smoking, drugs, teratogens, including diagnostic X-irradiation, etc.); (*b*) peculiar birth practices (e.g. hospitalisation, sedation of mother and incidentally of child, so limiting mother–child interaction post-natally; 'exposure'; circumcision); (*c*) peculiar nursing practices (e.g. use of cow's milk; nappies; 'wet-nursing'); (*d*) peculiar family practices (e.g. limiting family size, so restricting sibling interaction; separation (as a result of adoption, boarding schools)).

Maternal effects and plastid inheritance

R. A. E. TILNEY-BASSETT AND
O. A. L. ABDEL-WAHAB
Dept of Genetics, University College of Swansea, Singleton Park,
Swansea, SA2 8PP, UK

The genetic basis for maternal effects

Plant and animal cells are basically alike. They both possess a nucleus, containing the chromosomes, and a cytoplasm, containing the mitochondria, ribosomes, Golgi bodies and endoplasmic reticulum, all within a boundary membrane. In addition, plant cells contain another class of organelle, the plastids. The plastids, which are found in all plant cells, exist in a number of interconvertible forms – the undifferentiated proplastids of meristematic cells, the starch-filled amyloplasts of storage tissues, the chromoplasts of highly pigmented tissues (particularly some flowers and fruit), the all-important chloroplasts of photosynthetic tissues, and other less common types – all of which develop in their characteristic way to fulfil a particular function appropriate to the cells and tissues containing them. The plastids, which vary in shape and size, are all visible within the magnification of the light microscope, but electron microscopy is required to study their internal organization. The number of plastids varies from 1 per cell in many algae up to 300–400 per cell in large leaf cells and very much higher numbers in the exceptionally large cells of *Acetabularia*. They appear to multiply independently within the cell, each plastid being derived by division from a pre-existing plastid. The plastids contain all the machinery for protein biosynthesis in which messenger RNA is transcribed on plastid DNA and translated on plastid ribosomes.

Within the plastid the genetic information is restricted to one plastid chromosome, but this is often highly polyploid with as many as 40 or more copies. Within higher plants the many copies of the plastid genome appear to be localized within 1 to as many as 16 nucleoids, and within the large chloroplasts of some algae many more nucleoids are found; the actual number of nucleoids seems to increase with increasing plastid size as, for example, when a small proplastid develops into a large chloroplast. Estimates of the kinetic complexity for chloroplast DNA vary from 94 to 230 million, of which the lower value may be nearer the truth. Indeed, the normal physical state of the chloroplast DNA in higher plants is to exist as circles with a length of 30–46 μm, corresponding to a molecular weight range of 83–96 million; the DNA of algae may be more variable. About 90 million of DNA contains 272000 deoxynucleotides capable of coding for polypeptide chains containing a total of about 45000 amino acid residues, or about 150 proteins, each with polypeptide chains 300 amino acids long and having a molecular weight of about 40000. Some of this DNA is taken up by the genes for ribosomal and transfer RNA, but this leaves enough information to code for quite a large number of structural genes. The protein products of a few of these genes are now known, for example the large subunit of ribulose diphosphate (RuDP) carboxylase, the carbon dioxide fixation cycle enzyme, and the existence of others clearly demonstrated (Kirk & Tilney-Bassett, 1978). We have, therefore, the potentiality for considerable genetic variation between plastids, and when the plastids differ between parents, as is found to be the case in certain species of tobacco with respect to the exact composition of the large subunit of RuDP carboxylase, or are unequally transmitted in sexual reproduction, as again in tobacco, we have the genetic basis for maternal effects.

Restrictions upon plastid inheritance

Unequal plastid transmission at sexual reproduction appears to be the rule in plants. The unicellular green alga, *Chlamydomonas reinhardtii*, forms isogametes of positive and negative mating type which fuse with one another to form a zygote. Yet, although each cell contributes a chloroplast, genetic analysis has proved that the transmission of the chloroplast genome is predominantly maternal, and rarely biparental or paternal (Sager, 1972). According to Chiang (1976), the mechanism by which this result is achieved is rather complex, and still incompletely understood. The chloro-

plast DNA of both parents appears to be degraded after gametogenesis by endonuclease restriction enzymes, which become even
more active after gametic fusion. Sometime after zygote formation
there are at least two rounds of semiconservative chloroplast DNA
replication, in which neither parental chloroplast DNA is
favoured, and some of the newly formed DNA is restricted too.
In addition, there is extensive recombination between the DNA
strands which probably permits some genes to be rescued from
DNA molecules most of which are subsequently degraded.
Throughout these events the paternal chloroplast DNA is typically degraded faster than the maternal DNA so that, as a rule,
the DNA strands that survive, and which are transmitted to the
zoospores at meiosis, are descendants of the maternal chloroplast
genome. Occasionally the mature zygote still retains some paternal
chloroplast genes, owing to marker rescue or to the survival of a
paternally derived chloroplast genome, which accounts for the
observed 0.1–3.0 % biparental zygotes, or, very exceptionally,
only the paternal chloroplast DNA survives to account for the very
rare paternal zygote. What advantage the alga gains from the
severe restriction of plastid DNA from one parent is not clear, and
we also do not know if the laboratory observations are a true
representation of what happens in nature. We can only speculate
that there is an advantage in reducing the number of copies of the
chloroplast genome either for the genetic reason that many of the
copies may have imperfections that should not be allowed to
accumulate, or that they are not all capable of acting as templates
for future rounds of replication, or possibly that there is a
physiological advantage in reducing the load of chloroplast DNA
when the alga enters into a dormant stage of its life cycle as the
tough, resistant zygote. If, for whatever reason, the number of
copies of chloroplast DNA is severely reduced, it is imperative that
at least one or two copies survive. The cell may be answering this
problem by modifying some of the DNA and so protecting it
against degradation (Sager, 1975; Sager & Kitchin, 1975).

For plants in which there is a differentiation between the motile
male gamete and the non-motile female gamete, Paolillo (1974) has
drawn attention to the difference that exists between, on the one
hand, the algae *Chara* and *Nitella*, the ferns *Equisetum*, *Marsilea*
and *Pteridium*, and the gymnosperm *Zamia*, all of which have
numerous plastids in the male gamete, and on the other hand, the

mosses and liverworts *Bryum, Marchantia, Pellia, Phaeoceros* and *Polytrichum*, which have only one plastid per male gamete. This does not mean, however, that the plastid contribution by the male parent will be less in the latter than in the former group. The existence of plastids in the male gamete is not a guarantee that they will be transmitted. At fertilization the sperms of the liverwort *Sphaerocarpos* and the fern *Pteridium* lose all of their components except the headpiece and nucleus as they descend the neck canal of the archegonium, whereas the plastid of the *Marchantia* sperm enters the egg with the nucleus. It does not seem to follow, therefore, that the reduction in plastid number is a necessary step in eliminating male plastid transmission, as even one plastid may succeed where many fail, and neither one nor many is a guarantee of transmission.

Fine structural studies of the gymnosperms *Biota orientalis* (Chesnoy, 1969), *Chaemaecyparis lawsoniana* (Chesnoy, 1973) and *Larix decidua* (Camefort, 1969) have shown that, owing to the peculiarities of zygote development and not to any numerical supremacy, the embryo plastids are derived largely, if not entirely, from the male parents. Moreover, this startling conclusion is supported by the genetical analysis of another gymnosperm, *Cryptomeria japonica*, in which about 97 % of the progeny of a cross between two dissimilar cultivars received some, and often all, of their plastids from the male parent (Ohba, Iwakawa, Ohada & Murai, 1971). Here then the behaviour of the plastids is making possible not a maternal, but a paternal effect.

The flowering plants have been studied both by fine structural and by genetical analysis. Electron microscopy has shown that for many plants a critical step in plastid transmission by the male parent occurs in the first division of the haploid, unicellular pollen grain when it divides mitotically into a vegetative and a generative cell. At this division the plastids are reported to remain in the vegetative cell and to be excluded from the generative cell in *Mirabilis jalapa* (Lombardo & Gerola, 1968), *Endymion non-scriptus* and *Tradescantia bractiata* (Jensen, 1974). If the generative cell lacks plastids then when it divides into the two male gametes these will likewise contain no plastids and so there can be no male plastid transmission. The absence of plastids from the male gametes has been reported for the genera *Beta, Gossypium, Hordeum* and *Petunia* (Jensen, 1974), and independent genetical analysis has confirmed in each that the plastid inheritance is purely maternal (Kirk & Tilney-Bassett, 1978). Sometimes, however, the exclusion mechanism is imperfect, and the generative cells of *Antirrhinum*, *Beta* and *Lycopersicon* occasionally receive a single plastid, which would account for the rare biparental plastid transmission observed in the first-named genus (Hagemann, 1976). Finally, there are genera like *Oenothera* and *Pelargonium*, with regular biparental plastid inheritance, in which the generative cells are observed to contain many plastids, as is also the case for *Castilleia wightii* and *Lobelia exinus* (Hagemann, 1976), which have not been tested genetically. Hence there appears, on present evidence, to be a close correlation between the exclusion of plastids from the generative cell of the pollen grain and a purely maternal plastid inheritance, and between inclusion of plastids in the generative cell and a subsequent biparental plastid inheritance.

The distribution of maternal and biparental plastid inheritance in flowering plants

Various kinds of genetical analysis (Kirk & Tilney-Bassett, 1978) have now been employed to demonstrate a maternal or biparental plastid inheritance in a sufficient number of genera of flowering plants for us to begin to look at the distribution of this character (Table 1). Although some of the original data are not above criticism, and some genera should be analysed further, it is becoming clear that both maternal and biparental plastid inheritance is widespread. Examples of both types are found in monocots and dicots, and in both the Archichlamydeae and the Sympetalae. All the orders represented by two or more families include families in which there are genera with biparental plastid inheritance as well as maternal. Most families are represented by only one genus, but in families with two or more genera listed, the Cruciferae, Scrophulariaceae, Compositae and Liliaceae are uniformly maternal, the Caryophyllaceae, Leguminosae, Onagraceae, Solanaceae and Gramineae are mixed, and the Geraniaceae are uniformly biparental. The picture is therefore one of considerable heterogeneity, in which it seems probable that the evolution of a purely maternal plastid inheritance from a more primitive biparental inheritance has occurred on many occasions, and may even be reversible. It also follows that we cannot yet make any firm

Table 1. *Classification of the genera of flowering plants according to whether their plastids are transmitted at fertilization by the maternal parent alone (M) or by both parents (B)*

Major divisions	Orders	Families	Genera
DICOTYLEDONS Archichlamydeae	Urticales	Moraceae	*Humulus* (M)
	Polygonales	Polygonaceae	*Fagopyrum* (B)
	Centro-spermae	Aizoaceae	*Mesembryanthemum* (M)
		Caryophyllaceae	*Silene* (B), *Stellaria* (M)
		Chenopodiaceae	*Beta* (M)
		Nyctaginaceae	*Mirabilis* (M)
	Rhoedales	Cruciferae	*Arabidopsis* (M), *Arabis* (M), *Aubretia* (M)
	Rosales	Leguminosae	*Acacia* (B), *Medicago* (B), *Phaseolus* (B), *Pisum* (M), *Trifolium* (M)
		Saxifragaceae	*Hydrangea* (M)
	Geraniales	Geraniaceae	*Geranium* (B), *Pelargonium* (B)
	Malvales	Malvaceae	*Gossypium* (M)
	Parietales	Guttiferae	*Hypericum* (B)
		Violaceae	*Viola* (M)
	Myrtiflorae	Onagraceae	*Oenothera* (B), *Epilobium* (M)
Sympetalae	Ericales	Ericaceae	*Rhododendron* (B)
	Primulales	Primulaceae	*Primula* (M)
	Tubiflorae	Boraginaceae	*Borrago* (B)
		Convolvulaceae	*Pharbitis* (M)
		Labiatae	*Nepeta* (B)
		Scrophulariaceae	*Antirrhinum* (M), *Mimulus* (M)
		Solanaceae	*Browallia* (B), *Capsicum* (M), *Lycopersicon* (M), *Nicotiana* (M), *Petunia* (M), *Solanum* (M)
	Plantaginales	Plantaginaceae	*Plantago* (M)
	Cucurbitales	Curcurbitaceae	*Curcurbita* (M)
	Campanulatae	Compositae	*Helianthus* (M), *Lactuca* (M)

Table 1 (*cont.*)

Major divisions	Orders	Families	Genera
MONOCOTYLEDONS			
	Glumiflorae	Graminae	*Secale* (B), *Avena* (M), *Coix* (M), *Hordeum* (M), *Oryza* (M), *Sorghum* (M), *Triticum* (M), *Zea* (M)
	Liliflorae	Liliaceae	*Allium* (M), *Chlorophytum*(M), *Hosta*(M)

The table is based on information from a variety of genetical crosses within and between species and described by Kirk & Tilney-Bassett (1978). Genera in which the species exhibited only traces of biparental plastid inheritance are treated as maternal.

predictions as to the expected behaviour of any other genera. In other words, we cannot safely predict which new genera are likely to exhibit plastid-related maternal effects. All we can do is to take note of the trends and predict that in the Cruciferae, Solanaceae and Gramineae, for example, a purely maternal transmission is more probable than a biparental one. Even in those genera in which the plastids are excluded from the male germ line, we cannot be sure that this will apply to the mitochondria and other cytoplasmic constituents, so there always remains the likelihood of some contribution by the male, albeit a small and perhaps insignificant one.

The genetic control of plastid inheritance in *Oenothera* and *Pelargonium*

In *Chlamydomonas* we saw that the transmission of the male genome was normally blocked even though the chloroplast DNA was regularly introduced into the zygote; there is no reason why a similar genetic control should not operate in some higher plants. Further analysis of zygotic control has in fact concentrated on interspecific crosses within Oenothera and on intercultivar crosses within *Pelargonium* × *Hortorum*, making use of plants in which one parent contains normal green (G) plastids and the other mutant white (W) plastids in their respective germ layers. In both cases electron microscopy has shown that many plastids regularly enter the generative cells (Diers, 1963; Lombardo & Gerola, 1968), and Meyer & Stubbe (1974) have shown that the male gamete contributes 8–13 plastids, compared with the contribution of the mature egg of up to 32 plastids, to the zygote of *Oenothera erythrosepala*. Plastid inheritance, although nominally classified as biparental, is really mixed, that is to say some of the progeny receive all their plastids from the mother, some from both parents and, in *Pelargonium*, some progeny, often the majority, receive all their plastids from the male parent.

Schötz (1954) found that after G × W plastid crosses between *Oenothera* species there were green and variegated offspring, and after the reciprocal W × G plastid crosses there were white and variegated offspring. This result shows that there is a major sex effect which ensures that all offspring receive some, and often all, plastids from the female parent, while the male parent is a minor, and never the sole, contributor. Thus the relative importance of the two parents correlates closely with the three- to fourfold greater number of plastids contributed by the egg than by the male gamete. Some plastid replication occurs within the zygote before the first cell division to give a total population of 54–64 plastids (Meyer & Stubbe, 1974). At the asymmetrical, flask-shaped, cell division a high proportion of these plastids are probably retained in the large suspensor cell while a few enter the terminal cell so that, through chance alone, male plastids are often left out, and even if they do enter the terminal cell, as a small fraction of the total plastid population, they are again liable to be left out of the main embryo cell lineage at the next division. The numerical limitation of the male plastids explains the predominantly maternal plastid inheritance, but this is not the complete story. Schötz (1954, 1958, 1968, 1974, 1975) found that within 28 species or hybrids of the subgenus *Euoenothera* he could distinguish three different plastid types with respect to their relative speeds of multiplication; these he called fast, medium and slow. In any hybrid in which two of these plastid types were brought together, the faster-multiplying, or stronger, plastid had a competitive edge over the slower-multiplying, or weaker, plastid, which therefore modified the sex advantage. Hence the combination of strong female plastids and weak male ones virtually inhibits any effective male transmission, whereas male transmission is at its best in the combination of a weak female and a strong male plastid. In addition to this plastid effect, by comparing a large number of interspecific crosses in which the plastids were of the same kind and the hybrid genome variable, Schötz (1974, 1975) showed that the output ratio of green to while plastids was also modified by the nuclear genome. He found that the same white male plastids do better or worse in G × W crosses according to whether their competitiveness is encouraged or discouraged by the hybrid nucleus. The overall effect of the three levels of control within *Oenothera* is that, for any cross, a variable portion of the progeny receive all their plastids from the female parent and the remaining progeny receive plastids from both parents. The range of the plastid and nuclear effects is sufficiently great to enable us to make crosses with up to 70 % biparental progeny, or to make crosses in which the plastid inheritance is wholly maternal – the male plastids that enter the zygote simply cannot overcome the triple disadvantage of a numerical and multiplicative inferiority coupled with an unsympathetic nucleus.

The behaviour of the plastids in *Pelargonium* is quite different. Reciprocal crosses between plants with green and plants with white plastids in their germ layers generally give rise to varying proportions of green, variegated and white progeny, showing that the embryos receive their plastids either all from the female parent, or all from the male parent, or from both parents. When G × W and W × G crosses are compared, the majority of offspring do not automatically take after the female parent, as in *Oenothera* crosses, but instead they usually take after the parent that contributes the green plastids. This non-reciprocal behaviour is not attributed to the effects of selection or environment but is under genetic control. This control, which is predominantly by the nuclear genotype of the female parent, is modified by the poorer transmission of the mutant plastids compared with the normal ones. Thus, depending on their nuclear genotype, the six *Pelargonium* cultivars studied so far (Tilney-Bassett, 1976) fit into a series progressing from very strong females to weak ones in the following order:

Miss Burdette-Coutts > Lass O'Gowrie > Dolly Varden > J. C. Mapping > Flower of Spring ≈ Foster's Seedling

The two end cultivars well illustrate the combined effect of the maternal strength of the cultivar and the difference between green and white plastids. When Miss Burdette-Coutts is the female in G × W crosses the combination of a nuclear genotype that makes it a very strong female, and green plastids, produces an average segregation of 83 % green, 17 % variegated and no white embryos. In the reciprocal W × G cross, the average segregation is 19 % white, 77 % variegated and 4 % green embryos so that, in spite of the disadvantage of white plastids, the inheritance remains predominantly maternal. By contrast, when Foster's Seedling is the female parent in G × W crosses it is such a weak female that, with an average segregation of 40 % green, 28 % variegated and 32 % white embryos, the white male plastids are transmitted almost as well as green female plastids. With the reciprocal W × G cross, however, the average segregation of 2 % white, 21 % variegated and 77 % green embryos shows that the weak female is still further weakened by its white plastids and the net result is an overwhelming paternal plastid inheritance.

The above two cultivars are representatives of two distinct segregation patterns following G × W crosses. These are called type I (G > V > W), in which most offspring are green, rather less are variegated and usually a few are white, or occasionally the variegated are more frequent than green; and type II (G > V < W), in which green and white offspring are both more frequent than variegated which are usually few or absent. Breeding evidence suggests that the two patterns are controlled by alleles of a nuclear gene, named Pr_1 and Pr_2, with an effect on plastid replication at the time of zygote formation. Females, like Miss Burdette-Coutts, Lass O'Gowrie and Dolly Varden, whose off-spring segregate with a type I pattern are homozygous Pr_1Pr_1, and females, like J. C. Mapping, Flower of Spring and Foster's Seedling, whose offspring segregate with a type II pattern, are heterozygous Pr_1Pr_2. In the reciprocal W × G crosses the two types are less easily separable as, with the exception of Miss Burdette-Coutts, they all exhibit a predominantly paternal in-heritance; however, the cultivars are graded in the same order as for G × W crosses with the type II plants all showing a stronger paternal effect than the type I plants. As at least five, and probably all six, cultivars are significantly different from each other, there must be additional genes that modify the basic patterns.

In selfing and backcross experiments the heterozygotes behave as if there is gametic lethality of the Pr_2 allele on the female side.

Thus, when heterozygous Flower of Spring was selfed, instead of the expected $1 \ Pr_1Pr_1 : 2 \ Pr_1Pr_2 : 1 \ Pr_2Pr_2$ segregation a ratio of $1 \ Pr_1Pr_1 : 1 \ Pr_1Pr_2$ was observed. Similarly, when the cultivar was crossed with Dolly Varden (Pr_1Pr_1), the segregation ratio only approached $1 : 1$ when Dolly Varden was the female parent, and the progeny were almost all Pr_1Pr_1 when Flower of Spring was the female parent (Tilney-Bassett, 1973). From this finding it may be deduced that the disparity between the segregation patterns from type I and type II families after $G \times W$ crosses occurs despite the fact that the vast majority of the female eggs that are fertilized must be of uniform genotype Pr_1. Hence it appears that the outcome of the segregation pattern is predetermined by the female parent at or before meiosis when still heterozygous and diploid, and therefore still different from the homozygote (Tilney-Bassett, 1974).

A possible explanation of the gene action is to suppose that Pr_1 and Pr_2 behave as codominant alleles in which the enzyme product of Pr_1 encourages the replication of green female plastids, and the product of Pr_2 the replication of mutant male plastids. This would neatly account for the approximately $1 : 1$ ratio of green:white embryos found in the progeny of $G \times W$ crosses among many type II plants. The product from the Pr_2 allele, made in the Pr_1Pr_2 parent, would determine the behaviour of the eggs in which it entered even though their actual genotype was Pr_1. Such a system might include the switching off of one allele so that transcription takes place at the Pr_1 allele, or the Pr_2 allele, but not both together. If this is the case then we might expect to find mosaic tissues within which most embryos are controlled by Pr_1 and are green or by Pr_2 and are white. We have looked for evidence of such behaviour at three levels of development. First, we compared the frequency of green, variegated and white embryos after $G \times W$ crosses in six inflorescences from one plant of Flower of Spring, but there was no evidence of heterogeneity ($X_{10}^2 = 13.378$, $P = 0.5$–0.1). Secondly, we recorded the position of each flower in an inflorescence, scored its segregation behaviour after $G \times W$ crosses, and looked for evidence of the clustering of similar flowers within one inflorescence, but all appeared to be randomly distributed with respect to their segregation behaviour. Finally, we looked at the behaviour of individual flowers after $G \times W$ crosses with all three type II cultivars, to see if there was any tendency for embryos from the same flower to be alike more often than would be expected according to the binomial distribution. The results (Table 2) show no heterogeneity between cultivars and no significant departure from the binomial distribution for flowers with two and four embryos and only a slight departure for flowers with three embryos. We conclude that there is no clear correlation between the embryos of the same flower, and hence in the absence of an early switching mechanism the maternal effect does not begin to function until the development of the individual embryo sac.

Precisely how the maternal genotype transmits the information to determine the pattern of plastid inheritance remains unknown, but we can consider some clues as to how the plastids respond. After the type II G × W crosses there are approximately equal proportions of green and white embryos while variegated are few, or are often absent altogether. This observation is in conflict with the random sorting-out of even two plastids, and so Tilney-Bassett (1970) suggested that within a zygote frequently one plastid type replicates first, and the first to replicate effectively inhibits the replication of the other type. In fact the data are consistent with the possibility that all the plastids are degraded or fail to replicate except one, either green or white, apart from the occasional zygote in which two survive and from which variegated embryos are derived. The varying segregation ratios exhibited by all crosses can be accounted for by the assumption that the different maternal genotypes control the probability of the survival and replication of a green or white plastid, and of one or more plastids. Birky (1976) has called the phenomenon probabilistic inheritance, because there appears to be a decision made in each zygote whether to transmit maternal or paternal plastids or both in varying proportions. The choice may not be predicted in advance, but we can assign probabilities to it. The effect of the Pr_2 allele in heterozygous Pr_1Pr_2 females is to greatly increase the probability that plastids derived from the male will replicate in the zygotes as compared with their chances when the female parents are homozygous Pr_1Pr_1.

The maternally controlled probabilities appear to be modified, however, by an interaction between the plastids themselves, in which the green plastids are consistently superior to white plastids. We have investigated this phenomenon further by crossing two mutant plastids. One mutant, derived from the type II cultivars,

Table 2. *Tests of the goodness of fit between observed and expected frequencies of green and white embryos in individual flowers on the assumption of a binomial distribution*

		Between cultivars			
	Total	Heterogeneity X^2	P	Pooled X^2	P
Flowers with two embryos containing:					
2G+0W: 103 1G+1W: 163 0G+2W: 64	330	0.731	> 0.9	0.001	> 0.99
Flowers with three embryos containing:					
3G+0W: 58 2G+1W: 97 1G+2W: 116 0G+3W: 41	312	4.086	0.9–0.5	10.896	0.05–0.01
Flowers with four embryos containing:					
4G+0W: 19 3G+1W: 56 2G+2W: 92 1G+3W: 50 0G+4W: 12	229	6.981	0.9–0.5	1.450	0.9–0.5

The data are derived from G × W crosses using the type II plants Foster's Seedling, Flower of Spring and J. C. Mapping as the source of green plastids and six chimerical cultivars as the source of white plastids. In the table G is a green embryo and W is a white embryo, and the numbers express the frequency in a single flower.

is phenotypically highly stable, always producing pure white embryos on selfing, while the other mutant, derived from the type I cultivars, is phenotypically unstable, frequently producing white embryos bearing green flecks on selfing. By scoring the number of white embryos with green flecks, after crossing these two plastid types, we can estimate the extent of male and female plastid transmission. The results of the W × W crosses are compared with the W × G and G × W crosses using the same female cultivar for all three (Table 3); for the G × W and W × G crosses the male parent is the mean result from using all six cultivars as male (Kirk & Tilney-Bassett, 1978), and for the W × W crosses the mean result of using the three stable or unstable mutants as appropriate. We see that when the type I cultivars are the female parents (Table 3*a*) there is a considerable improvement in the performance of the white female plastids in competition with white male plastids as compared with their performance in competition with green male plastids. This improvement is greatest for Dolly Varden, the weakest of the three females as measured by W × G and G × W crosses, and least for Miss Burdette-Coutts, the strongest of the three females. When the type II cultivars are the female parents, there is no improvement in the performance of the white female plastids in competition with white male plastids as compared with their performance in competition with green male plastids (Table 3*b*). These contrasting results need to be explored more fully, but tentatively it looks as if the green plastids are superior to both mutant plastids, and the unstable mutant plastids superior to the stable mutant plastids. In the competition between the two mutant plastids it seems as if the unstable mutant plastid is comparable to a green plastid when it enters from the male parent (Table 3*b*), and not quite as strong as a green plastid when it enters from the female parent (Table 3*a*). Evidently the two mutants must differ in their replicative abilities as well as their stabilities, and also in the ability of the type II plastids to green up in winter time (Kirk & Tilney-Bassett, 1978). We are therefore witnessing in *Pelargonium* the property of mutations to alter the multiplicative competitiveness of plastids just as must have occurred during the evolution of *Oenothera* species, only without the associated loss of the ability to develop into a functional chloroplast.

If we are to draw one conclusion from these studies it must be to stress the fluidity of maternal effects. The relative weightings

Table 3. *Comparison of the estimated percentages of female plastid transmission after three sets of* Pelargonium *crosses*

(a) *The female plastids are derived from the type I cultivars listed*

Type I cultivars	$W^u \times G$ crosses	$W^u \times W^s$ crosses	$G \times W$ crosses
Miss Burdette-Coutts	62.7 %	80.7 %	96.4 %
Lass O'Gowrie	40.1 %	74.4 %	92.0 %
Dolly Varden	27.6 %	77.1 %	84.4 %

(b) *The female plastids are derived from the type II cultivars listed*

Type II cultivars	$W^s \times G$ crosses	$W^s \times W^u$ crosses	$G \times W$ crosses
Foster's Seedling	8.0 %	6.1 %	52.5 %
Flower of Spring	9.8 %	6.6 %	52.6 %
J. C. Mapping	17.9 %	18.4 %	54.0 %

The unstable mutant plastid is represented as W^u and the stable mutant plastid as W^s. In the $G \times W$ crosses both W^u and W^s mutants are equally represented in the mean result quoted.

of the various controls on plastid inheritance between *Oenothera* and *Pelargonium* differ significantly, but in both it is possible to obtain combinations of factors which in $G \times W$ crosses produce a wholly maternal inheritance. Neither biparental nor maternal inheritance is a fixed property in these two genera, rather their potential flexibility is a condition that probably also exists in at least some of the genera from families exhibiting both biparental and maternal inheritance (Table 1). Indeed, in many of the maternal genera only one or two individuals have ever been tested, so there is no reason to assume that the potentiality for biparental inheritance is lost for ever. On the contrary, the existence of individuals which permit an occasional paternal contribution suggests that the selection of suitable genotypes to return to a regular biparental transmission may still be feasible; moreover, such genotypes may already exist in large genetically variable populations. We should therefore treat the property of a purely maternal inheritance with caution. This is likely to be an important criterion for the localization of the genes of some chloroplast enzymes within the chloroplast DNA, but it will not work for all genera and may not work for all individuals within a species, so irregular results may sometimes occur even when there is a well-established record. There is a real need, both in breadth and depth, for the further evaluation of the role of the plastids in maternal effects.

References

BIRKY, C. W. JR (1976). The inheritance of genes in mitochondria and chloroplasts. *Bioscience*, **26**, 26–33.

CAMEFORT, H. (1969). Fécondation et proembryogénèse chez les Abiétacées (notion de néocytoplasme). *Revue de Cytologie et de Biologie Végétales*, **32**, 253–71.

CHESNOY, L. (1969). Sur la participation du gamète mâle à la constitution du cytoplasme de l'embryon chez le *Biota orientalis* Endl. *Revue de Cytologie et de Biologie Végétales*, **32**, 273–94.

CHESNOY, L. (1973). Sur l'origine paternelle des organites du proembryon du *Chamaecyparis lawsoniana* A. Murr (Cupressacées). *Caryologia*, **25**, supplement, 223–32.

CHIANG, K. S. (1976). On the search for a molecular mechanism of cytoplasmic inheritance: past controversy, present progress and future outlook. In *The Genetics and Biogenesis of Chloroplasts and Mitochondria*, ed. T. Bücher, W. Neupert, W. Sebald & S. Werner, pp. 305–12. Amsterdam: Elsevier/North-Holland Biomedical Press.

DIERS, L. (1963). Elektronenmikroskopische Beobachtungen an der generativen Zelle von *Oenothera hookeri* Torr et Gray. *Zeitschrift für Naturforschung*, **18B**, 562–6.

HAGEMANN, R. (1976). Plastid distribution and plastid competition in higher plants and the induction of plastom mutations by nitroso-urea compounds. In *The Genetics and Biogenesis of Chloroplasts and Mitochondria*, ed. T. Bücher, W. Neupert, W. Sebald & S. Werner, pp. 331–8. Amsterdam: Elsevier/North-Holland Biomedical Press.

JENSEN, W. A. (1974). Reproduction in flowering plants. In *Dynamic Aspects of Plant Ultrastructure*, ed. A. W. Robards, pp. 481–503. New York: McGraw-Hill.

KIRK, J. T. O. & TILNEY-BASSETT, R. A. E. (1978). *The Plastids: Their Chemistry, Structure, Growth, and Inheritance*, 2nd edn. Amsterdam: Elsevier/North-Holland Biomedical Press.

LOMBARDO, G. & GEROLA, F. M. (1968). Cytoplasmic inheritance and ultrastructure of the male generative cell of higher plants. *Planta, Berlin*, **82**, 105–10.

MEYER, B. & STUBBE, W. (1974). Das Zahlenverhältnis von mütterlichen und vaterlichen Plastiden in den Zygoten von *Oenothera erythrosepala* Borbas (syn. *Oe. lamarckiana*). *Berichten der Deutschen Botanischen Gelleschaft*, **87**, 29–38.

OHBA, K., IWAKAWA, M., OHADA, Y. & MURAI, M. (1971). Paternal transmission of a plastid anomaly in some reciprocal crosses of Suzi, *Cryptomeria japonica* D. Don. *Silvae Genetica*, **20**, 101–7.

PAOLILLO, D. J. (1974). Motile male gametes of plants. In *Dynamic Aspects of Plant Ultrastructure*, ed. A. W. Robards, pp. 504–31. New York: McGraw-Hill.

SAGER, R. (1972). *Cytoplasmic Genes and Organelles*. New York & London: Academic Press.

SAGER, R. (1975). Patterns of inheritance of organelle genomes: molecular basis and evolutionary significance. In *Genetics and Biogenesis of Mitochondria and Chloroplasts*, ed. C. W. Birky Jr, P. S. Perlman & T. J. Byers, pp. 252–67. Columbus, Ohio: Ohio State University Press.

SAGER, R. & KITCHEN, R. (1975). Selective silencing of eukaryotic DNA. *Science*, **189**, 426–33.

SCHÖTZ, F. (1954). Über Plastidenkonkurrenz bei *Oenothera*. *Planta, Berlin*, **43**, 182–240.

SCHÖTZ, F. (1958). Beobachtungen zur Plastidenkonkurrenz bei *Oenothera* und Beiträge zum Problem der Plastidenvererbung. *Planta, Berlin*, **51**, 173–85.

SCHÖTZ, F. (1968). Über Plastidenkonkurrenz bei *Oenothera*. II. *Biologisches Zentralblatt*, **87**, 33–61.

SCHÖTZ, F. (1974). Untersuchungen über die Plastidenkonkurrenz bei *Oenothera*. IV. *Biologisches Zentralblatt*, **93**, 41–64.

SCHÖTZ, F. (1975). Untersuchungen über die Plastidenkonkurrenz bei *Oenothera*. V. *Biologisches Zentralblatt*, **94**, 17–26.

TILNEY-BASSETT, R. A. E. (1970). The control of plastid inheritance in *Pelargonium*. *Genetical Research, Cambridge*, **16**, 49–61.

TILNEY-BASSETT, R. A. E. (1973). The control of plastid inheritance in *Pelargonium*. II. *Heredity*, **30**, 1–13.

TILNEY-BASSETT, R. A. E. (1974). The control of plastid inheritance in *Pelargonium*. III. *Heredity*, **33**, 353–60.

TILNEY-BASSETT, R. A. E. (1976). The control of plastid inheritance in *Pelargonium*. IV. *Heredity*, **37**, 95–107.

Transcription during amphibian oogenesis

JOHN SOMMERVILLE

Dept of Zoology, University of St Andrews, St Andrews, Fife, UK

Levels of transcriptional activity in oogenesis

Growing oocytes accumulate gene products, in the form of either RNA or protein, or, more typically, as ribonucleoprotein (RNP) complexes, in massive amounts. Thus, after several months of synthetic activity, a mature oocyte of *Xenopus laevis* contains, for example, about 2×10^{11} polyadenylated RNA molecules (Cabada, Darnbrough, Ford & Turner, 1977), 190 ng of histone protein (Woodland & Adamson, 1977) and about 10^{12} ribosomes (Perkowska, Macgregor & Birnstiel, 1968). The vast majority of these, and other, products are not involved directly in the metabolism of the oocyte; rather they are stored as maternal components, to be utilized during the rapid cleavage stages of early embryogenesis when there is little gene activity.

The formation of large amounts of material during oogenesis requires not only a long time but also intense transcriptional activity and the involvement of many genes. At the chromosomal level this is represented by a high percentage of transcriptionally active chromatin (see Table 1), which can be visualized directly as the loops of lampbrush chromosomes. They are formed by a periodic looping-out of DNA from regions of condensed chromatin and are fairly evenly distributed throughout the chromosome complement. Lateral loops are particularly rich in transcription products (Table 1), which in the native state exist as linear aggregates of RNP beads (Malcolm & Sommerville, 1977).

In addition to being isolated in the native state, lampbrush chromosomes can be spread in low salt at pH 9 and viewed by electron microscopy as transcriptional matrices, each matrix consisting of a series of RNP fibrils which extend laterally from

47

Table 1. *Composition of chromatin[a] of* Triturus *oocyte nuclei and of somatic cell nuclei*

		Oocytes[c]	
	Somatic cells[b]	Total chromosomal	Lampbrush loops[d]
DNA	1	1	1
RNA	< 0.1	2	40
Protein	1–2	25	500

[a] Ratio of mass of DNA to RNA and to protein.
[b] Values from a variety of sources.
[c] Sommerville (unpublished results).
[d] Assuming that about 5 % of the DNA of lampbrush chromosomes is in lateral loops. In *Triturus* there are approximately 10^4 lateral loops each with an average axial length of 20–30 μm. The total length of genomic DNA (haploid amount) is about 7 m.

points of attachment at RNA polymerase molecules on the chromatin strand (for definitions and references, see Franke & Scheer, 1978). The high level of transcriptional activity in oocyte nuclei is confirmed by the observations that spread lampbrush chromosomes contain transcriptional units that are almost maximally packed with polymerase molecules and that the DNA is probably extended in its B-configuration and not compacted into nucleosomes (Scheer, 1978).

The course of transcriptional activity in amphibian oocytes is largely correlated with the formation of lampbrush chromosomes, a process which occurs in the extended diplotene of first meiotic prophase. During this period, which may last up to several months in some Amphibia, the chromosomes decondense, the lateral loops extend in length and the looped-out chromatin loses its nucleosomal condition and becomes saturated with polymerase molecules (see Franke & Scheer, 1978). Peak activity occurs generally about early vitellogenesis. Later, as the oocytes grow to maximum size, the loops retract, polymerase density decreases, the chromatin becomes progressively more beaded as nucleosomes re-form and eventually the chromosomes condense in preparation for metaphase.

There are, however, important variations in the timing of transcription of some sequences. For instance, the onset of synthesis and accumulation of the stable small RNA species, transfer RNA and ribosomal 5S RNA, occurs in pre-vitellogenic oocytes and precedes the synthesis of both messenger RNA (mRNA) and ribosomal 18S and 28S RNA (rRNA). These three major RNA classes are independent to the extent that they are transcribed by different forms of RNA polymerase. Furthermore, the formation of rRNA is under separate control in the compartment of the extrachromosomal nuclei, and synthesis of rRNA does not reach its peak until mid-vitellogenesis. Also, it has been reported that different patterns of loop labelling can be obtained by hybridizing middle-repetitive DNA to RNA transcripts on *Triturus* chromosomes from slightly different oogenic stages (Macgregor & Andrews, 1977). However, cell-free translation studies have shown no differences in the coding properties of mRNA extracted from *Xenopus* oocytes at different stages of development from pre-vitellogenesis to maturity (Darnbrough & Ford, 1976), so the possibility of stage-specific transcription of mRNA sequences due to lampbrush loops that are extended and retracted for different oogenic periods remains uncertain.

Gene numbers and the number of genes active in oogenesis

In considering the functional significance of the extensive transcription found in oocytes, it is an important requisite to know something of the number of different genes and the number of repeats of each gene that are active in lampbrush chromosomes, and also, where possible, to identify the sequences in terms of the protein or RNA encoded. The following aspects should be taken into account.

Genome size

An evolutionary complication, which turns out to be quite useful in the analysis of gene numbers and their functional significance, is that large differences are found in the genome size (C-value) of different amphibian taxa. A few examples are shown in Table 2, ranging from *Xenopus* with 2.9 pg DNA per haploid chromosome complement, which is a value typical of vertebrate organisms, to *Necturus* with 78 pg DNA, which is, by any criterion, a vastly excessive amount. There is no reason to suppose that in high

C-value organisms there is any increase in number of different kinds of genetic sequence, in the sense that more biochemical functions can be performed. On the other hand, there is evidence to suggest an evolutionary increase in the frequency of many individual sequences as a consequence of unequal crossing-over and chromatid exchange (Smith, 1973). By considering the karyo-types of related Amphibia with different C-values, it appears that the relative dimensions of homologous chromosomes remain unaltered, favouring the view that genetic sequences have been duplicated fairly uniformly through the genome (see Macgregor, 1978).

Degree of reiteration

The maintenance and utilization of tandemly duplicated sequences will of course depend on the selective advantage that more genes of that one type will confer. Oogenesis is a particularly critical period in the development of an organism, for adequate supplies of gene products must be accumulated within a reasonable length of time (for values and discussion, see Sommerville, 1977). For instance, it is a direct consequence of an increase in genome size that proportionately more histone proteins must be synthesized to complex the total amount of DNA replicated during early embryogenesis. Also, if the storage product is RNA, there is a requirement for more efficient transcription and this is most easily obtained by an increase in gene repeat frequency. Although mRNA transcription products appear to be fairly stable over a period of several months (Ford, Mathieson & Rosbash, 1977) it is obviously advantageous to keep to a minimum the period between onset of lampbrush chromosome activity and potential fertilization. For these reasons it might be expected that any fortuitous increase in the number of genes encoding a product required in vast amount for early development will be conserved. If we consider the data derived from molecular hybridization experiments, it is significant that the percentage of the genome devoted to the formation of rRNA, or to 5S RNA, or to histones, remains remarkably constant despite large differences in the genome size of different amphibian genera (Table 2).

There is a lesser requirement for multiple copies of most protein coding sequences. Hybridization data suggest that most of the mRNA sequences found in the oocytes of both *Xenopus* and

Table 2. *Proportion of the genome homologous to ribosomal RNA, 5S RNA and histone-coding sequences and the corresponding gene repeat frequency in amphibians with vastly different genome sizes*

Organism	C-value (pg)[b]	Percentage of genome (number of copies per haploid genome)[a]		
		rRNA[c]	5S RNA[d]	Histone[e]
Xenopus	3	0.12 (450)	0.1 (24000)	0.01 (50)
Triturus	23	0.1 (3200)	0.08 (140000)	0.01 (350)
Notophthalmus	35	0.1 (5000)	0.09 (300000)	—
Siredon	57	0.1 (9000)	—	—
Necturus	78	0.02 (5000)	—	0.01 (1200)

In general these are average values, derived from the pooled DNA of several animals, and do not take into account any heterogeneity in gene numbers between individuals of the same species.
From data compiled by Sommerville (1977).
From Rosbash, Ford & Bishop (1974) and data compiled by Birnstiel, Chipchase & Spiers (1971).
For references see Pukkila (1975).
Results from Sommerville (1979).

Triturus are transcribed preferentially from DNA sequences that are present once, or only a few times, in the genome (Rosbash, Ford & Bishop, 1974), whereas thermal stability data of hybrid molecules indicate that for each gene of *Triturus* several additional DNA sequences are present, but that these are only partially homologous to the mRNA sequences (Sommerville & Malcolm, 1976). It is possible that organisms with large genomes retain, and even transcribe, sequences that have had an origin common with functional coding sequences, but that have diverged due to mutation to become non-functional. In many circumstances only one, and any one, complete sequence from a tandemly replicated series need be conserved.

Chromosomal location of active genes

By definition, genes active in oocytes are located on the loops of lampbrush chromosomes (apart, that is, from the rRNA genes in nucleoli). The transcription of specified sequences, such as 5S RNA and histone-coding sequences, has been detected by hybridizing labelled, purified DNA probes to DNA transcripts *in situ*. The pattern of labelling with 5S DNA shows homologous sequences to be contained in transcripts on four loops, each of which is located near the centromere of four different chromosomes of *Notophthalmus* (Pukkila, 1975). Histone-coding sequences are found to be contained in transcripts on five to seven loops of *Triturus* chromosomes (Old, Callan & Gross, 1977). Whereas some of the reactive loops are labelled over their whole length, others are labelled over only part of their length.

Thus at least some of the transcribed reiterated sequences are seen to be contained within a restricted number of loops. The fact that they tend not to be held in one continuous block might facilitate the expression of stage-specific or tissue-specific sequences (e.g. 'somatic' as opposed to 'oocyte' types of 5S RNA: Ford & Southern, 1973) by means of localized condensation of the chromatin (see Ford & Mathieson, 1976).

The cytological and biochemical data on the extent of sequence reiteration are, to a certain degree, compatible. For instance, the estimated repeat frequency of histone genes in *Triturus* is about 350 copies per haploid DNA amount (Sommerville, 1979). If the sequence organization is similar to that in sea urchins (i.e. as 6 kb units containing one each of five histone genes: see Kedes, 1976), then the total length of DNA containing histone-coding sequences is about 700 μm. This value is consistent with the summated length of loops labelled by hybridization *in situ*, which is about 500 μm. However, it is more difficult to account for the accommodation of the estimated 300000 5S genes in four loops of average length 60 μm (Pukkila, 1975).

Extent of transcription

Saturation values obtained by hybridizing oocyte total nuclear RNA with *Triturus* DNA indicate that the maximum level of transcription is from no more than 5 % of the genomic DNA (Sommerville & Malcolm, 1976). Because this is the amount of DNA estimated to be present in the loops at any one time, and because the RNA was extracted from a complete range of oocyte sizes, there is no evidence from this type of experiment to suggest that many additional DNA sequences are looped-out and transcribed at different oogenic stages. Even 5 % of the genome of *Triturus* is equivalent to a genetic complexity of 10^9 nucleotides, a value which is about 50 times the sequence complexity of oocyte polyadenylated mRNA. It must be concluded that much of the transcribed RNA is not equivalent to stable mRNA sequences.

Loop number and transcriptional units

There are more than 5×10^3, and probably as many as 10^4, loops in every haploid set of *Triturus* lampbrush chromosomes. This number has been considered as a basis for the estimation of the number of different genes that are expressed during oogenesis, because loops are units in the sense that they are inherited structures (which occasionally have a distinctive morphology and so can be mapped on the appropriate chromosome) and that they are often transcribed as single units (for discussion see Sommerville, Malcolm & Callan, 1978). A major problem is that in *Triturus* the loops are longer than genes by about two orders of magnitude. At present only the chromosomal location of reiterated sequences is known, and we must await information on the transcription pattern of more typical protein-coding sequences. What is clear from loop morphology is that more than one transcriptional unit can be contained in a single loop. This is especially obvious in examining spread preparations of lampbrush chromosomes where various types of multiple unit (multi-gene?) loops, including some with reverse polarity, have been described (Scheer, Franke, Trendelenburg & Spring, 1976).

Sequence complexity of mRNA

The most direct estimates of the number of different genes transcribed during oogenesis come from sequence complexity studies on isolated polyadenylated DNA. The kinetics of hybridization of labelled complementary RNA, reverse transcribed from oocyte mRNA, with excess amounts of its template RNA, give values which can be interpreted in terms of the number of different types of mRNA present. The limitations of this kind of experiment are that only polyadenylated mRNA sequences are subject to analysis and that the sensitivity of the technique is such that mRNA sequences that are present as relatively few copies per cell are undetected. This former limitation may be less important with respect to amphibian oocytes than to other tissues and organisms, for the major characterized non-polyadenylated mRNA, histone mRNA, is polyadenylated in amphibian oocytes (Ruderman & Pardue, 1977).

The kinetic data indicate that in oocytes there are two abundance classes of mRNA molecules, one consisting of about 1000 different sequences which are each present as 3×10^7 copies per oocyte, and the other of 18000 different sequences which are each present as 10^6 copies per oocyte (see Perlman, Ford & Rosbash, 1977). *In toto* there are about 2×10^{11} mRNA molecules in the mature oocyte (Cabada *et al.*, 1977) consisting of 2×10^4 different types. This, then, is the nearest we come to the number of actively transcribed genes and this value is similar to complexity values for mRNA from other cell types (see Lewin, 1975b). Whether or not all of these sequences are essential for early development is not clear. It has already been stated that many more sequences are transcribed than are stabilized as mRNA and this may be due to non-regulated or 'leaky' transcription in oocytes. But are there also mRNA sequences accumulated which are not immediately required?

Sequences homologous to globin mRNA have been detected in RNA extracted from *Xenopus* ovaries, by hybridization with complementary DNA itself transcribed from adult globin mRNA (Perlman *et al.*, 1977). These sequences are present at approximately 200000 copies per mature oocyte, a concentration not markedly different from that of the majority of other polyadenylated sequences. Furthermore, tadpole globin mRNA, which has no sequence homology with adult globin mRNA, is detected in ovary RNA at the same concentration as the adult form, although there is much more (300-fold) transcription of tadpole globin sequences during the first three days after fertilization. Therefore, here is an example of a class of sequences normally associated with differentiated erythroid cells, which is present in oocytes and is probably transcribed from lampbrush chromosomes. As suggested by Perlman *et al.*, the globin sequences may play some undetermined role in oogenesis or early development, or else they represent a property of lampbrush chromosomes whereby all structural gene sequences expressed during the life-time of the organism are transcribed at a high rate. Preliminary experiments, involving the hybridization of *Xenopus* globin cDNA with *Triturus* lampbrush chromosomes, indicate that there may be RNA transcripts which contain globin mRNA sequences present on lampbrush loops (see Sommerville, 1977). It would be interesting to know if other sequences normally associated with terminally differentiated cells are also transcribed and accumulated in oocytes.

Structure and composition of oocyte transcription products

Details of the structure of RNP transcripts can be derived from cytological examination of native or spread lampbrush chromosomes and by analysis of isolated nuclear RNP. It is evident that immediately after transcription the RNA sequences associate with specific proteins, and that the RNP so formed extends to considerable lengths, although compacted in beaded structures, as the polymerase molecules proceed round the loop. In general the fate of transcription products is that the long sequences of RNA are processed into smaller molecules, some of which are stabilized as mRNA, and that the proteins which remain associated with the RNA influence the cleavage and nucleotide modification events and also determine the final cellular location and activity of the sequence. (Such functions for RNP protein are complicated by the fact that there may be exchange of proteins as the RNA passes through various stages of processing and from nucleus to cytoplasm.)

Primary transcript RNP in loop matrices of *Triturus* reaches lengths of 10–30 μm or even more, and in some instances is equivalent to the transcribed length of a whole loop. It is not known whether RNA molecules are contiguous throughout these fibrils. Although RNP fibrils of 2–20 μm can be isolated, RNA extracted from this material is somewhat shorter, in the range 2–4 μm, i.e. between 6000 and 12000 nucleotides (Sommerville & Malcolm, 1976). Nevertheless, these are high molecular weight RNA molecules and fall into the category of heterogenous nuclear RNA (hnRNA) described for other cell types (see Lewin, 1975*a*). Because genetic sequences are transcribed from loops and because loops invariably produce long transcripts, it seems highly likely that, in amphibian oocytes at least, there is a precursor–product relationship between high molecular weight transcripts and the generally lower molecular weight mRNA molecules; what is less clearly demonstrated is the presence of correspondingly large hnRNA molecules free from the chromatin in the nucleus. However it must be remembered that some molecular species, notably 5S RNA and transfer RNA, are probably derived from short transcripts.

Histone-coding sequences in *Triturus* are transcribed from long loops which have fairly average amounts of RNP matrix associated with them (Old *et al.*, 1977). In order to quantitate this observation use can be made of the fact that labelled *Triturus* culture cell RNA, after sedimentation on denaturing sucrose gradients, hybridizes with cloned sea urchin DNA sequences which are denatured and immobilized on nitrocellulose filters (Sommerville, 1979). The significant feature of these experiments is that most of the labelled transcripts which contain histone-coding sequences sediment at 60S–80S. Therefore it appears that there exist high molecular weight precursors to histone mRNA and that these precursors are polycistronic transcripts, a situation similar to that reported for HeLa cells (Melli, Spinelli, Wyssling & Arnold, 1977). Preliminary experiments indicate that there are similarly large histone precursor transcripts in *Triturus* oocytes (Sommerville, unpublished results).

Electron microscopic observations on transcriptional units, patterns of hybridization to transcripts *in situ*, and biochemical analyses of transcript molecules, all indicate that transcribed sequences are removed not only at the termini of loops but also from the nascent, growing RNP chains. It is interesting to note that features commonly seen in transcriptional units are, first, small circles of RNP either formed within the fibril or lying adjacent to the fibril and possibly excised from it, and second, regions of branching of the RNP fibril, particularly near the ends of the transcripts. These structures, which are shown in Fig. 1, could well be associated with processing events occurring within the nascent transcripts. As well as directly visualizing the act of transcription in spread loops, it might also be possible to witness the excision and splicing of transcribed sequences. The major problem is that, as yet, the genetic sequences contained in spread transcriptional units cannot be easily identified.

The RNA sequences contained in isolated nuclear RNP can be studied more easily. It has been shown by molecular hybridization that sequences homologous to total oocyte mRNA (Sommerville & Malcolm, 1976) and to histone-coding sequences (Sommerville,

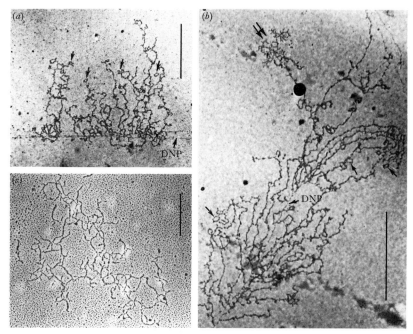

Fig. 1. Secondary structures in RNP transcripts from lampbrush chromosome loops of *Triturus cristatus* oocytes. In (*a*) and (*b*) circles of RNP can be seen (arrows) either as part of the main fibril or as structures lying adjacent to the fibril. Branching of the fibrils is also evident and a highly branched region is seen at the terminus of a long fibril (double arrows) in (*b*). The chromatin axis is indicated (DNP) and the bars represent 1 μm. Isolated oocyte nuclear RNP, when deproteinized, generates considerable secondary structure (*c*) which is reminiscent of the branched structure seen in (*b*). There may be more than one linear RNA molecule involved in this particular structure, the high degree of double-strandedness makes interpretation difficult. The bar represents 0.5 μm. All preparations were made as described by Malcolm & Sommerville (1977).

unpublished results) are present in oocyte nuclear RNA. However, mRNA sequences may account for only 3 % of the RNA transcribed from loops in *Triturus* oocytes, and although diverged redundant sequences may also be transcribed (see section on Degree of reiteration) the bulk of the RNA sequences remain unidentified.

An unusual feature of RNA extracted from oocyte nuclear RNP is that some preparations are extremely rich in uridylic acid, there being up to 60 mol% in some fractions (Sommerville, 1973). This disproportionate base composition is associated with the appearance, in summer months, of a second peak of RNP in

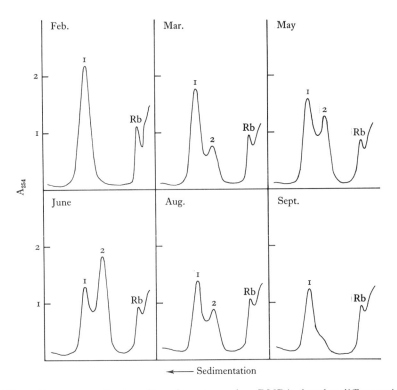

Fig. 2. Sucrose gradient profiles of oocyte nuclear RNP isolated at different times of year. In each preparation material was isolated from 12–16 ovaries of *Triturus cristatus* as described by Sommerville (1973) and banded on 30–70 % sucrose gradients. Peaks 1 and 2 as well as the contaminating 80S ribosomal peak (Rb) are indicated. From September to February most of the oocytes used were at early vitellogenic stages and from March to June there was a progressive increase in the proportion of fully grown oocytes.

sucrose gradients. Gradient profiles obtained at different times throughout the year reveal an annual pattern in RNP distribution (Fig. 2). This observation is not as strange as at first it seems, for the timing of the appearance, development and decline of peak 2 is correlated with the proportion of pre-vitellogenic and mature oocytes to vitellogenic stages found in the ovaries of the experimental animals. The size distribution of proteins associated with the two RNP peaks is similar but not identical (see Fig. 3*a* and *b*); only the base composition is distinctive (peak 1 has a more normal 31 mol % uridylic acid in its RNA). It is not known whether the U-rich sequences arise from selective and asymmetric transcription of AT-rich regions of the genome or whether they are formed post-transcriptionally by some special mechanism operating in either very small or fully grown oocytes. It is interesting to note that the early-synthesized RNA in germinating wheat and pea seeds is also very U-rich (up to 67.5 mol% : see Dobrzanska-Wiernikowska & Buchowicz, 1974). It is not known how common U-rich sequences are in eukaryotes and what function they may have in early development.

A more universally recognized property of transcribed RNA is that it immediately complexes with a fairly heterogeneous group of proteins to form linear arrays of RNP beads (for references, see Georgiev & Samarina, 1971; Beyer, Christensen, Walker & Le Stourgeon, 1977). In *Triturus* oocytes the RNP beads are consistently 20 nm in diameter and are connected, with fairly even spacing, by a thin fibril (Malcolm & Sommerville, 1974, 1977). Larger RNP structures, occasionally seen on chromosome loops, are formed by aggregation of 20 nm beads (Malcolm & Sommerville, 1974). Because of the regularity of RNP structure and its widespread reactivity with antibodies directed specifically against separated RNP proteins, in practically all transcripts in all loops (Sommerville, Crichton & Malcolm, 1978), it seems likely that beads are formed irrespective of RNA sequence and using the same set of packaging proteins. (Again 5S RNA and transfer DNA transcripts are notable exceptions, each complexing with its own specific protein.) The further observations that a major fraction of these proteins is basic (Sommerville, Crichton & Malcolm, 1978); that the protein 'cores' of beads are stabilized by protein–protein interaction (Malcolm & Sommerville, 1977); that there is considerable compaction of the RNA transcript length on assuming the beaded configuration; and that monomer particles can be produced after mild ribonuclease treatment (Malcolm & Sommerville, 1974, 1977), lead to the obvious analogy of RNP structural organization to that of nucleosomes in chromatin (e.g. Kornberg, 1974). A ribonucleosome interpretation has been favoured also for the structure of nuclear RNP derived from other sources (e.g. Pederson, 1976; Karn, Vidali, Boffa & Allfrey, 1977). The ease with which RNP beads can fuse into larger aggregates or melt into longer fibrils might explain variations in RNP structure seen in other cells.

The observations that distinctive loop morphologies are herit-

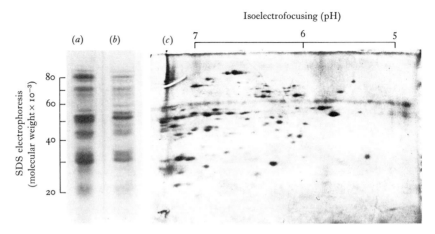

Fig. 3. Separation of nuclear RNP proteins isolated from oocytes of *Triturus cristatus*. (*a*) and (*b*) SDS–acrylamide gel electrophoresis of proteins derived from the peak 1 (*a*) and peak 2 (*b*) regions indicated in Fig. 2. For experimental details see Scott & Sommerville (1974). (*c*) Two-dimensional, isoelectrofocusing–SDS gel electrophoresis, analysis of proteins from a preparation similar to that used in (*a*). The procedure used was similar to that described by Gurdon, De Robertis & Partington (1976).

able characteristics of lampbrush chromosomes (see Sommerville, Malcolm & Callan, 1978) and that loop-specific proteins are found at some loci (Scott & Sommerville, 1974; Sommerville, Crichton & Malcolm, 1978) does not necessarily invalidate the ribonucleosomal interpretation given above. What they do mean is that additional (or alternative) proteins can bind to certain types of transcript by virtue of nucleotide sequence and that slightly different states of RNP aggregation can result from further (interbead) protein–protein interaction. In an attempt to examine in more detail the complexity in protein constitution of total nuclear RNP, two-dimensional separation analysis has been performed (D. Bell & J. Sommerville, unpublished results). What is seen is that the protein constitution of nuclear RNP is much simpler than that of oocyte nuclear sap, but more complex than is apparent from one-dimensional SDS–acrylamide electrophoresis (Fig. 3). However, much of the heterogeneity is due to charge modifications and probably loop-specific proteins are normally present at too low a concentration to be obvious over the background of common packaging proteins.

In conclusion, the favoured interpretation of protein–RNA

interaction is that the long RNA transcripts formed on lampbrush chromosomes are protected by binding basic proteins, and are made more manageable by being compacted into periodic 20 nm particles. Additional, specialized proteins for the modification, protection or cleavage of specific RNA sequences may also be bound at restricted sites. Processing events then occur on the nascent transcripts and RNA sequences required for oogenesis and early development are directed to specific regions of the cytoplasm where they are stored. The extent of polyadenylation may play a part in the latter processes (Rosbash & Ford, 1974; Cabada *et al.*, 1977) although protein association is undoubtedly important.

Some of the work reported here was supported by a grant from the Science Research Council of Great Britain.

References

BEYER, A. L., CHRISTENSEN, M. E., WALKER, B. W. & LE STOURGEON, W. W. (1977). Identification and characterization of the packaging proteins of core 40S hnRNP particles. *Cell*, **11**, 127–38.

BIRNSTIEL, M. L., CHIPCHASE, M. & SPEIRS, J. (1971). The ribosomal RNA cistrons. *Progress in Nucleic Acid Research and Molecular Biology*, **11**, 351–89.

CABADA, M. O., DARNBROUGH, C., FORD, P. J. & TURNER, P. C. (1977). Differential accumulation of two classes of poly(A) associated with messenger RNA during oogenesis in *Xenopus laevis*. *Developmental Biology*, **57**, 427–39.

DARNBROUGH, C. & FORD, P. J. (1976). Cell-free translation of messenger RNA from oocytes of *Xenopus laevis*. *Developmental Biology*, **50**, 285–301.

DOBRZANSKA-WIERNIKOWSKA, M. & BUCHOWICZ, J. (1974). Rapidly labelled very UMP-rich RNA in germinating wheat seeds. *Bulletin de l'Academie Polonaise des Sciences. Série des Science Biologique, Classe II*, **22**, 663–5.

FORD, P. J. & MATHIESON, T. (1976). Control of 5S RNA synthesis in *Xenopus laevis*. *Nature, London*, **261**, 433–5.

FORD, P. J., MATHIESON, T. & ROSBASH, M. (1977). Very long-lived messenger RNA in ovaries of *Xenopus laevis*. *Developmental Biology*, **57**, 417–26.

FORD, P. J. & SOUTHERN, E. M. (1973). Different sequences for 5S RNA in kidney cells and ovaries of *Xenopus laevis*. *Nature, New Biology*, **241**, 7–12.

FRANK, W. W. & SCHEER, U. (1978). Morphology of transcriptional units at different states of activity. *Philosophical Transactions of the Royal Society of London, Series B*, **283**, 333–42.

GEORGIEV, G. P. & SAMARINA, O. P. (1971). D-RNA containing ribonucleoprotein particles. *Advances in Cell Biology*, **2**, 47–110.

GURDON, J. B., DE ROBERTIS, E. M. & PARTINGTON, G. (1976). Injected nuclei in frog oocytes provide a living cell system for the study of transcriptional control. *Nature, London*, **260**, 116–24.

IZAWA, M., ALLFREY, V. G. & MIRSKY, A. E. (1963). Composition of the nucleus and chromosomes in the lampbrush stage of the newt oocyte. *Proceedings of the National Academy of Sciences of the USA*, **50**, 811–17.

KARN, J., VIDALI, G., BOFFA, L. C. & ALLFREY, V. G. (1977). Characterization of the non-histone nuclear proteins associated with rapidly labelled heterogeneous nuclear RNA. *Journal of Biological Chemistry*, **252**, 7307–22.

KEDES, L. H. (1976). Histone messengers and histone genes. *Cell*, **8**, 321–31.

KORNBERG, R. D. (1973). Chromatin structure: a repeating unit of histones and DNA. *Science*, **184**, 868–71.

LEWIN, B. (1975a). Units of transcription and translation: the relationship between heterogeneous nuclear RNA and messenger RNA. *Cell*, **4**, 11–20.

LEWIN, B. (1975b). Units of transcription and translation: sequence components of heterogeneous nuclear RNA and messenger RNA. *Cell*, **4**, 77–93.

MACGREGOR, H. C. (1978). Some trends in the evolution of very large chromosomes. *Philosophical Transactions of the Royal Society of London, Series B*, **283**, 309–18.

MACGREGOR, H. C. & ANDREWS, C. (1977). The arrangement and transcription of 'middle repetitive' DNA sequences on lampbrush chromosomes of *Triturus*. *Chromosoma, Berlin*, **63**, 109–26.

MALCOLM, D. B. & SOMMERVILLE, J. (1974). The structure of chromosome-derived ribonucleoprotein in oocytes of *Triturus cristatus carnifex*. *Chromosoma, Berlin*, **48**, 137–58.

MALCOLM, D. B. & SOMMERVILLE, J. (1977). The structure of nuclear ribonucleoprotein of amphibian oocytes. *Journal of Cell Science*, **24**, 143–65.

MELLI, M., SPINELLI, G., WYSSLING, H. & ARNOLD, E. (1977). Presence of histone mRNA sequences in high molecular weight RNA of HeLa cells. *Cell*, **11**, 651–61.

OLD, R. W., CALLAN, H. G. & GROSS, K. W. (1977). Localization of histone gene transcripts in newt lampbrush chromosomes by *in situ* hybridization. *Journal of Cell Science*, **27**, 57–97.

PEDERSON, T. (1976). A unifying concept for the molecular organization of chromatin and ribonucleoprotein. *Journal of Cell Biology*, **70**, 308a.

PERKOWSKA, E., MACGREGOR, H. C. & BIRNSTIEL, M. L. (1968). Gene amplification in the oocyte nucleus of mutant and wild-type *Xenopus laevis*. *Nature, London*, **217**, 649–50.

PERLMAN, S. M., FORD, P. J. & ROSBASH, M. (1977). Presence of tadpole and adult globin RNA sequences in oocytes of *Xenopus laevis*. *Proceedings of the National Academy of Sciences of the USA*, **74**, 3835–9.

PUKKILA, P. J. (1975). Identification of the lampbrush chromosome loops which transcribe 5S ribosomal RNA in *Notophthalmus (Triturus) viridescens*. *Chromosoma, Berlin*, **53**, 71–89.

ROSBASH, M. & FORD, P. J. (1974). Polyadenylic acid containing RNA in *Xenopus laevis* oocytes. *Journal of Molecular Biology*, **85**, 87–101.

ROSBASH, M., FORD, P. J. & BISHOP, J. O. (1974). Analysis of the C-value paradox by molecular hybridization. *Proceedings of the National Academy of Sciences of the USA*, **71**, 3746–50.

RUDERMAN, J. V. & PARDUE, M. L. (1977). Cell-free translation analysis of messenger RNA in echinoderm and amphibian early development. *Developmental Biology*, **60**, 48–68.

SCHEER, U. (1978). Changes of nucleosome frequency in nucleolar and non-nucleolar chromatin as a function of transcription: an electron microscopic study. *Cell*, **13**, 535–49.

SCHEER, U., FRANKE, W. W., TRENDELENBURG, M. F. & SPRING, H. (1976). Classification of loops of lampbrush chromosomes according to the arrangement of transcriptional complexes. *Journal of Cell Science*, **22**, 503–20.

SCOTT, S. E. M. & SOMMERVILLE, J. (1974). Location of nuclear proteins on the chromosomes of new oocytes. *Nature, London*, **250**, 680–2.

SMITH, G. P. (1973). Unequal crossing over and the evolution of multigene families. *Cold Spring Harbour Symposia on Quantitative Biology*, **38**, 507–14.

SOMMERVILLE, J. (1973). Ribonucleoprotein particles derived from the lampbrush chromosomes of newt oocytes. *Journal of Molecular Biology*, **78**, 487–503.

SOMMERVILLE, J. (1977). Gene activity in the lampbrush chromosomes of amphibian oocytes. In *International Review of Biochemistry, Biochemistry of Cell Differentiation* II, 15, ed. J. Paul, pp. 79–156. Baltimore: University Park Press.

SOMMERVILLE, J. (1979). Gene expression in lampbrush chromosomes. *Federation of European Biochemical Society Proceedings*, in press.

SOMMERVILLE, J., CRICHTON, C. & MALCOLM, D. B. (1978). Immunofluorescent localization of transcriptional activity on lampbrush chromosomes. *Chromosoma, Berlin*, **66**, 99–114.

SOMMERVILLE, J. & MALCOLM, D. B. (1976). Transcription of genetic information in amphibian oocytes. *Chromosoma, Berlin*, **55**, 183–208.

SOMMERVILLE, J., MALCOLM, D. B. & CALLAN, H. G. (1978). The organization of transcription on lampbrush chromosomes. *Philosophical Transactions of the Royal Society of London, Series B*, **283**, 359–66.

WOODLAND, H. R. & ADAMSON, E. D. (1977). The synthesis and storage of histones during the oogenesis of *Xenopus laevis*. *Developmental Biology*, **57**, 118–35.

Accumulation of materials involved in rapid chromosomal replication in early amphibian development

R. A. LASKEY, J. B. GURDON AND
M. TRENDELENBURG

MRC Laboratory of Molecular Biology, Hills Road,
Cambridge CB2 2QH, UK

After fertilization amphibian eggs cleave extremely rapidly to form 5000 cells in only 5 hours. Between 3 and 5 hours post-fertilization entire cell cycles average less than 15 minutes each (Graham & Morgan, 1966). Therefore at this stage the entire DNA content of a *Xenopus* nucleus can be replicated faster than the *E. coli* chromosome replicates in log phase. This paper considers the specializations which permit such accelerated chromosomal replication, and in particular the maternal contribution to this phenomenon.

The most obvious adaptation which permits rapid cell proliferation during cleavage is the uncoupling of growth from cell division throughout oogenesis. Thus a mature oocyte is at least 100000 times the volume of a somatic larval cell and it contains, for example, 100000 times more mitochondria (Chase & Dawid, 1972) or 200000 times more ribosomes (Brown & Littna, 1964; Gurdon & Brown, 1965). In this way oogenesis makes two essential contributions to the speed of cell proliferation in early embryogenesis. First, it provides such a large stockpile of cellular components that subsequent cell proliferation is simplified to a process of co-distributing pre-existing cell components with newly synthesized chromosomes and membranes. Second, it provides an enormous excess of the materials which are involved in the process of chromosomal replication itself. Obviously these two types of maternal contribution are intimately related; for example, maternal ribosomes are used to synthesize the histones which are assembled on to the embryo's newly replicated DNA. Therefore both aspects of the maternal contribution will be considered here.

Two experimental approaches which have been valuable in analysing this phenomenon will be emphasized. First, micro-injection of nucleic acid templates into intact eggs or oocytes allows particular enzyme activities and regulatory processes to be recognized and titrated in the intact cell. Second, the fractionation of in-vitro assay systems derived from egg homogenates allows the identification and quantitation of the molecular components involved.

Throughout this paper the term 'oocyte' is used to refer to full-grown female germ cells dissected from the ovary of adult frogs, and the term 'egg' refers to the mature gamete shed from female frogs after hormonally induced ovulation.

The capacity for DNA synthesis in amphibian eggs and oocytes

Fig. 1 illustrates the remarkable rate of cell proliferation which follows fertilization in *Xenopus laevis*. Graham & Morgan (1966) examined the individual phases of the cell cycle during this period of development and found that, not only was the DNA synthetic phase (S phase) of the cell cycle approximately 100 times shorter in mid-cleavage embryos than in larval cells, but that the non-synthetic phases, G_1 and G_2, were too short to be measured, and were probably completely absent. In addition, mitosis was shortened to only 2 minutes. Thus, unlike adult nuclei, the nuclei of mid-cleavage embryos appear to engage in uninterrupted alternating cycles of DNA replication and mitosis.

Callan (1973) has examined the mechanisms which shorten the S phase so much, to ask if the acceleration of DNA synthesis is achieved by increasing the rate of nucleotide chain elongation or by increasing the frequency of initiation. He examined this question by autoradiography of spread DNA preparations after pulse labelling cells with [³H]thymidine to determine the length of DNA synthesized in a given time and the distances between adjacent replication forks. The results showed that the shortening of S phase during embryogenesis is not achieved by accelerating the rate of chain elongation, but that it is accounted for solely by increasing the frequency of initiation and therefore by decreasing the length of DNA which must be replicated from an individual replication fork.

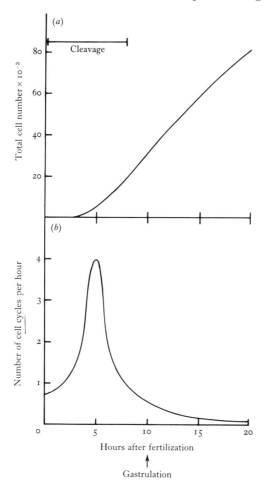

Fig. 1. Increase in (a) cell number and (b) rate of cell division following fertilization of *Xenopus laevis*. These schematic representations are compiled from unpublished data and from Graham & Morgan (1966) and Benbow, Pestell & Ford (1975).

An increase in the frequency of initiation of replication implies an abundant supply of DNA polymerases and precursor nucleotides. However, although these components are abundant in eggs and early embryos, the pattern of chromosomal replication cannot be explained simply by variations in polymerases or precursors. Thus the total DNA polymerase content of the embryo remains the same, at about 100 000 times the level of a larval somatic cell, throughout the period from ovulation to gastrulation, that is throughout the period of changes illustrated in Fig. 1 (Tato,

68 R. A. Laskey, J. B. Gurdon and M. Trendelenburg

Attardi Gandini & Tocchini-Valentini, 1974; Benbow, Pestell & Ford, 1975). Furthermore much of the polymerase activity is accumulated during oogenesis, so the oocyte's inability to replicate native DNA cannot be due to a lack of DNA polymerases. In contrast, the deoxynucleoside triphosphate pools of oocytes are very much smaller than those of unfertilized eggs, which are sufficient to synthesize 2500 diploid nuclei (Woodland & Pestell, 1972). Nevertheless, two types of microinjection experiment indicate that precursor concentrations do not have a regulatory role. First, when denatured (single-stranded) DNA is injected into oocytes it stimulates the incorporation of thymidine into DNA (Gurdon & Speight, 1969; Ford & Woodland, 1975). Second, artificial expansion of the oocyte's precursor pools, by direct precursor injection to achieve concentrations similar to those in eggs, does not stimulate the endogenous nucleus or injected native (i.e. double-stranded) DNA to replicate (Woodland, Ford & Gurdon, 1972). The general conclusion from this type of experiment is that the ability to replicate native DNA which appears when an oocyte matures into an egg is not related to the concentrations of precursors or DNA polymerases.

The same experimental approach allows us to titrate the egg's capacity for DNA synthesis. Gurdon, Birnstiel & Speight (1969) demonstrated that when native DNA is injected into an enucleated, unfertilized egg it serves as a template for thymidine incorporation. Subsequently Laskey & Gurdon (1973) and Ford & Woodland (1975) demonstrated that all of the incorporation proceeds by semiconservative synthesis of entire DNA strands. An additional reason for thinking that the DNA synthesis induced is biologically significant (i.e. regulated replication) is that the signal inducing synthesis of injected native DNA appears simultaneously with that for the endogenous nucleus (Gurdon, 1967; Ford & Woodland, 1975); moreover, as shown in Fig. 2, the amount of incorporation stimulated by injected DNA increases with the same kinetics as endogenous nuclear replication (compare Fig. 1 with Fig. 2). Ford & Woodland (1975) measured the amount of DNA synthesis stimulated by injected DNA and concluded that the enucleated egg could synthesize 120 pg DNA (i.e. the equivalent of 50 diploid nuclei) within the first 2 hours after injection. They pointed out that this would not be adequate for the observed rate of nuclear replication in the mid-cleavage stage. However, Fig. 2 illustrates that the capacity for DNA synthesis by enucleated unfertilized

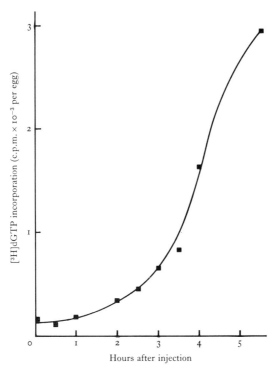

Fig. 2. Increase in rate of synthesis of injected DNA after injection of SV40 circular viral DNA and [³H]dGTP into unfertilized eggs of *Xenopus laevis*. Each egg received 2 ng SV40 DNA component I and 250 nCi of [³H]dGTP. Incubations of five eggs/sample were stopped by homogenizing in 1 ml 1 % SDS, 30 mM EDTA, 10 mM Tris pH 7.5, followed by extraction twice with a 24:1 vol.:vol. mixture of chloroform and isoamyl alcohol, and precipitation of the aqueous phase with 10 % trichloroacetic acid at 0 °C.

eggs rises sharply after 2 hours and keeps pace with the rate of DNA synthesis by intact embryos.

The conclusion that egg cytoplasm contains all of the materials required for exceptionally rapid replication of DNA templates is reinforced by two observations. First, a similar level of DNA synthesis is observed when the synthesis of new proteins is inhibited by puromycin. Second, when a non-replicating nucleus is injected into an enucleated egg it is induced to replicate its DNA (Graham, Arms & Gurdon, 1966). Furthermore since the injected quiescent nucleus can support development of the recipient egg to form a tadpole, it is clear that egg cytoplasm is able to induce it to replicate at the rapid rate characteristic of early embryos (Gurdon, Laskey & Reeves, 1975).

The capacity for chromatin assembly

If the speed of nuclear DNA replication is approximately 100 times faster in early embryos than in adults, then it must pose a serious problem of provision and assembly of other chromosomal components such as histones. Clearly the newly synthesized DNA must be complexed to histones rapidly, because normal mitotic chromosomes can be seen between each phase of DNA synthesis.

The rate of histone synthesis during early development of *Xenopus laevis* has been examined by Adamson & Woodland (1974). They found that each early embryo synthesizes histones at the remarkable rate of 2–5 ng/hour. Nevertheless the rate of DNA synthesis at this time is 15 ng/hour, so the high rate of histone synthesis is still inadequate to provide the 1 : 1 mass ratio of histones to DNA seen in chromatin. Therefore, although histone synthesis is coupled to DNA synthesis in somatic cells, Adamson & Woodland considered the possibility that a pool of histones is synthesized and accumulated in the oocyte before DNA synthesis starts. They demonstrated that each oocyte synthesizes histones at 40 pg/hour, i.e. sufficient to accumulate a pool of about 84 ng histones during oogenesis. They subsequently demonstrated the existence of this pool by direct extraction (Woodland & Adamson, 1977). Furthermore, they found that the rate of histone synthesis increased 20-fold during ovulation and that synthesis during the period of ovulation itself could provide a further 15 ng histones. Thus, by the time of fertilization each egg contains a pool of 100 ng histones, sufficient for 16000 diploid nuclei.

Although a stored histone pool can obviously simplify the problem of rapid chromosomal replication it does not provide a complete answer. Histones are not randomly scattered along DNA but organized into discrete repeating nucleosome subunits each of which contains an octamer of the core histones (i.e. two each of H2A, H2B, H3 and H4) complexed in a specific conformation to approximately 200 base pairs of DNA (see Felsenfeld, 1978, for review).

Between the third and fifth hours of development approximately 10^{11} nucleosomes must be precisely assembled from histones and DNA. The process of nucleosome assembly can be reproduced and analysed using egg homogenates *in vitro*. Thus a clarified supernatant from unfertilized eggs of *X. laevis* can convert purified DNA into a regularly spaced series of nucleosomes (Laskey, Mills & Morris, 1977). This observation directly confirms Adamson & Woodland's (1974) conclusion that eggs contain a stored histone pool, since not only are nucleosomes assembled without addition of extra histones, but assembly does not require new protein synthesis during the incubation (Laskey, Mills & Morris, 1977). Furthermore it demonstrates that most of the stored histone pool is available for immediate assembly with DNA, since the homogenate derived from each egg can assemble 40 ng DNA, sufficient for 6000 nuclei, into nucleosomes in only 1 hour (Laskey, Mills & Morris, 1977). When homogenates are diluted before centrifugation, the recovery of activity is increased twofold above this value. Therefore it agrees remarkably well with Adamson & Woodland's (1974) estimate of the size of the stored histone pool.

Nucleosome assembly is not achieved simply by mixing histones and DNA. If they are mixed at physiological ionic strength a precipitate forms. Egg homogenates contain a component which inhibits precipitation and allows an ordered interaction to occur. This component has been identified and purified (Laskey, Honda, Mills & Finch, 1978). It is an acidic protein which binds histones to form active complexes that then interact with DNA to form nucleosomes. The assembly protein is present in such large amounts that all of the stored histone pool could possibly be complexed to it in the egg.

No study has been made yet of the time of synthesis of the assembly protein during oogenesis. It would be interesting to know if it is synthesized co-ordinately with histones for example. However, we do know that a similar activity is present in the oocyte before ovulation. When purified DNA is injected into the oocyte nucleus *in situ* it is assembled into nucleosomes (Wyllie, Laskey, Finch & Gurdon, 1978). The oocyte's capacity for nucleosome assembly has not been titrated fully, but it appears to be at least 10 % of egg's capacity, though the rate of assembly of nucleosomes on injected DNA is probably slower in the oocyte than in the egg.

The capacity for protein synthesis

In the previous sections we have seen that early development makes extraordinary demands on certain proteins such as DNA polymerases and histones. Here we consider the capacity of the egg and oocyte for synthesizing these and other proteins. Unlike sea urchins, where total protein synthesis increases 15–30-fold suddenly at fertilization, Xenopus eggs show no sudden increase in protein synthesis during development. Woodland (1974) measured the proportion of ribosomes engaged in protein synthesis as judged by their presence in polyribosomes. He found that the proportion of active ribosomes increased gradually and progressively from full-grown oocytes to early larvae. His most striking finding was the observation that only 2 % of the ribosomes in the oocyte were engaged in protein synthesis.

The reason why 98 % of the oocyte's ribosomes are inactive has been investigated by microinjecting mRNA into intact oocytes to measure the translational capacity of the oocyte. The results obtained from this design of experiment can be misleading, depending on the choice of radioactive amino acid used to label the proteins synthesized. When this problem is overcome, however, the following results are obtained (Laskey, Mills, Gurdon & Partington, 1977). Increasing the available amount of mRNA does not increase the total amount of protein synthesis. The injected mRNA is translated in strict competition with the endogenous mRNA. The component which limits the oocyte's capacity for protein synthesis is present in crude polyribosome preparations. It has not been identified yet, though it is clear that both ribosomes and tRNA are present in the oocyte in excess. It remains to be seen if most of these are inactive or if the supply of another component such as initiation factors or tRNA charging enzymes is regulatory.

There is a conspicuous analogy between the over-provision of protein-synthesizing machinery in the oocyte and the over-provision of DNA-synthesizing machinery. In both cases a large excess of the major components is accumulated in the oocyte, but the overall system is switched off. Increasing the availability of the nucleic acid template has no effect on the level of activity. Additional regulatory components are required to make use of the maternal stockpile which is accumulated during oogenesis. The identities of the components which activate these two dormant systems remain to be determined.

The capacity for gene transcription

Roeder (1974) demonstrated that *Xenopus* oocytes contain exceptionally high levels of RNA polymerase activities, between 60000 and 100000 times as much as each larval somatic cell. Nevertheless a period of relative transcriptional quiescence follows oogenesis. Thus very little RNA synthesis occurs during the period of maximum DNA synthesis in mid-cleavage. Roeder (1974) demonstrated that the same high levels of each of the three main classes of RNA polymerase persist from the transcriptionally active oocyte, through the transcriptionally quiescent early embryo to the transcriptionally active late embryo. Thus the level of transcriptional activity does not reflect the levels of any of the three classes of RNA polymerase.

It might be expected that the persistent high levels of RNA polymerase activities simply reflect conservation of polymerase molecules which are fully active in the oocyte. While this argument could possibly be defended for RNA polymerase I, which synthesizes the precursor to 18S and 28S ribosomal RNA, it cannot explain the level of activity of RNA polymerase III, which synthesizes 4S and 5S RNA. When RNA polymerase III is titrated, by microinjecting excess template DNA into the oocyte nucleus, it is found to be present in a 200-fold excess over the amount which is used in the oocyte (Brown & Gurdon, 1977). Therefore there is an enormous spare capacity for RNA polymerase III, and activity is limited by the availability of the DNA template. It has not been possible to quantitate spare capacity with RNA polymerases I or II yet. Both polymerases I and II are engaged in active transcription of endogenous templates. Fig. 3(*a*) and (*b*) illustrates active transcription units characteristic of RNA polymerase class II (Fig. 3*b*) and class I (Fig. 3*a*). Nevertheless, as illustrated in Fig. 3(*c*), when additional DNA templates for these polymerases are supplied by microinjection into the oocyte nucleus they are transcribed (Mertz & Gurdon, 1977; Trendelenburg & Gurdon, 1978; Trendelenburg, Zentgraf, Franke & Gurdon, 1978; Gurdon, Melton & De Robertis, 1978).

The most striking results from experiments in which exogenous

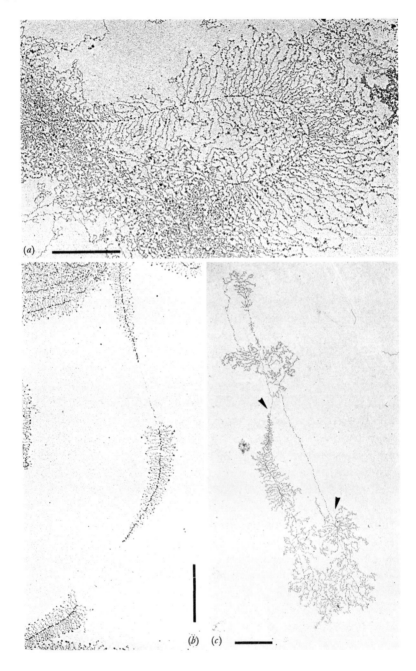

DNA templates are injected into oocyte nuclei concern the fidelity with which particular DNA sequences are selected for transcription. When a thousandfold excess of DNA coding for 5S RNA is injected into nuclei it is either transcribed selectively into a product of the correct size and sequence or it is scarcely transcribed at all (Brown & Gurdon, 1977). Perhaps surprisingly, the excess DNA injected does not serve as a template for random or aberrant transcription, even though it has completely saturated the oocyte's capacity for transcription by RNA polymerase III. The results are even more remarkable when the transcription of ribosomal DNA is studied. Trendelenburg & Gurdon (1978) used the electron microscope to examine transcription complexes formed after injection of ribosomal RNA genes which had been cloned in the bacterial plasmid PMB-9 by R. H. Reeder (Fig. 4*a*). The injected DNA can be recognized because it consists of circles of defined size. Most of the injected DNA which is recovered from an oocyte nucleus is simply found in the form of nucleosomes, but a small proportion of the recovered DNA molecules have the strikingly different appearance of fully active ribosomal transcription units (Fig. 4*b* and *c*). Such molecules have a characteristic length of DNA densely packed with polymerase molecules from which nascent RNA transcripts emanate. The length of nascent transcripts increases progressively with a strict polarity, indicating that all of the polymerases traversing this stretch of DNA initiated transcription at the same site. These structures can be unambiguously identified as ribosomal transcription units by several additional morphological criteria (Trendelenburg & Gurdon, 1978) and although they closely resemble endogenous ribosomal

---

Fig. 3. Electron microscopic visualization of transcription complexes from microinjected *Xenopus laevis* oocytes. (*a*) Lampbrush loop from endogenous oocyte chromatin. (*b*) Endogenous ribosomal RNA transcription complexes showing characteristic alternation of transcribed gene regions with untranscribed spacer regions. (*c*) Injected circular ribosomal DNA from the water-beetle *Dytiscus marginalis* showing the formation of regulated transcription complexes on the injected DNA in the oocyte nucleus. The contents of the oocyte nucleus were spread for microscopy 1 day (*b* and *c*) or 2 days (*a*) after injection of *Dytiscus* ribosomal DNA circles. All procedures and criteria for recognizing *Dytiscus* transcription complexes were those described previously (Trendelenburg, Zentgraf, Franke & Gurdon, 1978). Note the regularly transcribed region in (*c*) denoted by arrow heads. Scale bars indicate 1 μm.

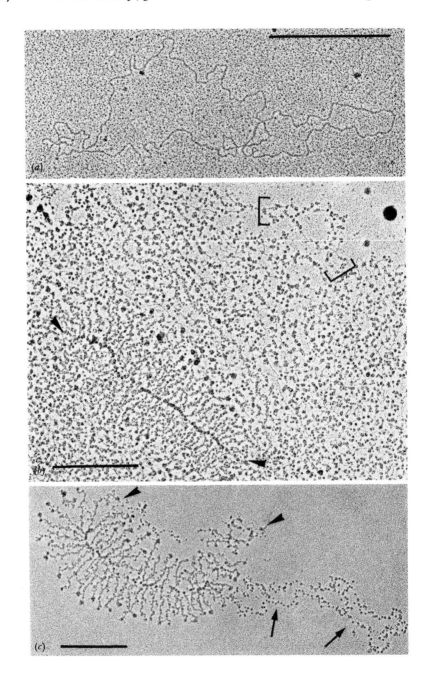

transcription complexes they can be distinguished from them by their characteristic small circular morphology.

The observation that more than 90 % of the injected DNA is assembled into transcriptionally inactive chromatin while a small proportion of injected molecules show maximum possible density of nascent transcripts initiating from a single site permits the following interesting conclusions. First, ribosomal gene transcription cannot be regulated by simple availability of RNA polymerase molecules since those genes which are transcribed are fully packed with RNA polymerase molecules. Second, even in the presence of a vast excess of ribosomal DNA the initiation of transcription is highly selective for a particular site in the DNA. Clearly the cellular components which confer transcriptional specificity and fidelity can function normally even in the presence of 100–1000 times more than the normal amount of DNA. So far this conclusion applies only to RNA polymerases I and III; a similar study for polymerase II which is responsible for transcribing mRNA and HnRNA might be particularly interesting. Finally the experiments described here suggest that the transcription of ribosomal RNA genes is perhaps regulated by a component which recruits a particular gene for full expression and it is this rather than RNA polymerase I which restricts the transcriptional capacity of the oocyte.

Fig. 4. A single repeat of *Xenopus laevis* ribosomal genes cloned in the bacterial plasmid PMB–9 (by Dr R. H. Reeder) observed before and after injection into *Xenopus* oocyte nuclei. (*a*) Free DNA circle before injection. (*b*) A dense cluster of chromatin formed from the plasmid containing ribosomal DNA after injection into an oocyte nucleus. Note that nearly all molecules are transcriptionally inactive except one fully active transcription complex (between arrow heads). Individual circles of untranscribed chromatin can be seen at the edges of the aggregate (brackets). (*c*) Chromatin formed from the injected plasmid containing ribosomal DNA after injection into an oocyte nucleus. Two molecules have been assembled into transcriptionally inactive chromatin (long arrows) and a third shows a characteristic fully active ribosomal transcription unit (between arrow heads). All procedures and criteria for distinguishing endogenous transcription complexes from those formed on injected DNA were those described previously (Trendelenburg & Gurdon, 1978). Scale bars indicate 0.5 μm.

Conclusion

The exceptionally rapid rate of cell division which occurs in early amphibian development is achieved by uncoupling cell growth from chromosomal replication. During the mid-cleavage stage DNA is replicated rapidly by a maternal stockpile of DNA polymerases initiating at close intervals along the DNA.

Newly replicated DNA is rapidly assembled into nucleosomes by an acidic protein which binds histones and transfers them to DNA. Histones are synthesized in large amounts during early embryogenesis, but to keep pace with the rate of DNA synthesis a stored pool of histones is provided by histone synthesis during oogenesis.

The cellular machinery for protein synthesis is accumulated during oogenesis but it remains relatively inactive in the oocyte and it is activated gradually as embryonic development proceeds.

The machinery for regulated gene transcription is also accumulated in excess in the oocyte. High levels of RNA polymerases are maintained throughout extreme changes in transcriptional activity. Microinjection of exogenous DNA templates indicates that gene transcription is regulated by components other than RNA polymerases.

References

ADAMSON, E. D. & WOODLAND, H. R. (1974). Histone synthesis in early amphibian development: histone and DNA syntheses are not coordinated. *Journal of Molecular Biology*, **86**, 263–85.

BENBOW, R. M., PESTELL, R. Q. W. & FORD, C. C. (1975). Appearance of DNA polymerase activities during early development of *Xenopus laevis*. *Developmental Biology*, **43**, 159–74.

BROWN, D. D. & GURDON, J. B.(1977). High-fidelity transcription of 5S DNA injected into *Xenopus* oocytes. *Proceedings of the National Academy of Sciences of the USA*, **74**, 2064–8.

BROWN, D. D. & LITTNA, E. (1964). RNA synthesis during the development of *Xenopus laevis*, the South African clawed toad. *Journal of Molecular Biology*, **8**, 669–87.

CALLAN, H. G. (1973). DNA replication in the chromosomes of eukaryotes. *Cold Spring Harbor Symposia on Quantitative Biology*, **38**, 195–203.

CHASE, J. W. & DAWID, I. (1972). Biogenesis of mitochondria during *Xenopus laevis* development. *Developmental Biology*, **27**, 504–18.

FELSENFELD, G. (1978). Chromatin. *Nature, London*, **271**, 115–22.

FORD, C. C. & WOODLAND, H. R. (1975). DNA synthesis in oocytes and eggs of *Xenopus laevis* injected with DNA. *Developmental Biology*, **43**, 189–99.

GRAHAM, C. F., ARMS, K. & GURDON, J. B. (1966). The induction of DNA synthesis by frog egg cytoplasm. *Developmental Biology*, **14**, 349–81.

GRAHAM, C. F. & MORGAN, R. W. (1966). Changes in the cell cycle during early amphibian development. *Developmental Biology*, **14**, 439–60.

GURDON, J. B. (1967). On the origin and persistence of a cytoplasmic state inducing nuclear DNA synthesis in frogs eggs. *Proceedings of the National Academy of Sciences of the USA*, **58**, 545–52.

GURDON, J. B., BIRNSTIEL, M. L. & SPEIGHT, V. R. (1969). The replication of purified DNA introduced into living egg cytoplasm. *Biochimica et Biophysica Acta*, **174**, 614–28.

GURDON, J. B. & BROWN, D. D. (1965). Cytoplasmic regulation of RNA synthesis and nucleolus formation in developing embryos of *Xenopus laevis*. *Journal of Molecular Biology*, **12**, 27–35.

GURDON, J. B., LASKEY, R. A. & REEVES, O. R. (1975). The developmental capacity of nuclei transplanted from keratinized skin cells of adult frogs. *Journal of Embryology and Experimental Morphology*, **34**, 91–112.

GURDON, J. B., MELTON, D. A. & DE ROBERTIS, E. M. (1978). Genetics in an oocyte. In *CIBA Symposium on Genetics and Human Biology : Possibilities and Realities*. Amsterdam: Associated Scientific Publishers (in press).

GURDON, J. B. & SPEIGHT, V. A. (1969). The appearance of cytoplasmic DNA polymerase activity during the maturation of amphibian oocytes into eggs. *Experimental Cell Research*, **55**, 253–6.

LASKEY, R. A. & GURDON, J. B. (1973). Induction of polyoma DNA synthesis by injection into frog egg cytoplasm. *European Journal of Biochemistry*, **37**, 467–71.

LASKEY, R. A., HONDA, B. M., MILLS, A. D. & FINCH, J. T. (1978). Nucleosomes are assembled by an acidic protein which binds histones and transfers them to DNA. *Nature, London*, **275**, 416–20.

LASKEY, R. A., MILLS, A. D., GURDON, J. B. & PARTINGTON, G. A. (1977). Protein synthesis in oocytes of *Xenopus laevis* is not regulated by the supply of messenger RNA. *Cell*, **11**, 345–51.

LASKEY, R. A., MILLS, A. D. & MORRIS, N. R. (1977). Assembly of SV40 chromatin in a cell-free system from *Xenopus* eggs. *Cell*, **10**, 237–43.

MERTZ, J. E. & GURDON, J. B. (1977). Purified DNAs are transcribed after microinjection into *Xenopus* oocytes. *Proceedings of the National Academy of Sciences of the USA*, **74**, 1502–6.

ROEDER, R. G. (1974). Multiple forms of deoxyribonucleic acid-dependent ribonucleic acid polymerase in *Xenopus laevis*. *Journal of Biological Chemistry*, **249**, 249–56.

TATO, F., ATTARDI GANDINI, D. & TOCCHINI-VALENTINI, G. P. (1974). Major DNA polymerases common to different *Xenopus laevis* cell types. *Proceedings of the National Academy of Sciences of the USA*, **71**, 3706–10.

TRENDELENBURG, M. F. & GURDON, J. B. (1978). Transcription of cloned *Xenopus* ribosomal genes visualized after injection into oocyte nuclei. *Nature, London*, **276**, 292–4.

TRENDELENBURG, M., ZENTGRAF, H., FRANKE, W. W. & GURDON, J. B. (1978). Transcription patterns of amplified *Dytiscus* ribosomal DNA after injection into *Xenopus* oocyte nuclei. *Proceedings of the National Academy of Sciences of the USA*, **75**, 3791–5.

WOODLAND, H. R. (1974). Changes in the polysome content of developing *Xenopus laevis* embryos. *Developmental Biology*, **40**, 90–101.

WOODLAND, H. R. & ADAMSON, E. D. (1977). The synthesis and storage of histones during the oogenesis of *Xenopus laevis*. *Developmental biology*, **57**, 118–35.

WOODLAND, H. R., FORD, C. C. & GURDON, J. B. (1972). Studies on genetic regulation utilizing microinjection of nuclei and DNA into living eggs and oocytes. *Advances in Biosciences*, **8**, 207–16.

WOODLAND, H. R. & PESTELL, R. Q. W. (1972). Determination of the nucleoside triphosphate contents of eggs and oocytes of *Xenopus laevis*. *Biochemical Journal*, **127**, 597–605.

WYLLIE, A. H., LASKEY, R. A., FINCH, J. & GURDON, J. B. (1978). Selective DNA conservation and chromatin assembly after injection of SV40 into *Xenopus* oocytes. *Developmental Biology*, **64**, 178–88.

Ribosomal RNA synthesis and ribosome production

PETER J. FORD

Dept of Molecular Biology, University of Edinburgh, Kings Buildings, Edinburgh EH9 3JR, UK

This review is concerned with the synthesis of ribosomes during oogenesis and development with special reference to maternal effects. Ribosomes are found in all living cells and are required for protein synthesis. In general, ribosome biosynthesis is linked to the demand for protein synthesis – cells with high growth and cell division rates or with high rates of protein synthesis produce more ribosomes than cells with low growth and cell division rates or with low rates of protein synthesis. There are exceptions to this rule, the most notable being the high rates of ribosome production and high ribosome content coupled with low protein synthesis rates that are found in the eggs and oocytes of animals and the ungerminated seeds and spores of plants and fungi. Systems relevant to the topic of this volume are thus unusual in escaping from the normally observed coupling between ribosome and protein biosynthesis.

The extent to which ribosome synthesis is controlled and, if it is, exactly how it is controlled, are questions of general biological interest. Attempts to answer these questions involve studies of the control and coordination of ribosomal ribonucleic acid (rRNA) and ribosomal protein (r protein) synthesis, as well as of the assembly of complex ribosomal ribonucleoprotein (rRNP) particles, and require knowledge of the specific interactions within and between different rRNA molecules, between rRNA and r protein and between different r proteins themselves. To illustrate the complexity of ribosome biosynthesis the reader has only to consider ribosome structure (see Brimacombe, Stofler & Whitman, 1978, for review). Ribosomes are composed of two subunits, one large and one small, which are present in cells in equal numbers. Each

81

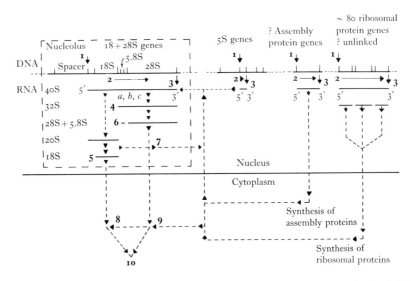

Fig. 1. Flow chart of main events and processes in ribosome biosynthesis. DNA: the genes involved are represented by those for *Xenopus laevis*. Ribosomal protein and assembly protein gene sequences have not yet been isolated from a eukaryote, and consequently it is not known how they are organized in the genome. RNA: the major events in RNA synthesis, processing, chemical modification and ribosome assembly are indicated by numbers. **1**, Initiation of transcription. **2**, Elongation of rRNA primary transcript, coupled with (*a*) specific methylations and base modifications, (*b*) sequential addition of ribosomal and assembly proteins, (*c*) addition of 5S rRNA. **3**, Termination of transcription and release of 80S ribonucleoprotein containing 40S RNA, 5S RNA and proteins. **4**, First processing step(s) cleaves 40S RNA to give 20S and 32S components. This is a slow process since a pool of 40S precursor is detectable in many cells. **5**, The 20S component is cleaved to give 18S RNA which is rapidly transported to cytoplasm. These are rapid processes in most cells since only very small amounts of 20S and 18S rRNA are detectable in nuclei. **6**, The 32S precursor is cleaved to give 5.8S + 28S rRNA which remain hydrogen-bonded to each other. This is a slow process since large amounts of 32S precursor are found in nuclei. The 28S/5.8S RNA is rapidly transported to the cytoplasm as only small amounts of 28S RNA are usually found in nuclei. **7**, Before leaving the nucleus any assembly proteins still remaining attached to the ribonucleo-proteins are removed and possibly recycled. **8**, After entry to cytoplasm late methylations occur on 18S rRNA and final protein components may be assembled. **9**, After entry to cytoplasm final protein components may be added to the large 60S ribosome subunit. **10**, Finally the newly synthesized ribosome subunits, together with specific protein factors, transfer RNA and messenger RNA form ribosomes active in protein synthesis.

subunit has at least one RNA molecule, several thousand nucleotides in length, folded in a defined manner and complexed with several dozen distinctly separate proteins in a reproducible way, such that each protein has a defined position relative to the other proteins and to the RNA. This is probably too static a picture of ribosome structure since the ribosome plays a dynamic role in protein synthesis, which must call for a variety of conformational states at different stages in the ribosome cycle, a cycle which includes events associated with binding to messenger ribonucleic acid (mRNA) and initiation of polypeptide synthesis, elongation of the polypeptide, and termination and release of the complete new protein.

It is possible to simplify discussion of ribosome production by considering each of three aspects in turn. The first deals with ribosome assembly, the second with r protein synthesis, and the third with rRNA synthesis. The main events and processes of ribosome biosynthesis are summarized in Fig. 1.

Ribosome assembly

In eukaryotes ribosomes are assembled in the nucleolus (see Warner, 1974, and Chooi, 1976, for reviews), a nuclear structure with a defined and heritable chromosomal locus, which is also the site of rRNA synthesis and thus contains the rRNA genes. Ribosome assembly involves the ordered sequential addition of r proteins or groups of proteins, which have been synthesized in the cytoplasm and transported to the nucleolus, to nascent rRNA precursor molecules still in the process of transcription (Kumar & Warner, 1972). The details of this assembly process are obscure but involve, besides the r proteins themselves, assembly proteins which act to facilitate assembly and never appear in the mature ribosome. Assembly proteins are presumably recycled through many assembly events, although it is not impossible that some might be used once or a few times only. Most r proteins are added in the nucleolus, but the final stages may take place in the cytoplasm, just as some chemical modifications to the RNA may take place late in the cytoplasm (Maden, Salim & Summers, 1972; Salim & Maden, 1973). Since the primary rRNA transcript is a single molecule containing the rRNA sequences of both ribosomal subunits, synthesized from contiguous regions of ribosomal deoxyribonucleic acid (rDNA), it may be oriented with respect to whether the small or large subunit sequences are transcribed first. It turns out in several organisms that the small subunit sequence appears first and is thus located towards the 5' end of the precursor (Dawid & Wellauer, 1976; Reeder, Higashinakagawa & Miller, 1976). It is not exactly at the 5' end in vertebrates, since a region of rDNA is transcribed first and subsequently destroyed by processing. Although there is no *a priori* reason why this should be so, it means that small subunit assembly precedes large subunit assembly on growing nascent precursor molecules. Availability of r proteins is a major control in ribosome production since rRNA is severely reduced in the absence of protein synthesis, achieved either by amino acid starvation (Maden, 1972) or by chemical inhibition (Willems, Penman & Penman, 1969). Under these conditions processing of rRNA precursors already present also stops. For instance, in HeLa cells treated with levels of puromycin which completely block protein synthesis, rRNA processing stops at the 32S stage, which thus accumulates (Soeiro, Vaughan & Darnell, 1968).

Besides the addition of r proteins in the correct manner ribosome assembly requires three other major processes: chemical modification of the rRNA, processing of the rRNA, and, finally, transport to the cytoplasm. Chemical modification involves the methylation of specific nucleotides (Maden *et al.*, 1972), either the base or the sugar, and the alteration of specific bases, for instance formation of pseudouridine in place of uridine. Correct methylation is essential for continued assembly since withdrawal of methionine, the methyl donor in methylation reactions, prevents the formation of mature rRNA, although some undermethylated 32S precursor does appear (Vaughan, Soeiro, Warner & Darnell, 1967). During processing the rRNA precursor is cleaved at specific sites to generate the RNAs of both small and large subunits (Weinberg & Penman, 1970). How these two phenomena are related to r protein assembly or to each other is not clear in detail, except that most chemical modifications occur rapidly before the first processing steps, while some of the modifications to the small subunit RNA are late events occurring in the cytoplasm (Maden *et al.*, 1972; Salim & Maden, 1973). Processing events take place after completion of transcription; the average lifetime of 45S rRNA precursor in HeLa cells is 20 minutes (Greenberg

& Penman, 1966), but whether all r proteins are assembled by this time is not clear. Both chemical modification and processing events are achieved by enzyme activities some of which are now being purified. Finally transport of completed subunits to the cytoplasm is an active process depending on continued rRNA synthesis, since 28S rRNA formed by processing of 45S rRNA precursor made prior to inhibition of rRNA synthesis by campto-thecin remains within the nucleus but is promptly transported when the drug is removed and rRNA synthesis resumes (Kumar & Wu, 1973).

Ribosomal protein synthesis

Ribosomal protein is a major component of total cell protein – 5 % in mammalian tissue culture cells, 15 % in exponentially growing yeast cells and 6 % in amphibian oocytes (excluding yolk proteins). Since there are approximately 80 different r proteins in eukaryotes, present in equimolar amounts, each protein itself represents 0.06–0.19 % of total cell protein.

Ribosomal proteins are, like all other proteins, synthesized on cytoplasmic polyribosomes (Craig, 1971). In yeast, r protein mRNA is found in the poly(A)-containing fraction and can be translated by an in-vitro cell-free protein synthesizing system to give at least 35 distinct r proteins which migrate with authentic r proteins in a two-dimensional electrophoretic assay (Warner & Gorenstein, 1977). This result suggests that r proteins are not made as precursor molecules which are subsequently cleaved (Warner & Gorenstein, 1977). Also in yeast it has been possible to show that transcription of r protein mRNAs is coordinately regulated (Gorenstein & Warner, 1976) and independent of the control of either rRNA synthesis or the synthesis of mRNA for other cell proteins. How the synthesis of a set of 80 proteins each present at an equivalent concentration may be regulated is an intriguing question. One possibility is wastage of proteins syn-thesized in excess of the cell's immediate requirement, the equi-molarity being imposed simply by the fact that there are equal numbers of assembly sites for each protein on maturing ribosomes; excess unused protein would be degraded. Such an explanation is consistent with the observation that ribosomal protein does not accumulate when rRNA synthesis is prevented. Another possibility for which there is also some evidence is that the various mRNAs are under coordinated transcriptional control, which together with identical average lifetimes would lead to similar intracellular mRNA concentrations and thus similar rates of synthesis for the individual proteins, providing also that each mRNA is equally efficient as a message. In yeast, single temperature-sensitive mutations may preferentially reduce or abolish r protein synthesis (Hartwell, McLaughlin & Warner, 1970), an effect which seems to involve the reduction of all r protein mRNA concentrations (Hereford & Rosbash, 1977). When wild-type yeast is shifted from low to high temperature there is a transitory inhibition of r protein synthesis without an effect on protein synthesis generally. The inhibition is apparently associated with the coordinated cessation of r protein mRNA transcription, for when r protein synthesis eventually resumes at the higher temperature all r protein mRNAs reappear together (Warner & Gorenstein, 1977).

During oogenesis or early development, where ribosome synthesis is clearly regulated, r protein synthesis has been shown to occur in a coordinated manner with rRNA synthesis. For instance, Hallberg & Smith (1975) have shown during oogenesis in *Xenopus laevis* that 30 % of total protein synthesis during early vitellogenic stages is of r protein, but that early in oogenesis, when there is very little rRNA synthesis, less than 5 % of total protein synthesis is of r protein. In development, r protein synthesis becomes detectable at the same time as rRNA synthesis itself becomes detectable; in *Xenopus* this is at gastrulation. Moreover r protein synthesis is not detectable in homozygous *o/o nu* (nucleolusless) mutant embryos of *Xenopus*, which do not synthesize rRNA either, suggesting that there is no accumulation of newly synthesized r protein in the absence of rRNA synthesis (Hallberg & Brown, 1969). However, in mouse L cells, when rRNA synthesis is inhibited by low concentrations of actinomycin, r protein synthesis may continue for several hours, but the r proteins made during this time are unable to enter ribosomes when the block is removed and rRNA synthesis resumed (Craig & Perry, 1971). Moreover r proteins do not accumulate in the nucleolus in the presence of actinomycin (Maisel & McConkey, 1971). Similar results have been obtained with camptothecin, a rapidly reversible inhibitor of RNA synthesis, and are interpreted as indicating that r protein is rapidly fixed in the nucleolus by attachment to nascent rRNA (Wu, Kumar & Warner, 1971); when new rRNA is absent unbound r protein passes rapidly out of the nucleolus and is destroyed. The estimated average lifetime of newly synthesized r protein in camptothecin-treated cells is less than 5 minutes. Finally, r protein potentially available for reassembly by turnover of old ribosomes is not re-utilized since protein and RNA moieties of ribosomes have the same average lifetimes (Tsurugi, Morita & Ogata, 1974). These observations are consistent with the idea that pools of free r proteins available for assembly are very small.

Ribosomal RNA synthesis

We have seen in Fig. 1 that 18, 28 and 5.8S rRNAs are synthesized as a single precursor to which are added proteins and 5S rRNA. 5S RNA is synthesized independently, usually in insects and vertebrates on genes not closely linked to 18 and 28S genes. In yeast, however, 5S RNA genes are intimately associated with 18 and 28S rRNA genes but are independently transcribed (Rubin & Sulston, 1973). Detailed structures and some complete nucle-otide sequences for these genes are now available for several organisms, notably yeast (Maxam, Tizard, Skryabin & Gilbert, 1977; Valenzuela *et al.*, 1977), *Drosophila* (Artavanis-Tsakanas *et al.*, 1977) and *Xenopus* (Federoff & Brown, 1978; Miller *et al.*, 1978). It is probably not too optimistic to say that within the next two to five years detailed knowledge of the control of transcription such as exists for many prokaryote systems will be achieved. The present review will consider some aspects of the control of rRNA synthesis especially relevant to development.

There are many points at which the rate of RNA synthesis may be regulated. Gene numbers themselves may be specifically altered, as in the case of rDNA amplification in many oocyte systems, to allow potentially greater rates of transcription, though whether this potential is realized depends on many other factors. Alternatively, large numbers of genes may be permanently main-tained in the genome for use at particular 'crisis' points in development, such as the large numbers of 5S rRNA genes most of which are normally active only during oogenesis in amphibians and fishes.

Transcription itself may be influenced at three stages –

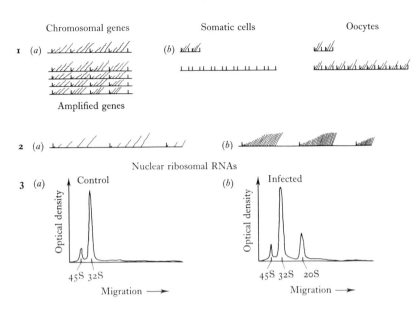

Fig. 2. Diagram of how ribosomal RNA synthesis may be regulated. 1, At the level of a gene cluster the rate of RNA synthesis may be regulated by the number of genes available for transcription. (a) Gene numbers may be specifically amplified in certain cells leading to an increase in number of potential templates for transcription. (b) Large numbers of genes may be maintained in the genome of all cells but only fully made use of by certain cells or at some points in development. 2, At the level of the individual gene RNA synthesis may be controlled by rate of initiation, which if limiting (a) will generate a gene which is not packed with polymerase, since elongation and termination events remove polymerases faster than initiation can replace them at the initiation site. If rate of initiation is not limiting (b), then elongation or termination will be limiting, and will generate a gene which is maximally packed with polymerase, since initiation can occur every time elongation reveals an exposed initiation site. 3, During RNA processing, changes at the rate-limiting steps will cause changes in rate of appearance of mature RNAs. For instance poliovirus infection of HeLa cells (b) causes a decrease in the production of mature 18S RNA by altering the rate of 20S → 18S processing. The result is that 20S RNA accumulates in the nucleolus.

initiation, elongation and termination – and factors specifically facilitating or inhibiting these steps may be important in the overall control of RNA synthesis rates. Increasing the rate of initiation will have an effect only if initiation is itself limiting transcription, and increased initiation will then serve to pack the DNA with polymerase; but if elongation or termination are limiting, the DNA is already maximally packed with polymerase.

Changes in the rate of rRNA processing will have an effect on ribosome synthesis rates only when the rate of processing is limiting, which may be the case since pools of ribosome precursor particles are seen in nuclei of cells synthesizing ribosomes. Finally, in order to ask meaningful questions about the nature of control in a multiple gene system it is an advantage to be able to study the population of genes at the individual gene level, since changes in the net rate of transcription will result from changes in both the proportion of active genes and the efficiency with which they are individually transcribed (Fig. 2). Techniques such as that developed by Miller and his colleagues (Miller & Beatty, 1969) enable this to be done.

Maternal effects and ribosome synthesis

The eggs of many organisms contain large numbers of ribosomes which provide a capacity for protein synthesis well into embryonic life. These ribosomes are made during oogenesis and are thus a specifically maternal contribution to early development. The point in development at which new embryonic ribosomes are required varies phylogenetically, but maternally contributed ribosomes are usually sufficient to sustain normal development into late embryonic life. Mutations which severely affect ribosome synthesis during oogenesis or the ability of those ribosomes to function normally in development might be expected to occur as recessive or temperature-sensitive female steriles, falling into a number of classes depending on the exact nature of the mutant lesion. Such mutations would have to be specifically in functions associated with ribosome synthesis in oocytes, since severe defects in ribosome synthesis generally in somatic cells will be embryonic zygotic lethals like the homozygous *o/o nu* mutant in *Xenopus* or severe *bobbed* mutants in *Drosophila*. The existence of maternal effect mutants for ribosome structure or synthesis is predicted from the knowledge that ribosome synthesis and accumulation in oocytes involve processes unique to oogenesis. For instance rDNA amplification, different sequences for 5S RNA and the feeding of ribosomes to oocytes by nurse cells are all examples of processes unique to oogenesis in a wide variety of organisms, which will be considered in more detail below.

In *Drosophila* there are, besides *bobbed* (Mohan & Ritossa,

1970), two other genetic loci which affect ribosome synthesis by nurse cells; they are *abnormal oocyte* (*abo*) (Sandler, 1970) and *daughterless* (*da*) (Mange & Sandler, 1973). Both genes map on the second chromosome and have an effect on expression of ribosomal RNA genes located in sex chromosome heterochromatin and on ribosome content in eggs. In *Xenopus* the effect of the *o nu* mutation on oogenesis has been studied in heterozygous + /*o nu* animals (Perkowska, Macgregor & Birnstiel, 1968). Such individuals compensate for lack of one nucleolar organizer by amplifying a normal amount of ribosomal DNA. There does not, however, seem to be an example of a maternal effect mutant which is traceable to altered ribosome function in eggs.

Ribosomal DNA amplification in oogenesis

The phenomenon of rDNA amplification, although phylogenetically widespread, was first discovered and has been most extensively studied in the amphibian *Xenopus laevis* (for review see Bird, 1979). A low level of rDNA amplification first appears during the very earliest stages of germ line development, when only a few primordial germ cells are present (Kalt & Gall, 1974). This 10–40-fold increased level of rDNA is maintained in both sexes during the gonial proliferative phase which occurs after migration of the primordial germ cells to the genital ridges. As meiosis begins these first-wave rDNA copies are lost and replaced, in females only, by a second 2500-fold increase in rDNA (Gall, 1968; Macgregor, 1968). Little is known about the first-wave amplification, but Bird (1977) has shown extrachromosomal circular DNA molecules of high buoyant density (Dawid, Brown & Reeder, 1970) which appear to replicate by Cairn's-type and rolling circle intermediates just like the second-wave amplification (Rochaix, Bird & Bakken, 1974). Although RNA synthesis does occur during the gonial divisions and in primordial germ cells (Dziadek & Dixon, 1977) it has not been proven that these early amplification products are active in transcription.

During pachytene rDNA of greater buoyant density in caesium chloride gradients than chromosomal rDNA is replicated primarily by rolling circle intermediates to a level 2500 times that for a diploid cell (Rochaix *et al.*, 1974). The amplified rDNA may be distinguished from chromosomal rDNA by its pattern of gene repeats (Wellauer *et al.*, 1974; Wellauer, Reeder, Dawid & Brown, 1976). Restriction enzyme analysis of chromosomal rDNA from a population of *Xenopus* reveals that rDNA repeats fall into many size classes (Wellauer, Dawid, Brown & Reeder, 1976). Length heterogeneity is due to variations in length of the non-transcribed spacer region (Botcham, Reeder & Dawid, 1977); the length of the transcription unit in *Xenopus* shows no detectable length variation. The number of length variants within an individual, or more precisely within a single chromosomal gene cluster, is restricted to a subset of the length variants available to the population as a whole, and the pattern of length variants is inherited in a simple Mendelian way (Reeder, Brown, Wellauer & Dawid, 1976). Unexpectedly, analysis of amplified rDNA from individual whole ovaries has shown that length variants rare in chromosomal rDNA from the same animal may predominate (Reeder *et al.*, 1976). Taking this analysis further Bird (1977) has shown (Fig. 3) that individual oocytes from one ovary may amplify quite different selections of the repeat units available to them; in 21 oocytes there were 18 different patterns of spacer variant. This demonstrates that oocytes are not initially dependent on excision of a *single* repeat unit for amplification, and further that selection of precisely which sequences are to be amplified is not rigorously controlled. If sequence variation within transcription units and not just length variation within spacers is also present in rDNA, then as a consequence of this highly selective but variable choice of genes for amplification it is predicted that individual oocytes within one ovary might eventually synthesize rRNA of predominantly different sequence. The consequences to the oocyte will depend precisely on the effects of the sequence variation on ribosome function. It has always been a problem to see how a point mutation occurring in one copy of a multiple gene cluster could be selected for or against in evolution given the diluting influence of the other gene copies, yet it is an observable fact that multiple genes do evolve in nucleotide sequence. Selective amplification would have the effect of allowing individual rRNA sequence variants a more rapid exposure to evolutionary selection pressure than would be possible if it did not occur. A special case of selective amplification which illustrates this last point is seen in interspecific hybrids between *Xenopus laevis* and *X. borealis* (this species was called *X. mulleri* but had been wrongly identified: see

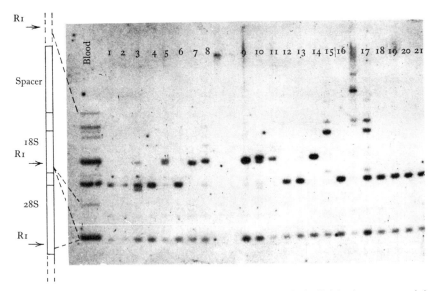

Fig. 3. *Eco*RI restriction analysis of amplified rDNA in individual oocyte nuclei, as compared with blood cell rDNA. All oocytes and blood DNA are from the same female frog. The figure is a 1-day radiograph of the blotted gel after hybridization to [^{32}P]rDNA. On the left is a schematic drawing of an rDNA unit, showing fragments generated by *Eco*RI (Wellauer *et al.*, 1974) and their positions on the gel. Slots containing no DNA or showing incomplete digestion have been omitted from the numbering. (From Bird (1977), with permission.)

Brown, Dawid & Reeder, 1977). Brown & Blackler (1972) have shown that in such hybrids, regardless of which species is the maternal parent, only *laevis* rDNA is amplified. Thus the rRNA synthesized during oogenesis will be of the *laevis* not *borealis* type.

Control of rRNA synthesis during amphibian oogenesis

Amphibian oogenesis has been used as a model system with which to study the control of specific gene activity. We have seen how rRNA genes are selectively amplified early in oogenesis, yet there is much evidence to show that they only become fully transcriptionally active during vitellogenesis, many months after the amplification events have taken place. Between rDNA amplification and activation 5S rRNA and transfer RNA (tRNA) are preferentially synthesized and accumulated (as we will discuss further later).

Different DNA-dependent RNA polymerases are responsible for transcription of 18 and 28S rRNA (polymerase I) and 5S and tRNA (polymerase III). It is possible therefore that the low level of rRNA and high level of 5S and tRNA synthesis during pre-vitellogenic stages is due to the lack of polymerase I and the enhanced presence of polymerase III. Although there is a much greater amount of polymerase III relative to polymerase I than is found in tissue culture or somatic cells, there is nevertheless sufficient polymerase I activity in stage I oocytes (Dumont, 1970) to allow considerable rates of rRNA transcription (Roeder, 1974). Since polymerase I is present some other factor must explain the low levels of rRNA synthesis in pre-vitellogenic oocytes.

Scheer, Trendelenburg & Franke (1976) have studied the patterns of rRNA transcription at the gene level during oogenesis in *Triturus alpestris* using Miller's spreading techniques. Ribosomal DNA spread from pre-vitellogenic oocytes shows very few regions of transcription, and those active genes which do occur may or may not be fully active and may or may not be closely linked to other active genes (Fig. 4). Pictures such as Fig. 4 clearly show that the low rate of transcription is due primarily to the total inactivity of the majority of transcription units (TUs), as well as to reduced activity of some of them, rather than to a low level of all potential TUs. They also indicate that rDNA activation may be a random effect at the level of individual TUs, rather than at the nucleolar or some intermediate level, which is dependent on a relatively rare event. Once a TU is activated for transcription it may, but not necessarily, become fully active. Because TUs that are fully active, that is maximally packed with polymerase, are seen, it seems likely that elongation or termination are limiting transcription rather than initiation. However the fact that submaximal packing of polymerase is also observed means that there may be an intermediate stage of activation of a TU during which initiation events limit transcription. Maximum rates of rRNA synthesis occur during vitellogenesis in *Triturus* and at this stage all available TUs are maximally packed with polymerase again, suggesting that elongation or termination events set the limits to transcription rate at this stage. In fully grown oocytes the net rate of RNA synthesis is about 10 % of the maximum level, and is achieved by reducing the number of active TUs as well as by limiting the number of polymerases initiating transcription on

P. J. Ford

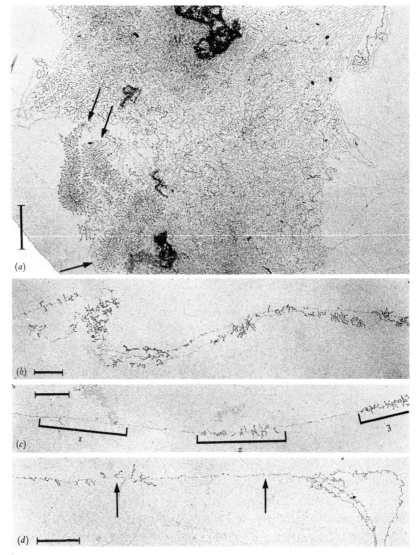

Fig. 4. Details of spread and positively stained nucleoli from pre-vitellogenic *T. alpestris* oocytes. Different types in the pattern of the onset of rDNA transcription are demonstrated. Large fibrillar aggregates (AF) with only a few but maximally active matrix units are recognized (arrows in *a*). This contrasts with the occurrence of many active adjacent matrix units, which, however, are incomplete in their fibril package (*b*; see also cistrons 2 and 3 in *c*, and the cistron denoted by the left-hand arrow in *d*). Occasionally, one notes cascades of increasing transcriptional activity along one axis (see, for example, *c*). In such cases, untranscribed cistrons (e.g. gene 1 in *c* and that denoted by the right-hand arrow in *d*) are clearly localized. Scale bars represent 1 μm. (From Scheer, Trendelenburg & Franke (1976), with permission.)

some TUs. The proportion of active TUs varies between nucleoli, which may indicate a level of control and a preference to shut down transcription in some nucleoli rather than others.

5S rRNA synthesis in oogenesis and development

In fish and amphibians large numbers of 5S rRNA genes (Brown & Weber, 1968) are maintained in the genome in order to accommodate the high demand for ribosomes during oogenesis and are not specifically amplified like the rRNA genes (Brown & Dawid, 1968). It turns out that only a subset of these genes are transcribed in somatic cells, while a second subset, distinguishable by nucleotide sequence, is active specifically in oocytes (Wegnez, Monier & Denis, 1972; Ford & Southern, 1973). Oocyte-type 5S genes are activated very early in oogenesis at or just after rDNA amplification in pachytene, and an amount of 5S RNA equivalent to 75 % of the final amount is accumulated before activation of rDNA takes place at the beginning of vitellogenesis (Dixon & Ford, 1979). Transfer RNA is also accumulated at the same time but there is no evidence to suggest oocyte-specific tRNA molecules (Denis, Mazabraud & Wegnez, 1975) or specific amplification of tRNA genes (Brown & Dawid, 1968).

During development 5S rRNA synthesis becomes detectable at the same time as 18 and 28S rRNA, and the detailed timing varies phylogenetically in the same way. However onset of 18 and 28S rRNA synthesis is not essential for 5S rRNA synthesis since Miller (1974) has shown that homozygous *o/o nu* mutant embryos of *Xenopus* start 5S RNA synthesis on time during late blastula stage, even though 18 and 28S rRNA synthesis is not occurring. These observations tell us that 5S rRNA synthesis is regulated completely independently of 18 and 28S rRNA synthesis – a conclusion further supported by the fact that almost normal rates of 5S RNA synthesis persist in metaphase-arrested cells although new 18 and 28S RNA is not detectable (Zylber & Penman, 1971).

Ribosomal RNA synthesis during oogenesis in *Drosophila*

The control of ribosome biosynthesis during oogenesis in *Drosophila* is very different and contrasts with the process in amphibians. The oocyte is attached by cytoplasmic bridges to and

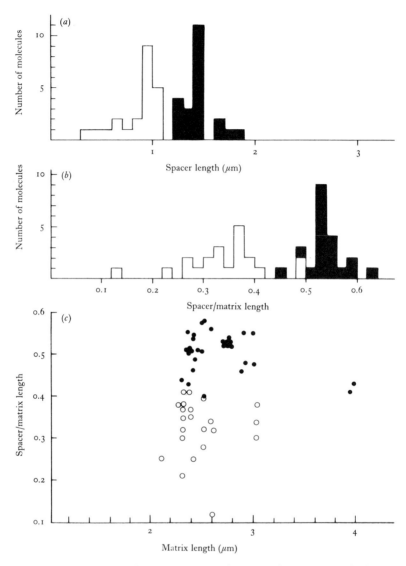

Fig. 5. (*a*) Histogram showing the distributions of spacer lengths for rDNAs from nurse cells (solid columns) and embryos (open columns) of *D. melanogaster*. For all the data presented, spacer lengths were measured only from preparations where at least the two adjacent rRNA cistrons were present. (*b*) Histogram showing the distributions of the ratios of spacer to matrix lengths from rRNA genes taken from nurse cells (solid columns) and embryos (open columns). (*c*) Plot of matrix lengths against ratios of spacer to matrix lengths for nurse cells (solid circles) and embryos (open circles). (From Chooi (1976), with permission.)

supplied with ribosomes by 15 nurse cells; the whole group of 16 cells is surrounded by follicle cells and forms the egg chamber (King, 1970). Although nurse cells replicate their DNA to 1000 times the haploid amount there is no evidence for increased differential amplification of rRNA genes. The chromosomes of nurse cells are similar to polytene chromosomes in salivary glands of larvae which are known to under-replicate rRNA genes (Spear & Gall, 1973). Nurse cells have single greatly enlarged nucleoli and at no stage are 5S and tRNA preferentially synthesized or accumulated. The rates of synthesis of all the major RNA species increases throughout oogenesis in parallel with the overall increase in DNA content of the nurse cells (Mermod, Jacobs-Lorena & Crippa, 1977).

A recent observation (Chooi, 1976; Fig. 5) is that nurse cells appear to transcribe rRNA from a subset of the available rRNA genes. Measurement of rRNA transcription unit and spacer lengths in nurse cells and embryo cells shows that in the former rRNA is transcribed from genes with relatively much longer spacer sequences. These observations may be related to sequence heterogeneity of rRNA genes in *Drosophila* (Glover & Hogness, 1977; Wellauer, Dawid & Tartof, 1978).

Ribosomal RNA synthesis during development

Ribosomal RNA synthesis is not detectable in mature unfertilized eggs but becomes detectable at some point in early embryogenesis. The exact point in development at which rRNA genes become active varies. In mammals, for instance, rRNA synthesis and nucleolus formation are first detected at the 2–4-cell stage, but in many other animals late blastula is the time when new rRNA is detected. Although many studies with phylogenetically very separate organisms have been made there is yet no truly adequate description in numerical terms of rRNA gene activation in development. Such questions as whether all rRNA genes are activated simultaneously, and if not whether the active genes are immediately fully active or whether there is a phase of gradual activation, cannot be answered by a purely biochemical approach. A start has been made to answer questions like these for the development of *Oncopeltus* (Foe, Wilkinson & Laird, 1976). In this insect rRNA synthesis is at a high level by 70–74 hours but is reduced at 120–140 hours. Analysis of Miller spreads at these two stages reveals that the change in rate is due both to a reduction of active rTUs and to a reduction in the number of transcripts on those TUs which are active.

Organisms are normally endowed with more rRNA genes than are required for development and individuals in a population may have quite different total numbers of rRNA genes (Miller & Brown, 1969; Miller & Knowland, 1970). For instance wild type ($+/+$ *nu*) *Xenopus laevis* have rRNA gene numbers varying between 900 and 1200 copies and two nucleolar organizers per diploid cell. However the anucleolate mutant (*o/o nu*), which is a recessive lethal, has less than 5 % of the wild-type rDNA and no true nucleoli. A second nucleolar mutant (p^1/p^1 *nu*) organizes two partial nucleoli and has just under half the wild-type gene number. The p^1 rRNA gene cluster (called partial lethal) is lethal when combined with an *o nu* chromosome (p^1/o *nu*) and such an organism has 23 % of the wild-type gene number (Miller & Knowland, 1970). Table 1 indicates the relative gene dosage and relative rates of RNA synthesis per embryo and per gene for various *Xenopus* nucleolar genotypes (Knowland & Miller, 1970). Clearly, reducing the rRNA gene number to less than 40 % of wild-type results in inadequate rates of rRNA synthesis for normal development.

A similar situation occurs in *Drosophila*, which has between 280 and 550 rRNA genes per diploid nucleus. These genes are located in regions of heterochromatin on the X and Y chromosomes. The

Table 1. *Average rate of rRNA synthesis per gene in various genotypes of* Xenopus laevis

	Nucleolar genotype					
	$+/+$	$+/p^1$	$+/o$	p^1/p^1	p^1/o	o/o
Relative amount of rDNA	100	73	50	46	23	6
Relative rate of rRNA synthesis	100	100	100	100	25–40	< 2
Relative rate of rRNA synthesis per gene	100	140	200	220	112–175	< 33

From data in Miller & Knowland (1970) and Knowland & Miller (1976).

locus *bobbed* (*bb*) has many alleles which are partial deletions of the nucleolar organizer and involve reductions in the number of rRNA genes. *Bobbed* embryos show a reduction of rRNA synthesis proportional to both the severity of the *bobbed* phenotype and the reduced rDNA. When a *bobbed* allele is placed opposite a deletion for the whole nucleolar organizer region the *bobbed* phenotype may revert over a few generations to wild-type. Correlated with reversion of phenotype is an accumulation of rDNA, a phenomenon called rDNA magnification. Magnification is generated in the male line and involves a true chromosomal change. However a magnified *bobbed* allele is initially unstable, since if combined with a wild-type allele it reverts to its previously *bobbed* condition. The instability of magnification is not permanent, and after about four generations a magnified *bobbed* locus will not return to *bobbed*. There is at present no convincing molecular explanation for these events (for review see Ritossa, 1976).

Ribosomal RNA synthesis in *X. laevis–X. borealis* interspecific hybrids

It is possible to distinguish the rRNA precursors of *X. laevis* and *X. borealis* (note that in some of the papers cited here *X. borealis* was called *X. mulleri* in error: see Brown, Dawid & Reeder, 1977) by nucleic acid hybridization procedures (Honjo & Reeder, 1973). The difference resides in the parts of the molecule eliminated by processing, since mature 18 and 28S rRNAs are indistinguishable by the same criteria. A study of the onset of rRNA synthesis in interspecific hybrids between these two species reveals that only one nucleolus develops at gastrulation (Blackler & Gecking, 1972) and that only the *laevis* rRNA genes are transcribed initially (Honjo & Reeder, 1973). Later in hybrid embryo development and in adult tissues a variable proportion of *borealis* rRNA genes may be transcribed. *Laevis* nucleolar dominance and repression of *borealis* rRNA genes is not absolute and gradually breaks down, but whether leakiness affects all cells or just a few is not known at present. Similar nucleolar dominance phenomena have been known for a long time (McClintock, 1934) and have widespread occurrence (Bicudo & Richardson, 1977), but only in *Xenopus* has the phenomenon been analysed in detail.

Providing *laevis* rDNA is present there is no paternal or

maternal effect since dominance of *laevis* genes is always observed. However by using appropriate $+/-$ *nu laevis* male or female animals it is possible to generate embryos which have only *borealis* rRNA genes donated either by the maternal or paternal parent. In this case the *borealis* genes are 'forced' to become active in transcription or the embryo will die. The time at which *borealis* genes activate is now dependent on the female parent species. When *laevis* is the maternal parent *borealis* genes activate very late in development, but when *borealis* is the maternal species rRNA synthesis is first detected at gastrulation as normal. Thus it appears that the presence of *laevis* rDNA is essential for repression of *borealis* rRNA genes either passively by competing more effectively for a rare essential component or actively by synthesizing an inhibitory substance. Furthermore maternally inherited components influence the time of rRNA gene activation, a conclusion which is consistent with the extraction of specific inhibitors of rRNA synthesis from eggs and embryos of *Xenopus laevis* (Laskey, Gerhart & Knowland, 1973).

5S rRNA synthesis in *X. laevis–X. borealis* interspecific hybrids

The 5S rRNA sequences in both species are distinguishable in both somatic cells and oocytes (Ford & Brown, 1976). In view of the nucleolar dominance effects (Honjo & Reeder, 1973) and the amplification of specifically *laevis* rDNA (Brown & Blackler, 1972) in hybrid oocytes, we have examined 5S rRNA synthesis in both somatic cells and oocytes from five separate batches of hybrids. Hybrids were produced by artificial fertilization of *laevis* eggs with *borealis* sperm; the reverse cross has not been investigated. Fig. 6 indicates the preparation of labelled 5S RNA from hybrid ovaries. It has been possible to detect the synthesis of 5S RNA from both parental species in both somatic cells (kidney) (Fig. 7*a*) and ovaries (Fig. 7*b*) of these hybrids. From an analysis of oligonucleotide yields it has been possible to conclude that the 5S genes of each species are approximately equally well transcribed in both somatic cells and pre-vitellogenic ovaries (Table 2). This conclusion holds for four hybrid matings between two different *laevis* females and two different *borealis* males. Individuals from a fifth mating between a *laevis* female and *borealis* male (both

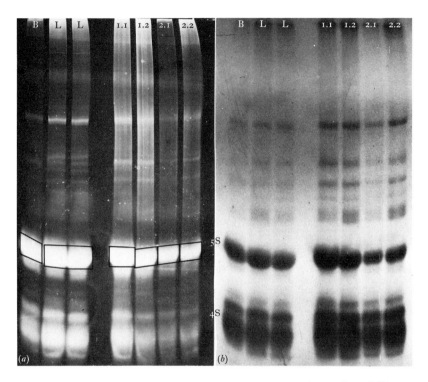

Fig. 6. Preparation of [^{32}P]5S rRNA from pre-vitellogenic ovaries of *Xenopus laevis* (L: seven ovaries), *Xenopus borealis* (B: two ovaries), and four different batches of *laevis/borealis* interspecies hybrids (H1.1, nine ovaries; H1.2, six ovaries; H2.1, five ovaries; H2.2, twelve ovaries). Ovaries were labelled by culturing with ^{32}PO$_4$ for 72 hours in modified HEPES buffered Barth X as described previously (Ford & Southern, 1973). RNA was extracted and low molecular weight (LMW) components separated from high molecular weight components by sucrose gradient centrifugation. The LMW RNAs were separated by electrophoresis in 5–15% acrylamide gel gradient using 50 mmol tris buffered to pH 8.6 with boric acid in 7 mol urea. Unlabelled RNA species were located by photographing the ethidium bromide stained gel in ultraviolet light (*A*) and labelled RNA species were located by radioautography (*B*). The 5S rRNA bands as indicated were cut from the gel, and the RNA eluted by extraction overnight and recovered by ethanol precipitation.

animals unrelated to the ones used for the other matings) synthesized both types of 5S RNA in their somatic cells, but we were unable to detect the synthesis or enhanced accumulation of any oocyte-type 5S RNA in their pre-vitellogenic ovaries, although somatic type sequences were detected in these ovaries (Fig. 7c).

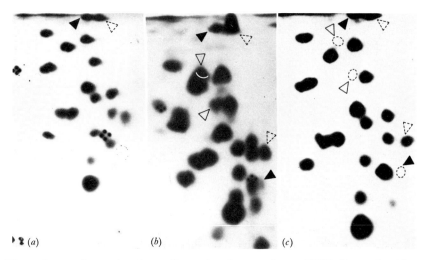

Fig. 7. Autoradiographs of two-dimensional separations of [^{32}P]oligonucleotides produced by digestion of (a) hybrid *laevis/borealis* somatic (kidney) cell 5S RNA; (b) hybrid (batch 1 H2.1) *laevis/borealis* total ovary 5S RNA; and (c) hybrid (batch 2) *laevis/borealis* total ovary 5S RNA. The procedures were exactly as described previously (Ford & Southern, 1973). Oligonucleotides specific to *laevis* (▽), specific to borealis (▼), and oligonucleotides with different yields in the two species (⌄⃝) are indicated (see Table 2).

It seems likely that variation in the number of oocytes entering the later stages of normal oogenesis between different hybrid matings can account for this anomaly. Examination of ovaries from the aberrant mating revealed very low numbers (50–200) of oocytes developing normally into vitellogenic stages (Fig. 8a) but large numbers of apparently normal gonial cells (Fig. 8b and d). Ovaries from the other four matings have shown much greater numbers of oocytes developing normally, although each mating is different in this respect and there is considerable variation between different ovaries from the same mating. In spite of this variation females from all five hybrid matings have shown the presence of abnormally large numbers of apparently pre-meiotic stages of oogenesis. The precise description of these cells will be given elsewhere but they have the apearance of primary oogonia with irregular nuclei and multiple nucleoli, or secondary oogonia with rounded nuclei and single centrally located nucleoli (Watson Coggins, 1973), and there are often apparently normal pachytene stages with amplified rDNA nuclear caps (Gall, 1968). Groups of

Table 2. *Yields of species-specific oligonucleotides from T₁ RNase digests of [³²P]5S RNA from laevis/borealis interspecies hybrids*

Oligonucleotide	Species		Hybrid expected	Hybrid observed	
	laevis	*borealis*		Experiment 1	Experiment 2
			(a) Somatic cells		
CCACACCACCCUG	1.00	0.20	0.60	0.44±0.03 (16)	0.48±0.04 (4)
CCAUACCACCCUG	0.00	0.80	0.40	0.56±0.04 (16)	0.50±0.04 (4)
			(b) Ovaries		
CCG	0.05	0.55	0.30	0.23±0.01 (16)	0.05±0.02 (4)
CCCG	0.10	1.00	0.55	0.59±0.06 (16)	0.90±0.04 (4)
AUACAG	0.75	0.00	0.37	0.44±0.04 (16)	0.02±0.01 (4)
AUCUCAG	0.70	0.00	0.35	0.40±0.18 (16)	0.01±0.01 (4)
CCACACCACCCUG	1.00	0.70	0.85	0.83±0.04 (16)	0.55±0.04 (4)
CCAUACCACCCUG	0.00	0.30	0.15	0.17±0.04 (16)	0.42±0.04 (4)

The mean yields are expressed in moles oligonucleotide/mole 5S RNA ± standard error of mean, with number of determinations in parentheses. Expected yields in hybrids were calculated on the assumption that both sets of parental genes are present in equal numbers, are equally active, and that there is no difference in stability of each type of 5S RNA. Oligonucleotide yields for *laevis* 5S RNA are taken from Ford & Southern (1973) and for *borealis* from Ford & Brown (1976). Control experiments with the parental species themselves, done at the same time as the hybrid experiments, were in agreement with previously published figures. Hybrid experiment 1 refers to four separate hybrid crosses between two *laevis* females and two *borealis* males made by artificial fertilization. Oligonucleotide yields were determined for four pools of ovaries from each cross and since there were no differences between these crosses the data was pooled. Hybrid experiment 2 refers to a completely independent set of hybrids produced by artificial fertilization. In this group of hybrids synthesis of ovary-specific sequences of either type was severely restricted. The results in this experiment are for four individual ovaries assayed on four separate occasions.

P. J. Ford

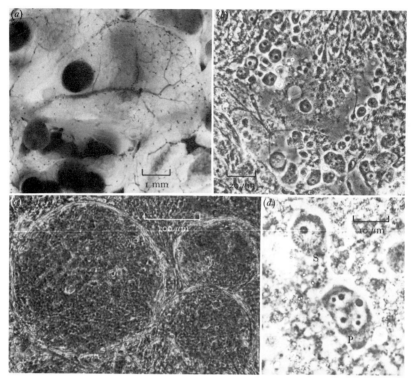

Fig. 8. (*a*) Macrophotograph of hybrid *laevis/borealis* ovary taken from an animal
1 year 2 months old showing very few fully developed oocytes and large numbers
of follicles (*c*) filled with normal looking oogonia (*b*). (Compare Watson Coggins,
1973.) (*d*) Primary (P) and secondary (S) oogonia from stage 66 *Xenopus borealis*
ovary.

several hundred such cells are found enclosed in follicles (Fig. 8*c*)
which are approximately the diameter (150–300 μm) of pre-
vitellogenic Dumont stage I oocytes, and to the untutored eye this
can give the ovary the superficial appearance of a normal pre-
vitellogenic ovary.

A maternal effect on the development of sterile ovaries in *Xenopus*

While examining large numbers of *laevis/borealis* interspecies
hybrids from four matings produced by artificial fertilization
using two *laevis* females and two *borealis* males in four possible
combinations, it became apparent that the eggs from one female

Table 3. *Possible maternal effect on sex ratio and development of one or two sterile gonads in* Xenopus

Parents	N	% males	% females	% indeterminate	% two ovaries	% one ovary
laevis 1 *borealis* 1	70	49	51	0	100	0
laevis 1 *borealis* 2	100	52	48	0	100	0
laevis 2 *borealis* 1	147	63	32	5	79	21
laevis 2 *borealis* 2	185	62	34	4	73	27

The figures recorded here were obtained from four hybrid crosses using two *laevis* females and two *borealis* males. Although some males also show only a single testis in crosses using *laevis* female 2, no record was kept of the frequency of single gonads, but subjectively it appeared to be the same as in females. Indeterminate animals had no distinguishable gonads. The four batches of hybrids were reared in separate tanks in the same room and were exact contemporaries treated and fed in the same way.

had given abnormal sexual development regardless of the male used. Abnormality is seen in two ways: one is sex ratio, with about two males for every female (Table 3), and the other is a high proportion of individuals with one or two sterile gonads (Table 3). These abnormalities are not related to the fact that the animals are hybrids since eggs from the other female gave two batches of hybrids with a normal sex ratio and no individuals with sterile gonads, while control crossing by artificial fertilization of the abnormal eggs with *laevis* sperm produced a group of good *laevis* animals which also showed the same effects.

Sterile gonads are probably produced by failure of the primordial germ cells to migrate and enter one or both of the genital ridges, though they may enter and fail to proliferate. If insufficient primordial germ cells are produced early in development there will be an increased probability that a genital ridge will not receive a population of primordial germ cells. Variation in number of primordial germ cells may result from several causes, the most likely being the amount of germinal plasm produced during oogenesis or the number of cell divisions prior to primordial germ cell migration.

I thank the Science Research Council for research support.

References

ARTAVANIS-TSAKANAS, S., SHEDL, P., TSCHUDI, C., PIRROTTA, V., STEWARD, R. & GEHRING, E. J. (1977). The 5S genes of *Drosophila melanogaster*. *Cell*, **12**, 1057–67.

BICUDO, H. E. M. C. & RICHARDSON, R. H. (1977). Gene regulation in *Drosophila mulleri*, *D.arizonensis*, and their hybrids: the nucleolar organiser. *Proceedings of the National Academy of Sciences of the USA*, **74**, 3498–502.

BIRD, A. P. (1977). A study of early events in ribosomal gene amplification. *Cold Spring Harbor Symposia on Quantitative Biology*, **42**, in press.

BIRD, A. P. (1979). Gene reiteration and gene amplification. In *Cell Biology : A Comprehensive Treatise*, vol. 3, in press.

BLACKLER, A. W. & GECKING, C. A. (1972). Transmission of sex cells of one species through the body of a second species in the genus *Xenopus*. II. Interspecific matings. *Developmental Biology*, **27**, 385–94.

BOTCHAN, P., REEDER, R. H. & DAWID, I. B. (1977). Restriction analysis of non-transcribed spacers in *Xenopus laevis* ribosomal DNA. *Cell*, **11**, 599–607.

BRIMACOMBE, R., STOFLER, G. & WHITMAN, H. G. (1978). Ribosome structure. *Annual Review of Biochemistry*, **47**, 217–49.

BROWN, D. D. & BLACKLER, A. W. (1972). Gene amplification proceeds by a chromosome copy mechanism. *Journal of Molecular Biology*, **63**, 75–83.

BROWN, D. D. & DAWID, I. B. (1968). Specific gene amplification in oocytes. *Science*, **160**, 272–80.

BROWN, D. D., DAWID, I. B. & REEDER, R. H. (1977). *Xenopus borealis* misidentified as *Xenopus mulleri*. *Developmental Biology*, **59**, 266–7.

BROWN, D. D. & WEBER, C. S. (1968). Gene linkage by RNA–DNA hybridisation. I. Unique DNA sequences homologous to 4S RNA, 5S RNA and ribosomal RNA. *Journal of Molecular Biology*, **34**, 661–80.

CHOOI, W. Y. (1976). RNA transcription and ribosomal protein assembly in *Drosophila melanogaster*. In *Handbook of Genetics*, ed. R. C. King, vol. 5, pp. 219–65. New York & London: Plenum Press.

CRAIG, N. C. (1971). On the regulation of the synthesis of ribosomal proteins in L cells. *Journal of Molecular Biology*, **55**, 129–34.

CRAIG, N. C. & PERRY, R. P. (1971). Persistent cytoplasmic synthesis of ribosomal proteins during the selective inhibition of ribosomal RNA synthesis. *Nature New Biology*, **229**, 493–6.

DAWID, I. B., BROWN, D. D. & REEDER, R. H. (1970). Composition and structure of chromosomal and amplified ribosomal DNAs of *Xenopus laevis*. *Journal of Molecular Biology*, **51**, 341–60.

DAWID, I. B. & WELLAUER, P. K. (1976). A reinvestigation of 5′→3′ polarity in 40S ribosomal RNA precursor of *Xenopus laevis*. *Cell*, **8**, 443–8.

DENIS, H., MAZABRAUD, A. & WEGNEZ, M. (1975). Biochemical research on oogenesis. Comparison between transfer RNAs from somatic cells and from oocytes in *Xenopus laevis*. *European Journal of Biochemistry*, **58**, 43–50.

DIXON, L. & FORD, P. J. (1979). Subcellular distribution and accumulation of 5S RNA and transfer RNA during oogenesis in *Xenopus laevis*. In preparation.

DUMONT, J. N. (1970). Oogenesis in *Xenopus laevis* (Dandin). I. Stages of oocyte development in laboratory maintained animals. *Journal of Morphology*, **136**, 153–80.

DZIADEK, M. & DIXON, K. E. (1977). An autoradiographic analysis of nucleic acid synthesis in presumptive primordial germ cells of *Xenopus laevis*. *Journal of Embryology and Experimental Morphology*, **37**, 13–31.

FEDOROFF, N. V. & BROWN, D. D. (1978). The nucleotide sequence of oocyte 5S DNA in *Xenopus laevis*. I. The AT-rich spacer. *Cell*, **13**, 701–16.

FOE, V. E., WILKINSON, L. E. & LAIRD, C. D. (1976). Comparative organisation of active transcription units in *Oncopeltus fasciatus*. *Cell*, **9**, 131–46.

FORD, P. J. & BROWN, R. D. (1976). 5S RNA sequences in *Xenopus mulleri* and the evolution of multigene families. *Cell*, **8**, 485–93.

FORD, P. J. & SOUTHERN, E. M. (1973). Different sequences for 5S RNA in kidney cells and ovaries of *Xenopus laevis*. *Nature New Biology*, **241**, 7–12.

GALL, J. G. (1968). Differential synthesis of genes for ribosomal RNA during amphibian oogenesis. *Proceedings of the National Academy of Sciences of the USA*, **60**, 553–60.

GLOVER, D. M. & HOGNESS, D. S. (1977). A novel arrangement of the 18S and 28S sequences in a repeating unit of *Drosophila melanogaster*. *Cell*, **10**, 167–76.

GORENSTEIN, C. & WARNER, J. R. (1976). Coordinate regulation of the synthesis of eukaryotic ribosomal proteins. *Proceedings of the National Academy of Sciences of the USA*, **73**, 1547–51.

GREENBERG, H. & PENMAN, S. (1966). Methylation and processing of ribosomal RNA in HeLa cells. *Journal of Molecular Biology*, **21**, 527–8.

HALLBERG, R. L. & BROWN, D. D. (1969). Coordinated synthesis of some ribosomal proteins and ribosomal RNA in embryos of *Xenopus laevis*. *Journal of Molecular Biology*, **46**, 393–402.

HALLBERG, R. L. & SMITH, D. C. (1975). Ribosomal protein synthesis in *Xenopus laevis* oocytes. *Developmental Biology*, **42**, 40–52.

HARTWELL, L. H., McLAUGHLIN, C. S. & WARNER, J. R. (1970). Identification of ten tenes that control ribosome formation in yeast. *Molecular and General Genetics*, **109**, 42–52.

HEREFORD, L. M. & ROSBASH, M. (1977). Regulation of a set of abundant mRNA sequences. *Cell*, **10**, 463–7.

HONJO, T. & REEDER, R. H. (1973). Preferential transcription of *Xenopus laevis* ribosomal RNA in interspecies hybrids between *Xenopus laevis* and *Xenopus mulleri*. *Journal of Molecular Biology*, **80**, 217–28.

KALT, M. R. & GALL, J. G. (1974). Observations on early germ cell development and premeiotic ribosomal DNA amplification in *Xenopus laevis*. *Journal of Cell Biology*, **62**, 460–72.

KING, R. C. (1970). *Ovarian Development in* Drosophila melanogaster. New York & London: Academic Press.

KNOWLAND, J. S. & GRAHAM, C. F. (1972). RNA synthesis at the two cell stage of mouse development. *Journal of Embryology and Experimental Morphology*, **27**, 167–76.

KNOWLAND, J. S. & MILLER, L. (1976). Reduction of ribosomal RNA synthesis and ribosomal RNA genes in a mutant *Xenopus laevis* which organises only a partial nucleolus. I. Ribosomal RNA synthesis in embryos of different molecular genotypes. *Journal of Molecular Biology*, **53**, 321–8.

KUMAR, A. & WARNER, J. R. (1972). Characterisation of ribosomal precursor particles from HeLa cell nucleoli. *Journal of Molecular Biology*, **63**, 233–46.

KUMAR, A. & WU, R. (1973). Role of ribosomal RNA transcription in ribosome processing in HeLa cells. *Journal of Molecular Biology*, **80**, 265–76.

LASKEY, R. A., GERHART, J. & KNOWLAND, J. S. (1973). Inhibition of ribosomal RNA synthesis in neurula cells by extracts from blastulae of *Xenopus laevis*. *Developmental Biology*, **33**, 241–7.

McCLINTOCK, B. (1934). The relationship of a particular chromosomal element to the development of the nucleoli in *Zea mays*. *Zeitschrift für Zellforschung und Mikorskopische Anatomie*, **21**, 294–302.

MACGREGOR, H. C. (1968). Nucleolar DNA in oocytes of *Xenopus laevis*. *Journal of Cell Science*, **3**, 437–53.

MADEN, B. E. H. (1972). Effects of amino acid starvation on ribosome formation in HeLa cells. *Biochimica et Biophysica Acta*, **281**, 396–408.

MADEN, B. E. H., SALIM, M. & SUMMERS, D. F. (1972). Maturation pathway for ribosomal RNA in the HeLa cell nucleolus. *Nature New Biology*, **237**, 5–9.

MAISEL, J. C. & McCONKEY, E. H. (1971). Nucleolar protein metabolism in Actinomycin D treated HeLa cells. *Journal of Molecular Biology*, **61**, 251–9.

MANGE, A. P. & SANDLER, L. (1973). A note on the maternal effect mutants of *daughterless* and *abnormal oocyte* in *Drosophila melanogaster*. *Genetics*, **73**, 73–86.

MAXAM, A. M., TIZARD, R., SKRYABIN, K. G. & GILBERT, W. (1977). Promoter region for yeast 5S ribosomal RNA. *Nature, London*, **267**, 643–5.

MERMOD, J.-J., JACOBS-LORENA, M. & CRIPPA, M. (1977). Changes in rate of RNA synthesis and ribosomal gene number during oogenesis of *Drosophila melanogaster*. *Developmental Biology*, **57**, 393–402.

MILLER, J. R., CARTWRIGHT, E. M., BROWNLEE, G. G., FEDEROFF, N. V. & BROWN, D. D. (1978). The nucleotide sequence of oocyte 5S DNA in *Xenopus laevis*. II. The GC-rich region. *Cell*, **13**, 717–25.

MILLER, L. (1973). Control of 5S RNA synthesis during early development of anucleolate and partial nucleolate mutants of *Xenopus laevis*. *Journal of Cell Biology*, **59**, 624–5.

MILLER, L. (1974). Metabolism of 5S RNA in the absence of ribosome production. *Cell*, **3**, 275–81.

MILLER, L. & BROWN, D. D. (1969). Variation in the activity of nucleolar organisers and their ribosomal gene content. *Chromosoma*, **28**, 430–44.

MILLER, L & KNOWLAND, J. S. (1970). Reduction of ribosomal RNA synthesis and ribosomal RNA genes in a mutant of *Xenopus laevis* which organises only a partial nucleolus. II. The number of ribosomal genes in animals of different nucleolar genotypes. *Journal of Molecular Biology*, **53**, 329–38.

MILLER, O. L. JR & BEATTY, B. R. (1969). Visualisation of nucleolar genes. *Science*, **164**, 955–7.

MOHAN, J. & RITOSSA, F. M. (1970). Regulation of RNA synthesis and its bearing on the bobbed phenotype of *Drosophila melanogaster*. *Developmental Biology*, **22**, 495–512.

PERKOWSKA, E., MACGREGOR, H. C. & BIRNSTIEL, M. L. (1968). Gene amplification in the oocyte nucleus of mutant and wild-type *Xenopus laevis. Nature, London*, **217**, 649–50.

REEDER, R. H., BROWN, D. D., WELLAUER, P. K. & DAWID, I. B. (1976). Patterns of ribosomal spacer lengths are inherited. *Journal of Molecular Biology*, **105**, 507–16.

REEDER, R. H., HIGASHINAKAGAWA, T. & MILLER, O. JR (1976). The 5′→3′ polarity of the *Xenopus* ribosomal RNA precursor molecule. *Cell*, **8**, 449–54.

RITOSSA, F. (1976). The *bobbed* locus. In *The Genetics and Biology of Drosophila*, ed. M. Ashburner & E. Novitski, vol. 1b, pp. 801–46. New York & London: Academic Press.

ROCHAIX, J. D., BIRD, A. P. & BAKKEN, A. (1974). Ribosomal RNA gene amplification by rolling circles. *Journal of Molecular Biology*, **87**, 473–87.

ROEDER, R. G. (1974). Multiple forms of deoxyribonucleic acid-dependent ribonucleic acid polymerase in *Xenopus laevis*. Levels of activity during oocyte and embryonic development. *Journal of Biological Chemistry*, **249**, 249–56.

RUBIN, G. M. & SULSTON, J. E. (1973). Physical linkage of the 5S cistrons to the 18 and 28S RNA cistrons in *Saccharomyces cerevisiae. Journal of Molecular Biology*, **79**, 521–7.

SALIM, M. & MADEN, B. E. H. (1973). Early and late methylations in HeLa cell ribosome maturation. *Nature, London*, **244**, 334–6.

SANDLER, L. (1970). The regulation of sex chromosome heterochromatic activity by an autosomal gene in *Drosophila melanogaster. Genetics*, **64**, 481–93.

SCHEER, U., TRENDELENBURG, M. F. & FRANKE, W. W. (1976). Regulation of transcription of genes of ribosomal RNA during amphibian oogenesis. A biochemical and morphological study. *Journal of Cell Biology*, **69**, 465–89.

SOEIRO, R., VAUGHAN, M. H. & DARNELL, J. E. (1968). The effect of puromycin on intranuclear steps in ribosome biosynthesis. *Journal of Cell Biology*, **36**, 91–104.

SPEAR, B. B. & GALL, J. G. (1973). Independent control of ribosome gene replication in polytene chromosomes of *Drosophila melanogaster. Proceedings of the National Academy of Sciences of the USA*, **70**, 1359–63.

TSURUGI, K., MORITA, T. & OGATA, K. (1974). Mode of degradation of ribosomes in regenerating rat liver *in vivo. European Journal of Biochemistry*, **45**, 119–24.

VALENZUELA, P., BELL, G. I., MASIARZ, F. R., DE GENNARO, L. J. & RUTTER, W. J. (1977). Nucleotide sequence of the yeast 5S ribosomal RNA gene and adjacent putative control regions. *Nature, London*, **267**, 641–3.

VAUGHAN, M. H., SOEIRO, R., WARNER, J. R. & DARNELL, J. E. (1967). The effects of methionine deprivation on ribosome synthesis in HeLa cells. *Proceedings of the National Academy of Sciences of the USA*, **58**, 1527–32.

WARNER, J. R. (1974). The assembly of ribosomes in eukaryotes. In *Ribosomes*, ed. M. Nomura, A. Tissieres & P. Lengyel, *Cold Spring Harbor Monograph Series*, pp. 461–88.

WARNER, J. R. & GORENSTEIN, C. (1977). The synthesis of eukaryotic ribosomal proteins *in vitro. Cell*, **11**, 201–11.

WATSON COGGINS, L. (1973). An ultrastructural and radioautographic study of early oogenesis in the toad *Xenopus laevis. Journal of Cell Science*, **12**, 71–93.

WEGNEZ, M., MONIER, R. & DENIS, H. (1972). Sequence heterogeneity of 5S RNA in *Xenopus laevis*. *FEBS Letters*, **25**, 13–20.

WEINBERG, R. A. & PENMAN, S. (1970). Processing of 45S nucleolar RNA. *Journal of Molecular Biology*, **47**, 169–78.

WELLAUER, P. K., DAWID, I. B., BROWN, D. D. & REEDER, R. H. (1976). The molecular basis for length heterogeneity in ribosomal DNA from *Xenopus laevis*. *Journal of Molecular Biology*, **105**, 461–86.

WELLAUER, P. K., DAWID, I. B. & TARTOF, K. D. (1978). X and Y chromosomal ribosomal DNA of *Drosophila*: comparison of spacers and insertions. *Cell*, **14**, 269–78.

WELLAUER, P. K., REEDER, R. H., CARROLL, D., BROWN, D. D., DEUTCH, A., HIGASHINAKAGWA, T. & DAWID, I. B. (1974). Amplified ribosomal DNA from *Xenopus laevis* has heterogeneous spacer lengths. *Proceedings of the National Academy of Sciences of the USA*, **71**, 2823–7.

WELLAUER, P. K., REEDER, R. H., DAWID, I. B. & BROWN, D. D. (1976). The arrangement of length of heterogeneity in repeating units of amplified and chromosomal ribosomal DNA from *Xenopus laevis*. *Journal of Molecular Biology*, **105**, 487–505.

WILLEMS, M., PENMAN, M. & PENMAN, S. (1969). The regulation of RNA synthesis and processing the nucleolus during inhibition of protein synthesis. *Journal of Cell Biology*, **51**, 177–91.

WU, R. S., KUMAR, A. & WARNER, J. R. (1971). Ribosome formation is blocked by camptothecin, a reversible inhibitor of RNA synthesis. *Proceedings of the National Academy of Sciences of the USA*, **68**, 3009–104.

ZYLBER, E. A. & PENMAN, S. (1971). Synthesis of 5S and 4S RNA in metaphase arrested HeLa cells. *Science*. **172**, 947–9.

The control of vitellogenin synthesis by oestrogen in *Xenopus laevis* and conversion of vitellogenin into yolk proteins

J. KNOWLAND AND B. WESTLEY

Dept of Biochemistry, University of Oxford, South Parks Road,
Oxford OX1 3QU, UK

Vitellogenin is the name given to a protein, made in female liver, which is secreted into the blood-stream, absorbed by the growing oocytes, and there converted into the egg-yolk proteins lipovitellin and phosvitin. The synthesis of vitellogenin and its ultimate conversion to yolk proteins have been studied most intensively in the frog *Xenopus laevis*. Since the original observation that oestrogen induces vitellogenin synthesis in male frogs, which normally make no vitellogenin at all (Wallace, 1967), it has been shown that vitellogenin synthesis can also be induced in male liver *in vitro* (Wangh & Knowland, 1975). The response *in vitro* is very fast, specific to oestrogen, and does not require cytodifferentation. For these reasons the system is unusually attractive for the study of how a particular gene is controlled by a steroid hormone. In *Xenopus* there is the added advantage that oocytes, eggs, embryos at all stages of development and adults are readily available, so that the activity of the vitellogenin gene and its control can be studied throughout development, with a view to understanding how the control systems present in adult liver are built up during development.

Some of the reasons for studying vitellogenesis, and the main findings, have recently been reviewed (Tata, 1976). The purpose of this article is to summarize our current knowledge of vitellogenesis in *Xenopus laevis*, especially the more recent advances, and to suggest future work.

Structure and properties of vitellogenin

Native *Xenopus* vitellogenin contains two polypeptides, each with a molecular weight approaching 200000 (Bergink & Wallace, 1974), but recent work suggests that the size of the chains may vary. At least three species, with molecular weights of 182000, 188000 and 197000, have been detected (Wiley & Wallace, 1978), which means that vitellogenin may be heterogeneous. The core polypeptides carry about 0.7 % of carbohydrate and 12 % of lipid, increasing the overall native molecular weight to about 450000. The carbohydrate seems to be covalently attached, but the lipid is almost certainly non-covalently bound, because it can be removed by extraction with chloroform–methanol (Bligh & Dyer, 1959; Wallace, 1965). The carbohydrate and lipid that form part of vitellogenin enter the oocyte, and are presumably as important as the polypeptide in providing the oocyte and egg with material that can support the growth of the developing embryo until it can feed for itself. Another interesting feature is that one region of vitellogenin, comprising about 35000 daltons, consists largely of serine. Fifty-six per cent of the amino acids in this region, which is released in the oocyte as phosvitin, are serine, and over 70 % of them are phosphorylated. Each phosphate residue is neutralized by a bound calcium ion (Follett & Redshaw, 1968), so that vitellogenin seems to transport calcium into the oocyte as well as providing a food store for the growing embryo. In view of the possible involvement of calcium ions in the conversion of the oocyte into an egg (Maller, Wu & Gerhart, 1977), this aspect of vitellogenin may also be important.

Vitellogenin is an extremely soluble protein, and serum levels can reach as much as 150 mg/ml if animals are heavily stimulated with oestrogen. Under normal circumstances, however, the output of vitellogenin from the liver seems to match the uptake by the oocytes, so that the serum concentration remains steady (Wallace & Jared, 1969).

Entry of vitellogenin into oocytes
and its subsequent fate

Vitellogenin appears to enter oocytes by pinocytosis (Wallace & Jared, 1969) and it seems that the innermost of the three layers of follicle cells is needed for entry (Wallace, Jared & Nelson, 1970).

It is generally believed that vitellogenin is the only protein to enter the oocyte from the blood, but the evidence is inconclusive because the proteins that are absorbed by the oocytes have not been analysed using the powerful techniques that are now available. Once inside the oocyte vitellogenin is cleaved, releasing the mature yolk proteins lipovitellin and phosvitin. Unfortunately, our knowledge of the pathway that yields the yolk proteins is rather scanty. The evidence indicates (Jared, Dumont & Wallace, 1973; Dehn & Wallace, 1973) that vitellogenin quickly becomes associated with material from the oocyte surface membrane, which slowly discharges vitellogenin to the yolk platelets, where the actual conversion to lipovitellin and phosvitin takes place. Vitellogenin that is injected into the oocyte rather than absorbed by it is not processed normally but merely degraded (Dehn & Wallace, 1973; Berridge & Lane, 1976), while vitellogenin synthesized inside the oocyte on either injected polysomes (Lanclos & Hamilton, 1975) or injected mRNA (Berridge & Lane, 1976) is correctly converted into lipovitellin and phosvitin. These somewhat paradoxical observations suggest that in normal circumstances the oocyte surface plays an important part in delivering vitellogenin to the proper compartment of the cell for processing, but that this route is not compulsory, and can be by-passed if the vitellogenin is synthesized inside the oocyte. This opens the possibility that some yolk proteins may actually be synthesized by the oocyte itself. Very little if any radioactivity is found in the yolk proteins of oocytes that have been incubated in radioactive amino acids, but the sensitivity of the measurements does not entirely rule out the possibility of a residual amount of synthesis of either yolk protein or vitellogenin (Berridge & Lane, 1976).

The products of the cleavage of vitellogenin inside the oocyte are lipovitellin and phosvitin. In 6 M guanidine hydrochloride, lipovitellin dissociates into two subunits, and the most recent evidence (Ohlendorf *et al.*, 1977) suggests that each subunit contains three strongly but non-covalently associated polypeptides, with molecular weights of 105 500, 35 500 and 32 000. Native lipovitellin therefore contains six polypeptides with a combined molecular weight of 346 000, which is increased to about 420 000 by association with the lipid that enters the oocyte with vitellogenin. As in vitellogenin, this lipid is non-covalently bound, and is removed by treatment with guanidine hydrochloride or sodium dodecyl sulphate (SDS) as well as by extraction with chloroform–

methanol (Bergink & Wallace, 1974; Bligh & Dyer, 1959). Phosvitin appears to form dimers in aqueous solution and in the yolk platelets, but has a monomer molecular weight of about 17000. Each vitellogenin monomer therefore seems to release three polypeptides contributing to lipovitellin and one contributing to phosvitin, accounting for nearly all the vitellogenin monomer, although the exact relationship between the components of lipovitellin and phosvitin and the apparent heterogeneity in vitellogenin (Wiley & Wallace, 1978) remains uncertain.

Lipovitellin and phosvitin are, in contrast to vitellogenin, insoluble; and one lipovitellin molecule containing six subunits combines with one dimeric phosvitin molecule in the yolk to form a crystalline lattice (Ohlendorf *et al.*, 1975). This enables the oocyte to accumulate a large store of yolk without creating osmotic problems, and indeed about 90 % of the protein in a mature oocyte (about 200 μg) is yolk (Benbow, Pestell & Ford, 1975).

Induction of vitellogenin synthesis
Primary response

It was shown more than 10 years ago (Wallace, 1967) that male frogs, which never normally make any vitellogenin, produce a great deal if they are injected with oestrogen, and the original observations have been confirmed and extended many times (Clemens, 1974). It was also shown that vitellogenin is made only in the liver (Rudack & Wallace, 1968). Further analysis of the mechanism of induction clearly required an in-vitro system, but until recently attempts to induce vitellogenin synthesis in male liver *in vitro* failed. Carinci, Locci, Bodo & Caruso (1974) cultured embryonic chicken liver cells and showed that proteins which reacted with antibodies to phosvitin appeared when oestrogen was added. Although the sex of the chick embryos which provided the liver was not established, and some of the tests suggested that the control cultures might also have synthesized some reactive material, it seems quite possible that vitellogenin synthesis was induced *in vitro*. (Note: the protein apparently induced was called phosvitin in this paper, but was almost certainly vitellogenin.)

The first clear demonstration that vitellogenin synthesis can be induced in male *Xenopus* liver *in vitro* used [35S]methionine to label liver cultures and SDS gel electrophoresis to identify the vitellogenin synthesized (Wangh & Knowland, 1975). Previous attempts had not met with success partly because small amounts of high specific activity [^{35}S]methionine were added to methionine-free medium, giving a methionine concentration that was too low to support protein synthesis in the cultures used. However, if the medium used to label the cultures contains enough methionine to allow protein synthesis to continue, the direct induction of vitellogenin synthesis by oestradiol in cultures of male *Xenopus* liver can be demonstrated (Wangh & Knowland, 1975; Wangh, Longthorne & Knowland, 1976), and the importance of replacing methionine-free medium (Clemens, Lofthouse & Tata, 1975) with methionine-depleted medium in such experiments has been confirmed (Green & Tata, 1976).

It is interesting to note in this connection that severe amino acid starvation of both bacterial and mammalian cells can sometimes result in extensive mis-translation (Parker, Pollard, Friesen & Stanners, 1978), which is an added reason for ensuring that cell or organ cultures are labelled with radioactive amino acids under conditions which allow protein synthesis to continue at a reasonable rate.

The work outlined above was concerned with the primary induction of vitellogenin synthesis, i.e. the response obtained with male liver that had never previously been exposed to oestradiol. The main conclusions to be drawn (Wangh & Knowland, 1975) are:

1. The response is specific to oestrogens, no other factor being required, and can be obtained at physiological concentrations of oestradiol.

2. The response is very fast, being detectable within 6 hours, and possibly as early as 2 hours (not at 1 hour as quoted by Baker & Shapiro, 1977), showing that cytodifferentiation is not required for the induction. It has since been shown (Green & Tata, 1976) that DNA synthesis is not needed for induction.

3. Oestrogen is required continuously. If it is withdrawn, vitellogenin synthesis stops, but starts again when the oestrogen is replaced.

More recent work (Baker & Shapiro, 1977) suggests that when male frogs are injected with oestrogen, vitellogen mRNA cannot be detected in the liver until 3–4.5 hours after the injection. This response appears to be somewhat slower than that found for the induction of vitellogenin *in vitro* (Wangh & Knowland, 1975). It may reflect a slower application of oestrogen to the liver cells as a consequence of injecting the hormone into the lymph sac rather than adding it directly to liver cultures. The speed of the primary response may also depend upon other variables; e.g. the temperature at which oestrogen-stimulated frogs are kept before their livers are removed, whether pieces of liver are labelled at 25 °C (Wangh & Knowland, 1975) or at room temperature (Baker & Shapiro, 1977), and the size and age of the animals used. We have found that the primary response is generally more consistent when small (up to 25 g), young (up to 1-year-old), laboratory-reared frogs are used, and it may be important to choose such animals to obtain the fastest possible primary response.

These findings have been extended in studies of the secondary response, where animals are first injected with oestrogen, but the response of the liver to oestrogen is examined only after the first wave of vitellogenin synthesis has declined. Some of the salient features of the secondary response to oestrogen are discussed below.

Secondary response

The distinctive characteristic of secondary induction of vitellogenin is that it is faster than primary induction. This so-called anamnestic response has been the subject of a certain amount of interest, presumably because comparison of untreated and oestrogen-treated livers should reveal factors that increase the efficiency of vitellogenin induction.

It seems fairly certain that the anamnestic response is not due to an unmasking of stored vitellogenin mRNA transcribed during primary stimulation, because the levels of vitellogenin mRNA are no higher prior to secondary stimulation than they are before primary stimulation (0–5 molecules per cell, compared with more than 10000 after full stimulation: Baker & Shapiro, 1977; Ryffel, Wahli & Weber, 1977). Most studies agree that there is a correlation between the level of vitellogenin mRNA and the rate of vitellogenin production (Shapiro, Baker & Stitt, 1976), and that the initial rate of vitellogenin mRNA synthesis is greater after secondary stimulation than after primary stimulation (Baker &

Shapiro, 1977; Ryffel *et al.*, 1977). This indicates that the effect is predominantly on the transcription of the vitellogenin gene into mRNA, rather than on the translation of vitellogenin mRNA, although there is some evidence to suggest that the latter might also occur (Farmer, Henshaw, Berridge & Tata, 1978). There have not yet been any suggestions as to which factors might be involved, but one possibility that obviously requires serious consideration is that there is more oestrogen receptor present after primary stimulation, and our preliminary results suggest that this is the case (Westley & Knowland, 1979).

The mechanism by which vitellogenin synthesis is induced

Perhaps the most striking features of vitellogenin synthesis are that the primary response in male liver is very fast, increasing rapidly from zero without the need for cytodifferentiation, that the secondary response is faster, and that oestrogen is needed continuously. Any mechanism proposed must account for these facts, but so far only limited progress has been made. The more recent findings are outlined below.

The rapid primary response in male liver (Wangh & Knowland, 1975) strongly suggests that only oestrogen is needed to trigger the induction, and that all other factors required must already exist in the liver cells, even though males normally never synthesize any vitellogenin. This focuses attention on how the adult characteristic of inducibility is acquired, a problem related to differentiation of the liver. A non-inducible mutant would clearly be invaluable here, but the chance of finding one seems remote. A possible alternative approach is to test liver from early embryos for ability to synthesize vitellogenin. Then, by comparing non-inducible with inducible embryos, it may be possible to identify the factors that confer inducibility on liver.

Tadpoles injected with oestradiol before metamorphosis do not synthesize vitellogenin, but newly metamorphosed frogs do (Follett & Redshaw, 1974). These findings have been extended by testing the inducibility of tadpole liver *in vitro* as well as *in vivo* (Knowland, 1978). Vitellogenin synthesis can first be induced by oestradiol in liver taken from tadpoles at the end of prometamorphosis (stage 57) when the front legs break through, and the response increases during the completion of metamorphosis. If tadpoles are reared from an early stage with oestradiol always present in their water, inducibility appears at the same stage and to the same degree, which shows that oestrogen alone is not sufficient to establish the inducible state. These findings suggest that inducibility appears at about stage 57 as part of the normal development of the liver, independently of the supply of oestrogen.

Although inducibility is established during metamorphosis it is not clear whether inducibility and metamorphosis are compulsorily linked. *Xenopus* tadpoles eventually synthesize adult haemoglobin even if metamorphosis is inhibited (Maclean & Turner, 1976), so that the expression of at least one adult gene can be disconnected from metamorphosis. The ability to synthesize vitellogenin can also be considered an adult function, and may appear independently of metamorphosis. In any case, a comparison of non-inducible with inducible liver may help in the search for factors involved in controlling the vitellogenin gene.

The oestrogen receptor from *Xenopus* liver

One factor that almost certainly plays a central part in activating the vitellogenin gene is a receptor protein that binds oestrogens with high affinity, and which we have recently identified in extracts of *Xenopus* liver (Westley & Knowland, 1978). So far there have been surprisingly few studies of oestrogen receptors in *Xenopus* liver, even though the induction of vitellogenin synthesis clearly provides a good system for studying the mechanism of steroid hormone action.

Steroid receptors are universally identified by virtue of their binding to radioactive steroid, but the following difficulties are encountered when using this procedure in *Xenopus* liver. First, binding of [^{3}H]oestradiol to the receptor has to be distinguished from binding to other steroid-binding components which might be present, such as contaminating steroid-binding proteins from the blood. Secondly, the rate of dissociation of the *Xenopus* [^{3}H]oestradiol– receptor complex is rather high, which means that modified procedures have to be adopted if the receptor is to be identified using lengthy techniques such as sucrose gradient ultracentrifugation. Thirdly, the liver is an important site of

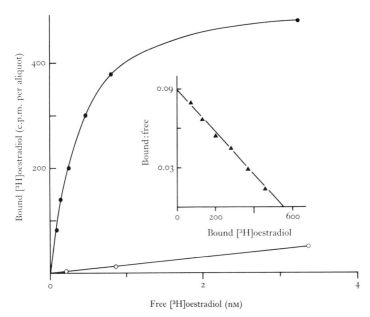

Fig. 1. Affinity of [³H]oestradiol binding in nuclear extracts from male *Xenopus* liver. Nuclei were prepared from male *Xenopus* liver and extracted with 0.5 M potassium chloride as described by Westley & Knowland (1978). Aliquots (200 μl) of nuclear extracts were incubated with various concentrations of [³H]oestradiol at 0 °C for 9 hours, and then bound [³H]oestradiol was determined as described by Westley & Knowland (1978). ●——●, the total [³H]oestradiol bound; ○——○, the [³H]oestradiol bound with low affinity. The insert shows the c.p.m. of tightly bound [³H]oestradiol plotted according to Scatchard, from which it was calculated that there were 99 ± 19 sites per nucleus having an affinity of 0.5×10^{-9} M. (From Westley & Knowland (1978). © MIT; published by MIT Press.)

steroid catabolism. This means that during incubations of steroids with liver *in vitro* or with subcellular fractions containing steroid-metabolizing enzymes the concentration of free steroid is continuously and sometimes very rapidly decreasing. Finally, the levels of oestrogen receptor in male liver are much lower than in most other steroid-responsive systems, so that it is important to use an assay system which is capable of reliably measuring these low levels.

We have found an oestrogen-binding component, which we believe is the oestrogen receptor involved in regulating vitellogenin synthesis, in salt extracts (0.5 M potassium chloride) of liver nuclei (Westley & Knowland, 1978). [³H]oestradiol bound to the

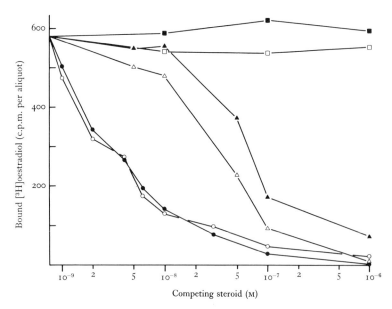

Fig. 2. Suppression of [³H]oestradiol binding in nuclear extracts by various unlabelled steroids. Aliquots (200 μl) of nuclear extracts prepared as described in Fig. 1 were incubated with [³H]oestradiol (5×10^{-9} M) in the presence of various concentrations of unlabelled oestradiol-17β (○——○), diethylstilboestrol (●——●), oestrone (△——△), oestriol (▲——▲), testosterone (□——□), or dexamethasone (■——■) at 20 °C for 1 hour. Bound [³H]oestradiol was determined as described in Fig. 1. (From Westley & Knowland (1978). © MIT; published by MIT Press.)

receiver is assayed using gel filtration on small columns of Sephadex LH-20, a method which has been used with great success to assay the low levels of steroid receptors found in brain tissue (Ginsburg *et al.*, 1974).

Fig. 1 shows that the male liver receptor has a high affinity ($K_D = 0.5 \times 10^{-9}$ M) for oestradiol, and this value is close to the minimum concentration of about 2×10^{-9} M oestradiol required to induce vitellogenin *in vitro* (Wangh & Knowland, 1975). From this data we have estimated that there are about 100 sites per nucleus, although the number increases dramatically after oestrogen treatment. Fig. 2 shows that the binding is specific for oestrogen. Oestradiol and diethylstilboestrol compete with high efficiency for [³H]oestradiol binding. The less potent oestrogens oestriol and oestrone compete less efficiently for [³H]oestradiol binding, while dexamethasone and testosterone do not compete at

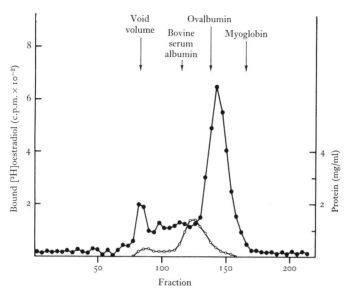

Fig. 3. Gel filtration of nuclear extract from male *Xenopus* liver. Nuclear extract (4–6 ml), prepared as described in Fig. 1, was chromatographed on a column (90 cm × 2.5 cm) of G100 Sephadex equilibrated with buffer containing 0.5 M potassium chloride at 20 °C. Fractions (2 ml) were collected and every fourth fraction was analysed for total protein (O——O) and high-affinity [³H]oestradiol binding activity (●——●) as described in Fig. 1. The major peak of high-affinity [³H]oestradiol binding activity eluted with a Stokes radius of 2.6 nm, estimated by comparing with the elution positions of various marker proteins as described by Westley & Knowland (1978). (From Westley & Knowland (1978). © MIT; published by MIT Press.)

all even at a concentration of 10^{-6} M. This ranking of affinities correlates well with the ability of steroids to induce vitellogenin synthesis both *in vivo* and *in vitro* (Redshaw, Follett & Nicholls, 1969; Wangh & Knowland, 1975).

We have also measured the sedimentation coefficient and Stokes radius of the receptor. Fig. 3 shows the elution profile of the receptor on Sephadex G100. The receptor elutes as one major peak with a Stokes radius of 2.6 nm. On sucrose gradients the receptor sediments at 3.5S and from these values we calculate a molecular weight of approximately 40000. This value is somewhat smaller than that for other oestrogen receptors measured in similar salt concentrations (generally around 60000: King & Mainwaring, 1974), but it is considerably higher than other estimates of steroid-binding components in *Xenopus* liver nuclei (10000: Klotz

& Rickwood, 1978) and cytoplasm (20000: Bergink & Wittliff, 1975).

We have also found a component in the cytoplasm of male *Xenopus* liver which resembles the nuclear receptor in its specificity and affinity for various steroids, but whether this represents true cytoplasmic receptor or simply nuclear receptor which has been released from the nucleus during homogenization is not yet certain.

Both the nuclear and cytoplasmic binding components can be distinguished from the steroid-binding proteins of the blood by their specificity, affinity and sedimentation properties, and therefore the nuclear and cytoplasmic oestrogen-binding activity is not due to contamination with blood.

Finally, it is interesting that a high proportion of the oestrogen receptor in male *Xenopus* liver is located in the nucleus; thus it has a rather different subcellular localization from the majority of other steroid receptors (King & Mainwaring, 1974) although similar to the oestrogen receptor of the chicken liver (Mester & Baulieu, 1972; Gschwendt & Kittstein, 1974) and certain steroid-responsive cell lines (Zava & McGuire, 1977).

Future work

It is clear from this outline of vitellogenin synthesis that a great deal of work is still required to reach a proper understanding of how the vitellogenin gene is regulated. Our knowledge of the vitellogenin molecule itself is still far from complete. Although the main properties of the oestrogen receptor from *Xenopus* liver have been established, the molecule itself has not been isolated, and the part that it plays in activating vitellogenin synthesis is obscure. For example, it is not clear how the receptor moves between nucleus and cytoplasm, or whether the appearance of inducibility during differentiation of the liver correlates with synthesis of the receptor. Nor do we know whether the receptor is the only molecule, apart from oestrogen, needed to activate the vitellogenin gene, or whether others may be required as well.

A possible direct test of the function of the receptor has been suggested (Wangh *et al.*, 1976). If the receptor is injected into *Xenopus* cells that do not normally synthesize vitellogenin, it may induce vitellogenin synthesis. Oocytes or embryonic cells of

Xenopus are suitable recipient cells for such an experiment, which, if successful, would provide direct evidence that the *Xenopus* liver oestrogen receptor is a regulatory protein.

Summary

In animals that lay eggs, the yolk is derived from a macromolecular precursor, vitellogenin. The polypeptide core of vitellogenin is synthesized in the mother's liver; carbohydrate and fat are added and the completed molecule is secreted into the blood. Vitellogenin is absorbed by the growing oocytes, where it is enzymatically cleaved, releasing the mature yolk proteins. Vitellogenin is synthesized in *Xenopus* liver only when oestrogen is supplied, and the induction is very interesting for two main reasons. First, vitellogenin synthesis can be induced in culture as well as *in vivo*, and rapidly increases from zero to high levels without any obvious cytodifferentiation, which makes the system very attractive for studying the control of gene expression by steroid hormones. Secondly, by establishing the stage at which vitellogenin synthesis can first be induced in embryos, and then analysing the transition from non-inducibility to inducibility, it may be possible to see how an adult function, namely the inducibility of vitellogenin synthesis, is acquired during differentiation.

We are grateful to the Medical Research Council for supporting this work, and for providing B.W. with a post-doctoral training fellowship.

References

BAKER, H. J. & SHAPIRO, D. J. (1977). Kinetics of estrogen induction of *Xenopus laevis* vitellogenin messenger RNA as measured by hybridization to complementary DNA. *Journal of Biological Chemistry*, **252**, 8428–34.

BENBOW, R. M., PESTELL, R. Q. W. & FORD, C. C. (1977). Appearance of DNA polymerase activities during early development of *Xenopus laevis*. *Developmental Biology*, **43**, 159–74.

BERGINK, E. W. & WALLACE, R. A. (1974). Precursor–product relationship between amphibian vitellogenin and the yolk proteins, lipovitellin and phosvitin. *Journal of Biological Chemistry*, **249**, 2897–903.

BERGINK, E. W. & WITTLIFF, J. L. (1975). Molecular weight of estrogen and androgen binding proteins in the liver of *Xenopus laevis*. *Biochemistry*, **14**, 3115–21.

BERRIDGE, M. V. & LANE, C. D. (1976). Translation of *Xenopus* liver messenger RNA in *Xenopus* oocytes: vitellogenin synthesis and conversion to yolk platelet proteins. *Cell*, **8**, 283–97.

BLIGH, E. G. & DYER, W. J. (1959). A rapid method of total lipid extraction and purification. *Canadian Journal of Biochemistry and Physiology*, **37**, 911–17.

CARINCI, P., LOCCI, P., BODO, M. A. & CARUSO, A. (1974). Estrogen-induced phosvitin [*sic*] synthesis in cultured chick embryo liver cells. *Experientia*, **30**, 88–9.

CLEMENS, M. J. (1974). The regulation of egg yolk protein synthesis by steroid hormones. *Progress in Biophysics and Molecular Biology*, **28**, 71–107.

CLEMENS, M. J., LOFTHOUSE, R. & TATA, J. R. (1975). Sequential changes in the protein synthetic activity of male *Xenopus laevis* liver following induction of egg-yolk proteins by estradiol-17β. *Journal of Biological Chemistry*, **250**, 2213–18.

DEHN, P. F. & WALLACE, R. A. (1973). Sequestered and injected vitellogenin. Alternative routes of protein processing in *Xenopus* oocytes. *Journal of Cell Biology*, **58**, 721–4.

FARMER, S. R., HENSHAW, E. C., BERRIDGE, M. V. & TATA, J. R. (1978). Translation of *Xenopus* vitellogenin mRNA during primary and secondary induction. *Nature, London*, **273**, 401–3.

FOLLETT, B. K. & REDSHAW, M. R. (1968). The effects of oestrogen and gonadotrophins on lipid and protein metabolism in *Xenopus laevis* Daudin. *Journal of Endocrinology*, **40**, 439–56.

FOLLETT, B. K. & REDSHAW, M. R. (1974). The physiology of vitellogenesis. In *Physiology of the Amphibia*, ed. B. Lofts, pp. 219–308. New York & London: Academic Press.

GINSBURG, M., GREENSTEIN, B. D., MacLUSKY, N. J., MORRIS, I. O. & THOMAS, P. J. (1974). An improved method for the study of high-activity steroid binding: oestradiol binding in brain and pituitary. *Steroids*, **23**, 773–92.

GREEN, C. D. & TATA, J. R. (1976). Direct induction by estradiol of vitellogenin synthesis in organ cultures of male *Xenopus laevis* liver. *Cell*, **7**, 131–9.

GSCHWENDT, M. & KITTSTEIN, W. (1974). Specific binding of estradiol to the liver chromatin of estrogenised roosters. *Biochimica et Biophysica Acta*, **301**, 84–96.

JARED, D. W., DUMONT, J. N. & WALLACE, R. A. (1973). Distribution of incorporated and synthesized protein among cell fractions of *Xenopus* oocytes. *Developmental Biology*, **35**, 19–28.

KING, R. J. B. & MAINWARING, W. I. P. (1974). *Steroid–Cell Interactions*. London: Butterworth.

KLOTZ, M. & RICKWOOD, D. (1978). Estradiol-binding protein in the nucleus of male *Xenopus laevis* liver. *Molecular and Cellular Biochemistry*, **19**, 125–9.

KNOWLAND, J. (1978). Induction of vitellogenin synthesis in *Xenopus laevis* tadpoles. *Differentiation*, **12**, 47–51.

LANCLOS, K. D. & HAMILTON, T. H. (1975). Translation of hormone-induced messenger RNA in amphibian oocytes. I. Induction by estrogen of messenger RNA encoded for vitellogenic protein in the liver of the male African clawed toad (*Xenopus laevis*). *Proceedings of the National Academy of Sciences of the USA*, **72**, 3934–8.

MACLEAN, N. & TURNER, S. (1976). Adult haemoglobin in developmentally retarded tadpoles of *Xenopus laevis*. *Journal of Embryology and Experimental Morphology*, **35**, 261–6.

MALLER, J., WU, M. & GERHART, J. C. (1977). Changes in protein phosphorylation accompanying maturation of *Xenopus laevis* oocytes. *Developmental Biology*, **58**, 295–312.

MESTER, J. & BAULIEU, E. E. (1972). Nuclear estrogen receptor of chick liver. *Biochimica et Biophysica Acta*, **261**, 236–44.

OHLENDORF, D. H., COLLINS, M. L., PURONEN, E. O., BANASZAK, L. J. & HARRISON, S. C. (1975). Crystalline lipoprotein–phosphoprotein complex in oocytes from *Xenopus laevis*: determination of lattice parameters by X-ray crystallography and electron microscopy. *Journal of Molecular Biology*, **99**, 153–65.

OHLENDORF, D. H., BARBARASH, G. R., TROUT, A., KENT, C. & BANASZAK, L. J. (1977). Lipid and polypeptide components of the crystalline yolk system from *Xenopus laevis*. *Journal of Biological Chemistry*, **252**, 7992–8001.

PARKER, J., POLLARD, J. W., FRIESEN, J. D. & STANNERS, C. P. (1978). Stuttering: high-level mistranslation in animal and bacterial cells. *Proceedings of the National Academy of Sciences of the USA*, **75**, 1091–5.

REDSHAW, M. R., FOLLETT, B. K. & NICHOLLS, T. J. (1969). Comparative effects of the estrogens and other steroid hormones on serum lipids and proteins in *Xenopus laevis* Daudin. *Journal of Endocrinology*, **43**, 47–53.

RUDACK, D. & WALLACE, R. A. (1968). On the site of phosvitin synthesis in *Xenopus laevis*. *Biochimica et Biophysica Acta*, **155**, 299–301.

RYFFEL, G. U., WAHLI, W. & WEBER, R. (1977). Quantitation of vitellogenin messenger RNA in the liver of male *Xenopus* toads during primary and secondary stimulation by estrogen. *Cell*, **11**, 213–21.

SHAPIRO, D. J., BAKER, H. J. & STITT, D. T. (1976). In vitro translation and estradiol-17β induction of *Xenopus laevis* vitellogenin messenger RNA. *Journal of Biological Chemistry*, **251**, 3105–11.

TATA, J. R. (1976). The expression of the vitellogenin gene. *Cell*, **9**, 1–14.

WALLACE, R. A. (1965). Resolution and isolation of avian and amphibian yolk-granule proteins using TEAE-cellulose. *Analytical Biochemistry*, **11**, 297–311.

WALLACE, R. A. (1967). A serum lipophosphoprotein produced by vitellogenic females and estrogenized males of *Xenopus laevis*. *Journal of Cell Biology*, **35**, 137A.

WALLACE, R. A. & JARED, D. W. (1969). Studies on amphibian yolk. VIII. The estrogen-induced hepatic synthesis of a serum lipophosphoprotein and its selective uptake by the ovary and transformation into yolk platelet protein in *Xenopus laevis*. *Developmental Biology*, **19**, 498–526.

WALLACE, R. A., JARED, D. W. & NELSON, B. L. (1970). Protein incorporation by isolated amphibian oocytes. I. Preliminary studies. *Journal of Experimental Zoology*, **175**, 259–70.

WANGH, L. J. & KNOWLAND, J. (1975). Synthesis of vitellogenin in cultures of male and female frog liver regulated by estradiol treatment *in vitro*. *Proceedings of the National Academy of Sciences of the USA*, **72**, 3172–5.

WANGH, L. J., LONGTHORNE, R. F. & KNOWLAND, J. (1976). Investigation of the mechanisms and consequences of steroid hormone action on vitellogenin synthesis in *Xenopus laevis*. In *The Molecular Biology of Hormone Action*, ed. J. Papaconstantinou, *Symposium of the Society for Developmental Biology 34*, pp. 151–69. New York & London: Academic Press.

WESTLEY, B. & KNOWLAND, J. (1978). An estrogen receptor from *Xenopus laevis* liver possibly connected with vitellogenin synthesis. *Cell*, **15**, 367–74.

WESTLEY, B. & KNOWLAND, J. (1979). Estrogen causes a rapid, large and prolonged rise in the level of nuclear estrogen receptor in *Xenopus laevis* liver. *Biochemical and Biophysical Research Communications*, in press.

WILEY, H. S. & WALLACE, R. A. (1978). Three different molecular weight forms of the vitellogenin peptide from *Xenopus laevis*. *Biochemical and Biophysical Research Communications*, **85**, 153–9.

ZAVA, D. T. & McGUIRE, W. L. (1977). Estrogen receptor. Unoccupied site in nuclei of the breast tumour cell line. *Journal of Biological Chemistry*, **252**, 3703–8.

Cytoplasmic localizations in mosaic eggs

M. R. DOHMEN and N. H. VERDONK

University of Utrecht, Zoological Laboratory, Transitorium 3, Padualaan 8,
Utrecht, The Netherlands

In most animal species the early stages of development are determined to a large extent by maternal factors that are synthesized during oogenesis. These factors are generally located in the cytoplasm or the cortex of the egg, and they are segregated into special regions of the egg by a process called 'ooplasmic segregation' (Costello, 1948) or 'cytoplasmic localisation' (Wilson, 1929). Cleavage then leads to a mosaic of cells with different developmental destinies. When these cells, after isolation, show a conspicuous capacity of self-differentiation, the egg is called a 'mosaic' or 'determinate' egg. When regulative processes intervene in the development of isolated blastomeres, thus causing a re-programming of their original destiny, the egg is called a 'regulative' egg (Heider, 1900). Many intermediate types of eggs are found between these extremes, and it was recognized long ago (Wilson, 1929, 1937) that the distinction between mosaic and regulative eggs does not reflect a fundamental difference.

Extreme mosaic eggs are in some respects very useful for the study of maternal determination of development via cytoplasmic factors. In many of these eggs special features indicate where important morphogenetic determinants are localized. In eggs with unequal cleavage, one of the two cells generally contains the basic determinants and the other cell is of minor importance. In eggs of certain annelids and molluscs part of the cytoplasm forms a so-called polar lobe at successive cleavages, and this lobe contains essential determinants. Unequal cleavage and polar lobe formation both seem to be mechanisms for segregating the vegetal pole region of the egg into one blastomere. The cytoplasmic regions which contain important developmental factors are generally

127

characterized by special localizations. These localizations may facilitate the study of early determination, as exemplified by the germ cell determinants in insect and amphibian eggs. In this review cytoplasmic and cortical localizations will be discussed mainly with respect to annelids, molluscs and ascidians.

Cytoplasmic and cortical localizations

In a large number of species special cytoplasms are segregated into specific cell lines. It is assumed, therefore, that these plasms contain morphogenetic determinants. In some cases experimental evidence supports this assumption. Irradiation or deletion of the plasm may prevent the appearance of the corresponding differ-entiations, and transplantation or abnormal segregation of the plasm may induce these differentiations in the cell lines in which the plasms become located as a result of the experiments. There are also cases, however, in which the presence of morphogenetic determinants in certain regions of an egg can be demonstrated experimentally without any cytoplasmic peculiarity to indicate the presence of a morphogenetic localization in that region. The structurally distinguishable morphogenetic cytoplasms can be divided into several classes, which are discussed below.

Vesicular aggregates

In a number of molluscs and annelids the plasm at the vegetal pole of the egg is constricted off at cleavage, forming a so-called polar lobe. This lobe fuses with one of the blastomeres, thereby shunting the vegetative pole plasm to this blastomere. In some

---

Fig. 1. (a) *Clepsine sexoculata* (Annelida, Hirudinea). Detail of animal pole plasm at the 2-cell stage. The plasm consists of a large number of mitochondria (m) and β-glycogen particles (arrow). The superficial layer of the pole plasm, from which this detail is taken, contains many vesicles (diameter 100–150 nm) with moderately electron dense contents (double arrow). No difference between animal and vegetal pole plasm has been observed. Scale bar represents 1.0 μm. (b) *Buccinum undatum* (Mollusca, Gastropoda). Germ-plasm-like structure, found exclusively in the polar lobe. The dense mass consists of extremely small vesicles (diameter c. 30 nm). Scale bar represents 0.5 μm. (c) *Crepidula fornicata* (Mollusca, Gastropoda). Germ-plasm-like structure, found exclusively in the polar lobe. The structure consists of vesicles and clusters of dense granules (diameter c. 7 nm). Scale bar represents 1.0 μm. (From Dohmen & Lok, 1975.)

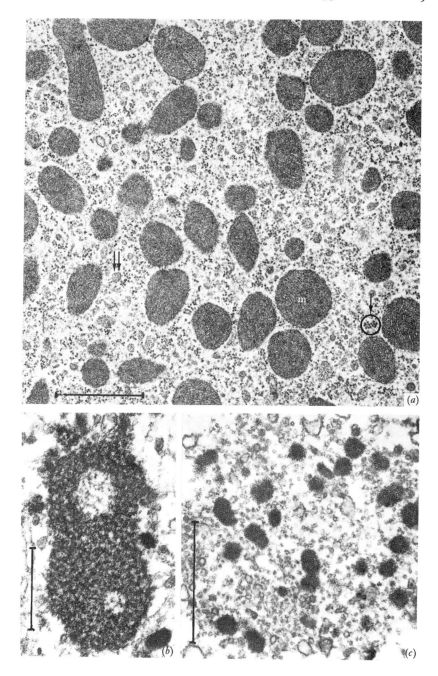

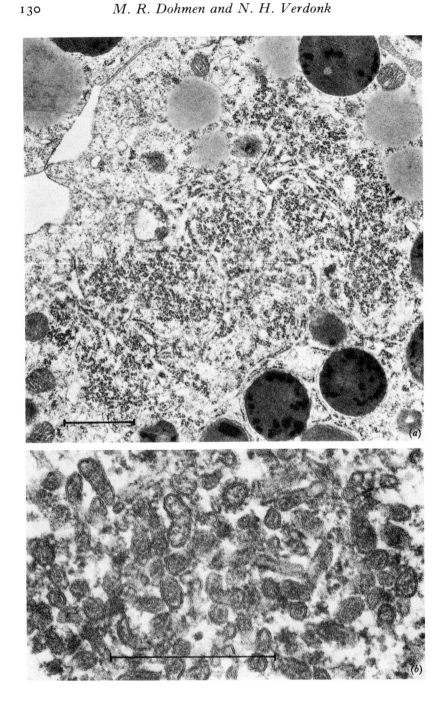

species polar lobes are as large as a blastomere; in other species they are much smaller, some of them containing less than 1 % of the total cytoplasm of the egg.

Experiments in which polar lobes were deleted prove that the lobes contain essential determinants. In the absence of a polar lobe the developing embryo shows important larval defects and most adult differentiations are absent. Small polar lobes appear to affect development no less than large lobes (Cather & Verdonk, 1974). Electron microscopical studies of gastropod eggs with small polar lobes indicate that several species contain typical aggregates of vesicles. The most conspicuous example is the cup-shaped aggregate, named the vegetal body, in the polar lobe of *Bithynia tentaculata* (Dohmen & Verdonk, 1974). A detail of a nascent vegetal body in an immature oocyte is shown in Fig. 2(b). The vesicles develop from the endoplasmic reticulum and during this process their contents become electron dense. Initially they are mostly elongated, but the mature egg contains spherical vesicles only.

Centrifugation experiments demonstrate that the vegetal body can be displaced from its original site. Subsequent deletion of the polar lobe, which is now deprived of the vegetal body, has no effect on development. Apparently the presence of the vegetal body in a blastomere suffices to induce normal development of a lobeless egg. It is concluded, therefore, that the morphogenetic determinants of the polar lobe of *Bithynia* are located in the vegetal body. This body is rich in RNA, as indicated by staining with methylgreen–pyronin and with the fluorescent dyes acridine orange and Hoechst 33258 (Dohmen & Verdonk, 1979). It would be premature, however, to attribute the morphogenetic effects to the RNA component.

The fate of the vegetal body is not yet clear. Just before second cleavage the cup-shape suddenly disappears and the body cannot be distinguished any longer in histological sections. In electron

Fig. 2. (a) *Lymnaea stagnalis* (Mollusca, Gastropoda). Ectosome at the inner end of a macromere at the late 24-cell stage. The ectosome consists of a dense aggregate of ribosomes, with endoplasmic reticulum and small vesicles (diameter *c.* 50 nm). The vesicles are not clearly visible at this magnification. Scale bar represents 1.0 μm. (From Dohmen & van de Mast, 1978.) (b) *Bithynia tentaculata* (Mollusca, Gastropoda). Detail of a nascent vegetal body in an oocyte during vitellogenesis. Scale bar represents 0.5 μm.

microscopical preparations the vesicles appear to remain intact and a large mass of them is segregated into the second polar lobe. During the 4-cell stage the vesicles disappear. The assumption of Verdonk & Cather (1973) that part of the vegetal body is segregated into the C-cell during second cleavage cannot yet be confirmed by actual observation.

Similar aggregates of electron dense vesicles have been observed in the polar lobes of the gastropods *Crepidula fornicata* (Dohmen & Lok, 1975), *Buccinum undatum* and *Littorina saxatilis*. In the last species the observed aggregate is very small (Dohmen & Verdonk, 1979) and its morphogenetic significance cannot be evaluated yet. In eggs of the gastropod *Lymnaea stagnalis* another type of vesiculate plasm is found, the so-called ectosomes, which are supposed to play a part in the induction of bilateral symmetry (see Raven, 1976). The ectosomes consist of a complex aggregate of small electron dense vesicles with ribosomes and endoplasmic reticulum (Dohmen & van de Mast, 1978). They appear for the first time at the 8-cell stage, in the same area where subcortical accumulations (SCA) were observed by Raven (1967, 1970). Probably the ectosomes are condensations of the SCA material. At the end of the 24-cell stage the ectosomes have become compact structures in which the ribosomes are predominant (Fig. 2a). The induction of bilateral symmetry is supposed to take place at this stage, but the role of the ectosomes in this process remains to be established.

Germ-plasm-like structures

In several groups of animals special cytoplasms, named germ plasms, are segregated into the germ cell line; in some species there is evidence that they are involved in the determinatioan of germ cells (see reviews by Beams & Kessel, 1974; Eddy, 1975). The so-called 'nuage' material found in the germ cells of numerous species is very similar to the germ plasms of insects and amphibians and the question has been raised whether the nuage represents germ-cell determinants.

Germ plasm or nuage material is reported in several species of annelids, molluscs and ascidians (see Eddy, 1975). In two gastropods, *Crepidula fornicata* (Dohmen & Lok, 1975) (Fig. 1c) and *Buccinum undatum* (Fig. 1b), germ-plasm-like structures are found exclusively in the polar lobe. They consist of a complex of small vesicles with aggregates of dense granules. In *Buccinum* some of these bodies have an electron translucent core (Fig. 1*b*). This type of core is characteristic for germinal granules in several insect species. Neither in *Crepidula* nor in *Buccinum* is the possible relation of these structures to primordial cells known. Their presence needs to be demonstrated in the 4d-cell, as in molluscs the gonads seem to originate from this cell (Raven, 1966).

Mitochondria

The eggs of many animals, especially annelids, molluscs and ascidians, contain concentrations of mitochondria (for references see Villa, 1976). The mitochondria may be concentrated in regions that contain essential morphogenetic determinants. Sometimes they are associated with special structures, such as the germinal granules in insects and amphibians or the vegetal body in the polar lobe of *Bithynia*, but they are also found in large concentrations without any structure to which a morphogenetic role can be attributed. The question arises, therefore, as to whether the mitochondria proper may function as morphogenetic agents. There is little evidence for this.

The finding of Sin (1975) that mitochondrial factors may induce puffs in chromosomes of the salivary glands of *Drosophila* demonstrates that mitochondria are to a certain extent capable of inducing differential gene activation. Bell & Holland (1974) showed that in *Ciona intestinalis* a minimum number of mitochondria are required to cause muscle cell differentiation, as indicated by a cytochemical reaction for cholinesterase. This result appears to be in conflict, however, with earlier reports by Tung, Ku & Tung (1941), who found that in the same species myofibrillae were often present in muscle cells although the mitochondria had been displaced into other cells by centrifugation of the unfertilized egg. The displaced mitochondria were not capable of inducing myofibrillae in the cells in which they came to be located. Also, in another species, *Styela*, larvae or normal form with contracting muscles can develop when the mitochondria are displaced to other parts of the body (Conklin, 1931). Interference with the energy-producing system of mitochondria, by inhibiting cytochrome oxidase and succinic dehydrogenase, does not prevent the differentiation of muscle cells, although their functioning is defective (Reverberi, 1971).

In the eggs of several oligochaetes and leeches the animal and vegetal pole plasms contain large numbers of mitochondria (cf. Fig. 1*a*). In *Tubifex* (Penners, 1922) and in *Clepsine* (Schleip, 1914) the pole plasms are segregated into the somatoblasts, which give rise to the ectodermal and mesodermal germ bands. Unfortunately, the centrifugation experiments on *Tubifex* eggs (Penners, 1924; Lehmann, 1948) do not permit conclusions to be drawn as to an eventual determinative significance of the mitochondria in the pole plasms.

Mitochondria-rich plasms are also found in the polar lobes of several species of annelids and molluscs, e.g. in *Myzostoma* (Pitotti, 1947), *Ilyanassa* (Geuskens & de Jonghe d'Ardoye, 1971) and *Dentalium* (Reverberi, 1970). In the last two species the contents of the polar lobes can be displaced by centrifugation without interfering with the developmental potential of this region (Verdonk, 1968; Clement, 1968). We have verified by electron microscopy that mitochondria are almost completely absent from polar lobes of eggs which were properly orientated during centrifugation. It is unlikely, therefore, that mitochondria play a determinative role in the development of these organisms. Also, in an insect species, *Smittia*, there is evidence that a concentration of mitochondria in the anterior region of the egg, which coincides with the localization of determinants for the head region of the embryo, does not play a determinative role (Kalthoff *et al.*, 1975). The available information thus leaves the impression that the mitochondria are not acting as morphogens, but are merely co-localized with morphogenetic determinants.

Cortical localizations

There is evidence that in some organisms morphogenetic determinants are not located in the cytoplasm but in the cortex of the egg. Centrifugation experiments with eggs of *Ilyanassa* (Clement, 1968) and *Dentalium* (Verdonk, 1968) showed that the displaceable components of the polar lobe cytoplasm do not contain the morphogens. Electron microscopical investigation of centrifuged eggs did not reveal any non-displaceable cytoplasmic component in the polar lobe. These data suggest that the cortex is the site where the morphogens of the lobe must be located. This conclusion is supported by recent unpublished results of van den Biggelaar, who found that after the cytoplasm has been sucked out of the polar lobe of *Dentalium*, development may proceed normally. Similar experiments were carried out with sea urchin eggs (Hörstadius, Lorch & Danielli, 1950). After removing half or more of the ooplasmic contents, normal plutei may develop. And in *Loligo* eggs it was demonstrated in various ways that a pattern of developmental information must be present in the egg cortex (Arnold, 1968; Arnold & Williams-Arnold, 1974, 1976).

Unfortunately, the concept of cortical localization is as poorly defined as the cortex itself. The cortex is generally considered to be the layer of the egg that is not displaced by ooplasmic streaming or other displacement of the cytoplasm. It is difficult at this time to present a definition which tries to single out the structure that imparts the rigidity to this layer and acts as a kind of cortical skeleton. Elbers (1959) concluded that, in this restricted sense, the cortex is probably identical with the plasma membrane, as no cortical gel layer could be observed. More recent observations show the plasma membrane to form a complex with extra- and intracellular structures. The whole complex probably serves as a 'cortical skeleton' to which cytoplasmic components may be more or less strongly bound. Depending on the binding strength and on the applied disruptive force, the results of centrifugation experiments and other mechanical interferences with the egg contents will vary. Centrifugation experiments with eggs of *Bithynia* show that relatively high forces (1400 g) may be required to displace cytoplasmic structures, such as the vegetal body, that are bound to the cortex. It is easily conceivable that in the polar lobes of *Ilyanassa* and *Dentalium*, which according to our electron microscopic studies do not contain non-displaceable cytoplasmic components, the morphogenetic determinants are inconspicuous structures that are bound to the cortex and cannot be displaced by the relatively low centrifugal force applied to the eggs (600 g). The same argument applies to the experiments in which a portion of the cytoplasm is sucked out of the egg. It has never been demonstrated that determinants are located in the plasma membrane. We feel that the distinction between cortical and cytoplasmic localization does not reflect two essentially different types of localization, as the tightly bound cytoplasmic vegetal body of *Bithynia* may be considered as an intermediate type.

Nature and mode of action of morphogens

Our present knowledge on the chemical nature of morphogens and their mode of action in specifying cell differentiation is very limited. The complexity of development certainly requires the operation of complicated regulative mechanisms. Two of these mechanisms rendering the analysis of morphogens and their functioning particularly difficult are (1) the occurrence of determinative intercellular interactions, e.g. in ascidians (Reverberi, 1971) and (2) the capacity of many cells to adjust their developmental programme after experimental injuries of the embryo, e.g. in *Ilyanassa* (Clement, 1971, 1976).

Probably there are several categories of morphogens. One distinction to be made is between morphogens with a narrow determinative histospecific capacity and organizer-like morphogens that are capable of inducing the development of more or less complete organisms. Germ plasms may be considered as examples of the first category, as they specify only the development of germ cells. Vegetal pole plasms of many molluscs and annelids may prove to be organizer-like determinants (Wilson, 1929; Clement, 1976). Morphogens can also be distinguished as regards their chemical nature. There are several obvious possibilities. Morphogens may be preformed histospecific molecules, such as mRNA or proteins, but also substances regulating translation or transcription, or substances controlling processes such as the cleavage rhythm, the position of cleavage planes, or the establishing of specific intercellular contacts.

Most research has been dedicated to determining the role of translation and transcription in development. In several experiments maternal mRNA has been demonstrated to support early development and differentiation. Morrill, Rubin & Grandi (1976) showed that in the gastropod *Lymnaea* early development through the trochophore stage is independent of transcription. In ascidian embryos there is preformed maternal mRNA for the development of larval palp structures (Minganti, 1959) and for the synthesis of endodermal alkaline phosphatase (Whittaker, 1977). For one species, *Ascidia callosa*, it is even claimed that RNA synthesis is not needed before the time of hatching (Lambert, 1971). In the annelid *Sabellaria*, Guerrier (1971) demonstrated that polar-lobe-dependent structures may develop in embryos in which transcription is inhibited. In the gastropod *Ilyanassa* the difference in protein synthesis between normal and lobeless embryos is independent of concomitant transcription (Raff, Newrock, Secrist & Turner, 1976). On the other hand, in ascidian embryos the appearance of the histospecific enzymes tyrosinase and acetylcholinesterase, restricted to certain cell-lines, is dependent on transcription (Whittaker, 1973). Indications for the control of transcription by polar lobes were found in *Ilyanassa* by Davidson *et al.* (1965) and Collier (1977).

The nature of the regulative substances that control translation, transcription, or other processes involved in morphogenesis, is unknown. Recently several regulative RNA species have been reported (see Deshpande *et al.*, 1977). RNA is present in large amounts in the vegetal body of *Bithynia*. Preliminary poly(U)–poly(A) hybridization experiments *in situ* indicate that this is not mRNA. Therefore, the vegetal body possibly controls development by means of regulative RNA.

Time and site of localization

Mosaic development refers to a specification of the blastomeres by means of morphogenetic determinants, inherited from the egg cell and differentially distributed during cleavage. A definite localization of morphogenetic determinants is not necessarily established, however, before a cleavage which separates two cells with different developmental potential. Therefore, localization may take place during oogenesis or as a consequence of the reorganization of the egg structure, during maturation or fertilization, or during cleavage.

Direct evidence for a localization of a morphogenetic potential during oogenesis is scarce. A clear-cut case of such an early localization is the vegetal body of *Bithynia*, which becomes located at the vegetal pole of the oocyte during vitellogenesis (Dohmen & Verdonk, 1979). Indirect evidence for a localization during oogenesis is provided by experiments in which freshly laid eggs of the scaphopod *Dentalium* were fragmented. Animal fragments form polar bodies but no polar lobe or lobe-dependent structures, whereas fragments that contain a major part of the vegetal region form a polar lobe and lobe-dependent structures (Wilson, 1904; Verdonk, Geilenkirchen & Timmermans, 1971). When unfertilized eggs of the annelid *Chaetopterus* are cut into animal and vegetal halves, only vegetal fragments form a polar lobe (Whitaker & Morgan, 1930). These data indicate that in both *Dentalium* and *Chaetopterus* the ability to form a polar lobe is already confined to the vegetal pole of the egg during oogenesis.

In other eggs maturation appears to be an important period for localization. In the eggs of the annelids *Clepsine* (Schleip, 1914) and *Tubifex* (Lehmann, 1948) the formation of the animal and vegetal pole plasm starts after the extrusion of the first polar body and continues during cleavage when both plasms are segregated into the CD-blastomere, where they fuse into a single plasm that becomes localized in the D-blastomere at the 4-cell stage.

In ascidians a rearrangement of the cytoplasmic constituents of the egg takes place after fertilization. In *Cynthia (Styela) partita* after fertilization five differentially pigmented regions become visible, which during cleavage are distributed over the blastomeres in a specific way. Cell lineage studies indicate that through this partitioning of the cytoplasmic materials the presumptive tissue areas are delineated (Conklin, 1905). Reverberi & Ortolani (1962) divided unfertilized eggs of *Ascidia malaca* into two equal parts by equatorial, meridional, or oblique sectioning. In each case the two fragments may develop into complete larvae. A few minutes after fertilization, however, the situation changes completely. When fertilized eggs are divided, vegetal fragments only give rise to normal larvae, whereas animal fragments form permanent blastulae (Reverberi, 1971).

The segregation of morphogenetic potential continues during cleavage, as is evident from experiments in which at successive cleavages blastomeres were isolated or specific blastomeres deleted, e.g. in *Ilyanassa* (Clement, 1962) and in *Tubifex* (Penners, 1926). It seems highly improbable that this segregation is merely a subdivision of potentials which are exactly localized already in the egg. That an extensive rearrangement of morphogenetic potential can take place during cleavage is shown by the localization of the potential for apical tuft formation in the scaphopod *Dentalium*. In the uncleaved egg this potential is located at the vegetal pole. During cleavage it is incorporated in the first but not in the second polar lobe. After second cleavage the potential for apical tuft formation moves towards the animal pole, where eventually the apical tuft is formed (Wilson, 1904; Geilenkirchen, Verdonk & Timmermans, 1970).

The above data indicate that localization is a dynamic process that may start already during oogenesis. It continues during cleavage, when the various morphogenetic determinants are parcelled out to special cells.

Mechanisms of localization

It is generally assumed that morphogenetic determinants are prelocalized already in the uncleaved egg according to a pre-existing pattern. Apart from the egg of the squid *Loligo pealei*, in which according to Arnold & Williams-Arnold (1976) an elaborate pattern of developmental information is present already in the uncleaved egg, a rather simple organization may explain the results of experiments on localization in mosaic eggs.

Removal of parts of the uncleaved egg does not affect development unless a special region is removed, which in nearly all cases is the vegetal region of the egg: e.g. in *Dentalium* (Wilson, 1904) and in fertilized eggs of *Ascidia malaca* (Reverberi, 1971). This means that the morphogenetic potential is accumulated in a special region of the egg definitely related to the animal–vegetal axis. This primary polarity is present already in the oocyte and is most probably determined by the position of the oocyte in the ovary (for a discussion see Kühn, 1965). As this polarity cannot be changed by centrifugation it is assumed to be bound to the cortex. During oogenesis the cytoplasmic components such as granules, yolk and pigment become arranged along the animal–vegetal axis under the influence of attracting or repulsing forces situated in the egg cortex (Raven, 1964). In this way also the morphogenetic determinants may become localized with respect to the original polarity of the egg, mostly at the vegetal pole (cf. the vegetal body of *Bithynia*). The existence of attractive forces is suggested by the fact that in eggs of *Tubifex* the pole plasms, displaced by centrifugation, reassemble under the cortex of the original poles (Lehmann, 1948).

The morphogenetic determinants, still concentrated in the uncleaved egg and often responsible for an extensive programme of differentiation, are parcelled out over the blastomeres during cleavage. Cleavage apparently plays an important role in establishing a definite localization. The question therefore arises as to whether the cleavage planes separate passively the already differently localized morphogenetic determinants or whether cleavage may play a causal role in setting up a localization. From his experiments with ctenophores Freeman (1976a, b, 1977) concluded a causal relation between cleavage and localization. In the eggs of these animals determinants for comb plates and photocytes are segregated into distinct cells. Prior to first cleavage these determinants are not strictly localized but localization is progressive during the first cleavages. A pre-existing fixed polarity seems to be absent, as the main axis, the oral–aboral axis of the future embryo, is established as a consequence of first cleavage and at this moment also the potential for comb plate formation begins to concentrate in the aboral region. By reversibly inhibiting selected cleavages Freeman found a correlation between the degree of localization in a blastomere and the cleavage that generates the blastomere. When second cleavage is blocked, either with 2,4-dinitrophenol or cytochalasin B, the next cleavage after removal of the drug may have the character of a second or third cleavage. In the first case the localization corresponds to the second cleavage, in the second case to the third cleavage. This indicates that cleavage in ctenophores determines the way in which localizations are set up.

Also in other systems, such as spiralians and ascidians, in which morphogenetic determinants responsible for an extensive programme of differentiation are localized in the uncleaved egg, cleavage may play a causal role in the final localization of the determinants in specific blastomeres, but convincing arguments for a causal relation between cleavage and localization cannot yet be presented. In both cortex- and cleavage-controlled localization the operating molecular mechanism is completely unknown. Recently, indications have been found that surface differentiations may be correlated with morphogenetic localizations (Dohmen & van der Mey, 1977). The relation between these two phenomena is still obscure. Microfilaments are involved in cortical binding of morphogens according to Arnold & Williams-Arnold (1976), who observed that local application of cytochalasin B in uncleaved eggs of *Loligo* causes specific defects. A correlation between ooplasmic segregation and a membrane-generated electrical current has been reported in the egg of the brown alga *Pelvetia* (Nuccitelli, 1978). As currents are probably generated by local changes in the plasma membrane, these findings support a cortical control of development.

Summary

Early development is to a large extent determined by maternal factors. The nature of the morphogenetic determinants, their mode of action, and the mechanisms that control their localization are largely unknown. In a few cases well-defined structures with demonstrable morphogenetic capacity have been identified, e.g. the germinal granules in insects and amphibians and the vegetal body in the gastropod *Bithynia*. In many eggs the morphogenetic cytoplasms, although clearly visible, are still poorly defined as regards the identification of the organelles which contain the morphogens, e.g. in eggs of ascidians and annelids. A cortical localization of determinants has been demonstrated, e.g. in the gastropod *Ilyanassa* and the scaphopod *Dentalium*. None of the visible localizations has as yet yielded conclusive information as to the nature of the morphogens. Biochemical experiments indicate that preformed maternal mRNA may support many types of differentiation, but there is also some evidence of morphogens possibly controlling development via differential embryonic transcription.

References

ARNOLD, J. M. (1968). The role of the egg cortex in cephalopod development. *Developmental Biology*, **18**, 180–97.

ARNOLD, J. M. & WILLIAMS-ARNOLD, L. D. (1974). Cortical–nuclear interactions in cephalopod development: cytochalasin B effects on the informational pattern in the cell surface. *Journal of Embryology and Experimental Morphology*, **31**, 1–25.

ARNOLD, J. M. & WILLIAMS-ARNOLD, L. D. (1976). The egg cortex problem as seen through the squid eye. *American Zoologist*, **16**, 421–46.

BEAMS, H. W. & KESSEL, R. G. (1974). The problem of germ cell determinants. *International Review of Cytology*, **39**, 413–79.

BELL, W. A. & HOLLAND, N. D. (1974). Cholinesterase in larvae of the ascidian, *Ciona intestinalis*, developing from fragments cut from centrifuged eggs. *Wilhelm Roux' Archiv für Entwicklungsmechanik der Organismen*, **175**, 91–102.

CATHER, J. N. & VERDONK, N. H. (1974). The development of *Bithynia tentaculata* (Prosobranchia, Gastropoda) after removal of the polar lobe. *Journal of Embryology and Experimental Morphology*, **31**, 415–22.

CLEMENT, A. C. (1962). Development of *Ilyanassa* following removal of the D macromere at successive cleavage stages. *Journal of Experimental Zoology*, **149**, 193–215.

CLEMENT, A. C. (1968). Development of the vegetal half of the *Ilyanassa* egg after removal of most of the yolk by centrifugal force, compared with the development of animal halves of similar visible composition. *Developmental Biology*, **17**, 165–86.

CLEMENT, A. C. (1971). *Ilyanassa*. In *Experimental Embryology of Marine and Freshwater Invertebrates*, ed. G. Reverberi, pp. 188–214. Amsterdam: North-Holland.

CLEMENT, A. C. (1976). Cell determination and organogenesis in molluscan development: a reappraisal based on deletion experiments in *Ilyanassa*. *American Zoologist*, **16**, 447–53.

COLLIER, J. R. (1977). Rates of RNA synthesis in the normal and lobeless embryo of *Ilyanassa obsoleta*. *Experimental Cell Research*, **106**, 390–4.

CONKLIN, E. G. (1905). The organization and cell-lineage of the ascidian egg. *Journal of the Academy of Natural Sciences of Philadelphia*, **13**, 1–119.

CONKLIN, E. G. (1931). The development of centrifuged eggs of ascidians. *Journal of Experimental Zoology*, **60**, 1–119.

COSTELLO, D. P. (1948). Ooplasmic segregation in relation to differentiation. *Annals of the New York Academy of Sciences*, **49**, 663–83.

DAVIDSON, E. H., HASLETT, G. W., FINNEY, R. J., ALLFREY, V. G. & MIRSKY, A. E. (1965). Evidence for prelocalization of cytoplasmic factors affecting gene activation in early embryogenesis. *Proceedings of the National Academy of Sciences of the USA*, **54**, 696–704.

DESHPANDE, A. K., JAKOWLEW, S. B., ARNOLD, H.-H., CRAWFORD, P. A. & SIDDIQUI, M. A. Q. (1977). A novel RNA affecting embryonic gene functions in early chick blastoderm. *Journal of Biological Chemistry*, **252**, 6521–7.

DOHMEN, M. R. & LOK, D. (1975). The ultrastructure of the polar lobe of *Crepidula fornicata* (Gastropoda, Prosobranchia). *Journal of Embryology and Experimental Morphology*, **34**, 419–28.

DOHMEN, M. R. & VAN DE MAST, J. M. A. (1978). Electron microscopical study of RNA-containing cytoplasmic localizations and intercellular contacts in early cleavage stages of eggs of *Lymnaea stagnalis* (Gastropoda, Pulmonata). *Proceedings of the Koninklijke Nederlandse Akademie van Wetenschappen, Series C*, **81**, 403–14.

DOHMEN, M. R. & VAN DER MEY, J. C. A. (1977). Local surface differentiations at the vegetal pole of the eggs of *Nassarius reticulatus*, *Buccinum undatum*, and *Crepidula fornicata* (Gastropoda, Prosobranchia). *Developmental Biology*, **61**, 104–13.

DOHMEN, M. R. & VERDONK, N. H. (1974). The structure of a morphogenetic cytoplasm, present in the polar lobe of *Bithynia tentaculata* (Gastropoda, Prosobranchia). *Journal of Embryology and Experimental Morphology*, **31**, 423–33.

DOHMEN, M. R. & VERDONK, N. H. (1979). The ultrastructure and role of the polar lobe in development of molluscs. In *Determinants of Spatial Organization*, ed. I. R. Konigsberg & S. Subtelny. New York & London: Academic Press (in press).

EDDY, E. M. (1975). Germ plasm and the differentiation of the germ cell line. *International Review of Cytology*, **43**, 229–80.

ELBERS, P. F. (1959). Over de beginoorzaak van het Li-effect in de morphogenese, een electronenmicroscopisch onderzoek aan eieren van *Limnaea stagnalis* en *Paracentrotus lividus*. Thesis, University of Utrecht.

FREEMAN, G. (1976*a*). The role of cleavage in the localization of developmental potential in the ctenophore *Mnemiopsis leidyi*. *Developmental Biology*, **49**, 143–77.

FREEMAN, G. (1976*b*). The effects of altering the position of cleavage planes on the process of localization of developmental potential in ctenophores. *Developmental Biology*, **51**, 332–7.

FREEMAN, G. (1977). The establishment of the oral–aboral axis in the ctenophore embryo. *Journal of Embryology and Experimental Morphology*, **42**, 237–60.

GEILENKIRCHEN, W. L. M., VERDONK, N. H. & TIMMERMANS, L. P. M. (1970). Experimental studies on morphogenetic factors localized in the first and the second polar lobe of *Dentalium* eggs. *Journal of Embryology and Experimental Morphology*, **23**, 237–43.

GEUSKENS, M. & DE JONGHE D'ARDOYE, V. (1971). Metabolic patterns in *Ilyanassa* polar lobes. *Experimental Cell Research*, **67**, 61–72.

GUERRIER, P. (1971). A possible mechanism of control of morphogenesis in the embryo of *Sabellaria alveolata* (Annelide polychaete). *Experimental Cell Research*, **67**, 215–18.

HEIDER, K. (1900). Das Determinationsproblem. *Verhandlungen der Deutschen Zoologischen Gesellschaft*, **10**, 45–97.

HÖRSTADIUS, S., LORCH, J. J. & DANIELLI, J. F. (1950). Differentiation of the sea urchin egg following reduction of the interior cytoplasm in relation to the cortex. *Experimental Cell Research*, **1**, 188–93.

KALTHOFF, K., KANDLER-SINGER, J., SCHMIDT, O., ZISSLER, D. & VERSEN, G. (1975). Mitochondria and polarity in the egg of *Smittia* spec. (Diptera, Chironomidae): UV irradiation, respiration measurements, ATP determinations and application of inhibitors. *Wilhelm Roux' Archives of Developmental Biology*, **178**, 99–121.

KÜHN, A. (1965). *Entwicklungsphysiologie*, 2nd edn. Berlin: Springer-Verlag.

LAMBERT, C. C. (1971). Genetic transcription during the development and metamorphosis of the tunicate *Ascidia callosa*. *Experimental Cell Research*, **66**, 401–9.

LEHMANN, F. E. (1948). Zur Entwicklungsphysiologie der Polplasmen des Eies von *Tubifex*. *Revue Suisse de Zoologie*, **55**, 1–43.

MINGANTI, A. (1959). Androgenetic hybrids in ascidians. I. *Ascidia malaca* ($\female$) × *Phallusia mamillata* ($\male$). *Acta Embryologiae et Morphologiae Experimentalis*, *244–56*.

MORRILL, J. B., RUBIN, R. W. & GRANDI, M. (1976). Protein synthesis and differentiation during pulmonate development. *American Zoologist*, **16**, 547–61.

NUCCITELLI, R. (1978). Ooplasmic segregation and secretion in the *Pelvetia* egg is accompanied by a membrane-generated electrical current. *Developmental Biology*, **62**, 13–33.

PENNERS, A. (1922). Die Furchung von *Tubifex rivulorum*. *Zoologische Jahrbücher. Abteilung für Anatomie und Ontogenie der Tiere*, **43**, 323–67.

PENNERS, A. (1924). Experimentelle Untersuchungen zum Determinationsproblem am Keim von *Tubifex rivulorum* Lam. I. Die Duplicitas cruciata und organbildende Substanzen. *Arvhiv für Mikroskopische Anatomie und Entwicklungsmechanik*, **102**, 51–100.

PENNERS, A. (1926). Experimentelle Untersuchungen zum Determinationsproblem am Keim von *Tubifex rivulorum* Lam. II. Die Entwicklung teilweise abgetöteter Keime. *Zeitschrift für Wissenschaftliche Zoologie*, **127**, 1–140.

PITOTTI, M. (1947). La distribuzione delle ossidasi e perossidasi nelle uova di *Myzostoma, Beroe e Nereis. Pubblicazioni della Stazione Zoologica di Napoli*, **21**, 93–100.

RAFF, R. A., NEWROCK, K. M., SECRIST, R. D. & TURNER, F. R. (1976). Regulation of protein synthesis in embryos of *Ilyanassa obsoleta*. *American Zoologist*, **16**, 529–45.

RAVEN, CHR. P. (1964). Mechanisms of determination in the development of gastropods. *Advances in Morphogenesis*, **3**, 1–32.

RAVEN, CHR. P. (1966). *Morphogenesis. The Analysis of Molluscan Development*, 2nd edn. Oxford: Pergamon Press.

RAVEN, CHR. P. (1967). The distribution of special cytoplasmic differentiations of the egg during early cleavage in *Limnaea stagnalis*. *Developmental Biology*, **16**, 407–37.

RAVEN, CHR. P. (1970). The cortical and subcortical cytoplasm of the *Lymnaea* egg. *International Review of Cytology*, **28**, 1–44.

RAVEN, CHR. P. (1976). Morphogenetic analysis of spiralian development. *American Zoologist*, **16**, 395–403.

REVERBERI, G. (1970). The ultrastructure of *Dentalium* egg at the trefoil stage. *Acta Embryologiae et Morphologiae Experimentalis*, 31–43.

REVERBERI, G. (1971). Ascidians. In *Experimental Embryology of Marine and Freshwater Invertebrates*, ed. G. Reverberi, pp. 507–50. Amsterdam: North-Holland.

REVERBERI, G. & ORTOLANI, G. (1962). Twin larvae from halves of the same egg in ascidians. *Developmental Biology*, **5**, 84–100.

SCHLEIP, W. (1914). Die Furchung des Eies der Rüsselegel. *Zoologische Jahrbücher. Abteilung für Anatomie und Ontogonie der Tiere*, **37**, 313–68.

SIN, Y. T. (1975). Induction of puffs in *Drosophila* salivary gland cells by mitochondrial factor(s). *Nature, London*, **258**, 159–60.

TUNG, T.-C., KU, S.-H. & TUNG, Y.-F.-Y. (1941). The development of the ascidian egg centrifuged before fertilization. *Biological Bulletin, Woods Hole*, **80**, 153–68.

VERDONK, N. H. (1968). The effect of removing the polar lobe in centrifuged eggs of *Dentalium*. *Journal of Embryology and Experimental Morphology*, **19**, 33–42.

VERDONK, N. H. & CATHER, J. N. (1973). The development of isolated blastomeres in *Bithynia tentaculata*. *Journal of Experimental Zoology*, **186**, 47–61.

VERDONK, N. H., GEILENKIRCHEN, W. L. M. & TIMMERMANS, L. P. M. (1971). The localization of morphogenetic factors in uncleaved eggs of *Dentalium*. *Journal of Embryology and Experimental Morphology*, **25**, 57–63.

VILLA, L. (1976). An ultrastructural investigation of the polar plasm of the egg of *Sternaspis* (Annelida, Polychaeta). *Acta Embryologiae et Morphologiae Experimentalis*, 153–65.

WHITAKER, D. & MORGAN, T. H. (1930). The cleavage of polar and antipolar halves of the egg of *Chaetopterus. Biological Bulletin, Woods Hole*, **58**, 145–9.

WHITTAKER, J. R. (1973). Segregation during ascidian embryogenesis of egg cytoplasmic information for tissue-specific enzyme development. *Proceedings of the National Academy of Sciences of the USA*, **70**, 2096–100.

WHITTAKER, J. R. (1977). Segregation during cleavage of a factor determining endodermal alkaline phosphatase development in ascidian embryos. *Journal of Experimental Zoology*, **202**, 139–53.

WILSON, E. B. (1904). Experimental studies in germinal localization. I. The germ-regions in the egg of *Dentalium. Journal of Experimental Zoology*, **1**, 1–74.

WILSON, E. B. (1929). The development of egg-fragments in annelids. *Wilhelm Roux' Archiv für Entwicklungsmechanik der Organismen*, **117**, 179–210.

WILSON, E. B. (1937). *The Cell in Development and Heredity*, 3rd edn. New York: Macmillan.

Oocyte mitochondria

F. S. BILLETT

Dept of Biology, Medical and Biological Sciences Building, University of Southampton, Southampton SO9 3TU, UK

The mature oocytes of many animals contain large numbers of mitochondria and it is well established that these mitochondria form an important maternal contribution to the early development of the embryo. In Amphibia, for instance, the mature oocyte contains sufficient mitochondria to populate many thousands of embryonic cells and the formation of new mitochondria does not take place in the embryo until a relatively late stage in development. Statements such as 'there are about 2×10^6 mitochondria in the mature *Rana pipiens* oocyte' and 'the number of mitochondria remain constant until hatching in both frogs and sea urchins' (Grant, 1978) appear to be a fair and adequate summary of the present state of knowledge. However, not all the evidence supports such a general view and in this contribution I would like to suggest reasons why it should be given only qualified support.

Oocyte mitochondria are usually assumed to bear a close resemblance to those found in other cells. The frequent representation of mitochondria as blunt-ended cylinders about 1–2 μm long and about 0.2 μm in diameter does not seem far removed from the reality revealed by the examination of living cells by light microscopy and of thin sections by electron microscopy. Such techniques also suggest that mitochondria may branch, divide and fuse. Characteristically the inner membrane is thrown into a number of parallel folds or cristae. Close packing of the cristae usually corresponds with cells having high energy demands, e.g. the flight muscle of insects; conversely relatively few cristae would indicate low activity as has been claimed for some oocytes. The molecular structure of the mitochondrion in relation to its role as an energy producer need not concern us here except to recall that

147

a key enzyme in this process, cytochrome oxidase, has often been used to monitor fractions prepared from oocytes (Chase & Dawid, 1972; Webb, LaMarca & Smith, 1975). Of more immediate interest is the presence of small amounts of DNA associated with a limited amount of genetic autonomy and intimately connected with mitochondriogenesis.

Following the observations of Chèvremont, Baeckland & Chèvrement-Comhaire (1960) and others, the significance of mitochondrial DNA was generally recognized by the mid 1960s and its basic structure and properties are now established (Borst & Kroon, 1969; Borst, 1970, 1972). The mitochondrial DNA of most metazoan cells appears to be strikingly uniform and it can be isolated as circular strands about 5 μm in total length with a molecular weight of about 1.1×10^7. The DNA can be detected in electron micrographs of mitochondria (Nass, Nass & Afzelius, 1965) as small electron dense masses and strands within the matrix.

The number of DNA molecules per 'standard' mitochondrion is usually said to vary from 2 to about 5, although as many as 10 strands have been reported in so-called 'giant' mitochondria (Mercker, Herbst & Kloss, 1967). It has also been suggested that the amount of DNA per mitochondrion may increase as the rate of replication increases. Bearing in mind the supposed, but sometimes disputed (Raff & Mahler, 1975), similarity between mitochondria and bacteria it is interesting to note that an upper limit of 5–6 genomes for a bacterial cell is suggested by the work of Cooper & Helmstetter (1968). However, unlike the bacterial genome, which is much larger, mitochondrial DNA has a very limited repertoire confined to the manufacture of the indigenous RNA, some structural protein and a small fraction of the total enzyme (see Beale & Knowles, 1978). Much of the genetic information necessary to manufacture the fully functional organelle must reside in the nuclear DNA. It is of particular interest to note that in *Xenopus*, for instance, only a small portion of the mitochondrial ribosomal proteins are encoded by the mitochondrial DNA (Leister & Dawid, 1974).

Preceding the discovery of mitochondrial DNA in other cells, and not without influence on the realization of its genetic significance, was the detection of large amounts of DNA in the cytoplasm of the eggs of Amphibia (Hoff-Jorgensen & Zeuthen,

1952; Sze, 1953). Obviously it was not long before the presence of this DNA was related to the large number of mitochondria known to exist in the eggs of many animals (Raven, 1961). Subsequently work on the oocytes themselves and on embryos, particularly Amphibia (Dawid, 1965, 1966, 1972; Dawid & Brown, 1970; Chase & Dawid, 1972) has contributed significantly to our general understanding of the role of mitochondrial DNA.

The purpose of the present paper is to look at some aspects of oocyte mitochondriogenesis which remain slightly obscure and to assess the value of the oocyte mitochondria as a maternal store for the future development of the embryo. The examples are confined to those oocytes which have a fairly well defined pre-vitellogenic phase during which most of the mitochondria appear to be formed within the oocyte itself; the extrinsic supply of mitochondria by nurse cells in insect ovarioles of the meroistic type is not considered (Mahawold, 1972).

The origin of oocyte mitochondria

In the absence of any convincing evidence that the mitochondria are either assembled from subunits found elsewhere in the cytoplasm or that they arise from a specific mitochondrial precursor, equivalent to the pro-mitochondria of yeast (Jayraman, Cotman, Mahler & Sharp, 1966) it must be assumed that oocyte mitochondria are the linear descendants of those found in primordial germ cells. The simplest model is that mitochondria arise by the division of existing mitochondria and that this leads to their passage from primordial germ cells, first to oogonia, and then to oocytes. The cytological evidence for this is reasonably strong but not conclusive. It should be emphasized that the division model need only imply the reproduction of a basic mitochondrial structure and not necessarily the complete replication of a fully functional organelle.

Although variations in structure have been suggested in other animals, e.g. mammals (Weakley, 1976), our own experience with Amphibia and fish (Al-Mukhtar & Webb, 1971; Kanobdee, 1975; Billett & Adam, 1976; Singhih, 1977) suggests the mitochondrial structure, as evidenced by the examination of electron microscope profiles, is very similar in primordial germ cells, oogonia and oocytes. In all three types there is a similar degree of inner membrane complexity, which in other cells would be associated with normal mitochondrial activity. There is some variation in the size and shape of the profiles and their relation to other structures in the cytoplasm. A detailed examination of the supposed antecedents of oocyte mitochondria is beyond the scope of this article (see Nørrevang, 1968, for review) but one or two features of primordial germ cell and oogonial mitochondria will be mentioned which appear to be relevant to the main theme.

A consistent feature of both primordial germ cells and oogonia is the association of small groups of mitochondria with electron dense material in the vicinity of the nucleus. There is some evidence to suggest that this material, sometimes referred to as 'nuage' is derived from the nucleoli and passaged through the nuclear pores. However, neither its composition nor its function is clear. Speculation that it is somehow involved in mitochondriogenesis seems unfounded. Although apparently similar material is found at a later stage in fish oocytes (Kanobdee, 1975) (Fig. 1*a*), it is certainly not associated with mitochondria in the previtellogenic oocytes of *Xenopus* when mitochondrial replication is at a peak. The relation of this substance to germ cell determination is dealt with elsewhere in the symposium (Smith & Williams, this volume).

The number of mitochondria in a primordial germ cell and the way in which they are distributed within the cell vary considerably from one animal to another, but it is interesting to note that in *Xenopus* they form a distinct aggregate in the vicinity of the nuclear envelope (Al-Mukhtar & Webb, 1971). This aggregate closely resembles the much larger cloud seen in the pre-vitellogenic oocytes; it contains the equivalent of 300–500 mitochondria. In *Carassius* there are far fewer mitochondria in the primordial germ cells and they form a perinuclear array around the nuclear envelope (Kanobdee, 1975).

In oogonia the mitochondria are frequently arranged around the nuclear envelope but neither in these cells nor in oocytes is there any firm evidence to suggest that the nuclear membrane is in any way directly involved in the formation of new mitochondria.

Replication and differentiation of mitochondria in oocytes

Soon after the onset of meiosis, marking the transformation of oogonium to oocyte, the number of mitochondria begins to increase rapidly. The size and rate of the increase obviously depend on the size of the final product, the mature oocyte. Starting with a few hundred or a few thousand mitochondria the oocyte may eventually contain tens of thousands, as in *Priapulus* (Nørrevang, 1968) or well over ten million, as in the stone loach *Misgurnis fossilis* (Ozernyuk & Palmback, 1975). Three general principles seem to apply: there is a rough correlation between the amount of yolk and the number of mitochondria at the end of oogenesis, most of the mitochondria appear to be formed during the pre-vitellogenic phase, and the amount of cytoplasmic DNA in the oocyte is largely accounted for by the number of mitochondria.

As material for the study of mitochondriogenesis the oocyte seems to offer a number of advantages, for instance: replication of mitochondrial DNA in the absence of nuclear DNA synthesis, accurate staging of events in relation to oocyte size, large amounts of tissue, the possibility of in-vitro culture of the immature ovary. There are however a number of disadvantages: large amounts of yolk and closely adhering follicle cells make bulk biochemical analysis difficult, the possibility of genetic analysis is very limited, and above all mitochondriogenesis in oocytes may have special features which could make it misleading as a general model. It should be remembered that the mitochondria of oocytes are needed for two purposes, namely to provide a large number of functional mitochondria in relation to the metabolic demands of the growing oocyte and also to provide a store, either actual or potential, to meet the initial requirements of the developing embryo. Thus the possibility must be envisaged that oocytes contain a mixture of mitochondria differing both in structure and in function, some fully differentiated, others not. Unless there were appreciable differences in density bulk biochemistry would not be able to distinguish between the members of a homogeneous and a heterogeneous population. For instance, low cytochrome oxidase activity in a mitochondrial fraction could mean either that all mitochondria possess small amounts of the enzyme, or that some possess high activity and others none at all. For oocyte

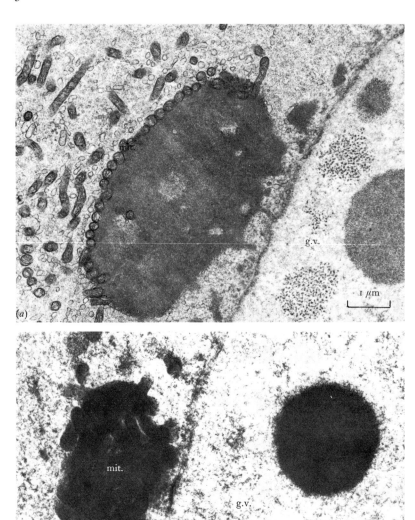

mitochondria the nature of the problem can be appreciated if identifiable aspects of replication, such as an increase in mitochondrial number and mitochondrial DNA synthesis, are considered separately from marks of differentiation, such as cytochrome oxidase activity.

Numbers of mitochondria and DNA content

In this section two significant indices of replication, namely mitochondrial number and mitochondrial DNA, are examined in more detail. Particular emphasis is placed on Amphibia, especially *Xenopus*. It is only in this group that significant comparisons between the various aspects of mitochondriogenesis can be made.

The obvious, but certainly naive, approach to mitochondrial replication in oocytes, is to estimate the numerical increase as the oocyte grows in size. Counts of mitochondrial profiles per unit area from electron micrographs, such as that shown in Fig. 2(c), should be translatable into numbers of mitochondria per oocyte. An attempt has been made to do this with *Xenopus*. In this species the mitochondria in the young oocyte initially form a perinuclear array. At this early stage thick sections reveal that the mitochondria area unusually long, occasionally branched, and sometimes form very dense aggregates (Fig. 1b). When the oocyte reaches about 100 μm in diameter a distinct mitochondrial cloud is formed in the region of the germinal vesicle (Fig. 2a and b). The cloud grows in size with the growth of the oocyte and begins to disperse when the oocyte is between 300 and 400 μm in diameter. The earlier work of Balinsky & Devis (1963) showed that this cloud consisted of many thousands of mitochondria, but a combination of electron microscope techniques (Billett & Adam, 1976) revealed the presence of continuous and branched regions of mitochondrial structure. Thus the cloud appears to be a large and much tangled mass of mitochondrial material rather than a

Fig. 1. (*a*) Mitochondria (mit.) and associated material in vicinity of germinal vesicle (g.v.) of oocyte (*c.* 80 μm diameter) of *Carassius auratus*. Electron microscope preparation, fixed in osmium and stained with uranyl acetate and lead citrate. (From Kanobdee, 1975.) (*b*) Dense aggregate of mitochondria (mit.) in vicinity of germinal vesicle (g.v.) of oocyte (*c.* 40 μm diameter) of *Xenopus*. Electron microscope preparation, fixed in osmium, 1 μm section, unstained.

Table 1. *Maximum number of mitochondria* (0.2 μm × 2 μm) *in clouds of* Xenopus *oocytes*

Diameter of mitochondrial cloud (μm)	Maximum no. of mitochondria ($\times 10^3$)	Corresponding oocyte diameter (μm)
10	12	60
20	95	100
30	263	140
40	788	170
50	1550	200
60	2673	225
70	4272	250

collection of discrete mitochondria and may resemble the mito- chondrial structures described in other cells (Brandt, Martin, Lucas & Vorbeck, 1971; Rancourt, McKee & Pollack, 1975), but be on a very much larger scale.

Despite the apparent complexity of the mitochondrial cloud it is still possible, and instructive, to work out a mitochondrial number in terms of a stereotyped cylindrical mitochondrion measuring 2 μm × 0.2 μm. The spherical nature of the cloud allows the calculation, in terms of volume, of the *maximum* number of standard mitochondria which clouds of different diameter could accommodate. This relation between cloud size and maximum mitochondrial number is given in Table 1. No allowance is made for any space between mitochondria. The space occupied by the mitochondrial material in the cloud can be determined by examination of electron micrographs of the cloud (Fig. 2c). The procedure is valid because the profiles themselves are randomly and evenly distributed and the number of profiles per unit area between clouds of different sizes is fairly constant. An estimate of

Fig. 2. (*a*) Stereoscan of surface of 25 μm section of *Xenopus* (*c.* 100 μm diameter) oocyte fixed in osmium and embedded in araldite. Note prominent mitochondrial cloud (mit.) adjacent to germinal vesicle (g.v.). (*b*) Autoradiograph of 1 μm section of *Xenopus* oocyte *c.* 70 μm in diameter. Tritiated thymidine labelling of mitochondrial cloud (mit.) in vicinity of germinal vesicle (g.v.) (From Al-Mukhtar, 1971.) (*c*) Electron micrograph showing part of the mitochondrial cloud of *c.* 200 μm oocyte of *Xenopus*. Fixed in osmium and stained with uranyl acetate and lead citrate.

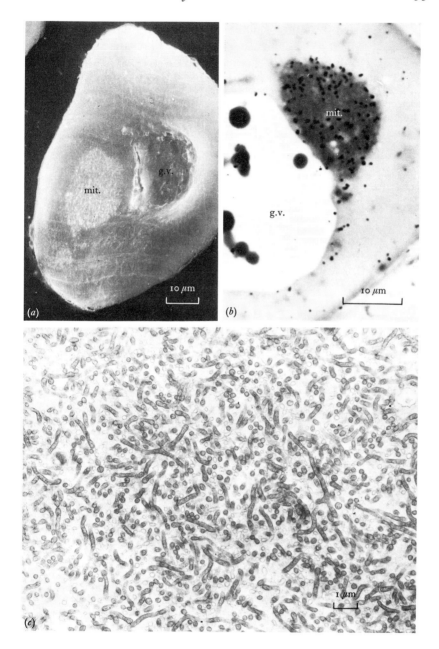

F. S. Billett

the number of mitochondria in clouds of different sizes has been made by Marinos (1978a) using the methods of morphometric analysis (Weibel, 1969). The relation between cloud size and oocyte diameter was also determined. According to these results the number of mitochondria range from about 5000 in a 50 μm diameter oocyte to just under 400000 in a 250 μm oocyte. This covers the range over which the cloud exists and where most of the mitochondria appear to be during this phase of oogenesis. In the light of these calculations the figure of about two million for the mitochondrial population of an amphibian oocyte does not seem to be unreasonable (see p. 147), although it must be considered a little on the high side.

Assuming the figure of two million mitochondria per amphibian egg is of the right order an interesting consequence follows, for although this number is strikingly large it seems quite inadequate to populate all the cells of the embryo at the hatching stage. Estimated numbers of cells in amphibian embryos at this stage are 440000 for *Rana pipiens* (Sze, 1953) and 250000 for *Xenopus laevis* (Bristow & Deuchar, 1964). If the mitochondrial population of *Xenopus* did not increase from the mature oocyte until hatching there would only be the equivalent of eight mitochondria per cell. A calculation based on the average amount of mitochondrial DNA in somatic cells (from 0.1 to 1.0 % of total DNA), assuming up to five mitochondrial genomes per organelle, suggests that on average cells contain between 100 and 1000 mitochondria. On this basis the amphibian embryo would run out of its maternal store by the end of gastrulation (*c*. 20000 cells); after this the synthesis of new mitochondrial material (but *not* mitochondrial DNA) would appear to be necessary. It is interesting to note that mitochondrial RNA synthesis can be detected in amphibian embryos at this stage but disappointing to learn that the amount of mitochondrial protein remains constant until a much later stage (Chase & Dawid, 1972). To possess a reasonable number of mitochondria per cell an amphibian embryo at the hatching stage would need a maternal store of between 2×10^7 and 2×10^8 mitochondria; figures which are not incompatible with the amount of mitochondrial DNA in the mature oocyte.

Not all the cytoplasmic DNA of eggs is mitochondrial, some of it is associated with yolk (see p. 161), but the amount is always substantial (over 75 % of the total) and it usually exceeds the quantity of nuclear DNA many times. In *Xenopus* the mitochondrial DNA content of the egg is firmly established at around 4 ng (Dawid, 1966; Webb & Smith, 1977). This represents about 2×10^8 mitochondrial genomes. At the start of oogenesis the newly formed oocyte would have a mitochondrial DNA content roughly equivalent to that of a somatic cell, about 1×10^3 mitochondrial genomes, similar to that found per cell in the tail-bud stage (Chase & Dawid, 1972). During the growth of the oocyte there is therefore a massive (200000-fold) increase in the amount of mitochondrial DNA.

By using tritiated thymidine the replication of oocyte mitochondrial DNA can be detected autoradiographically in the oocytes of Amphibia and fish (Al-Mukhtar, 1971; Kanobdee, 1975). Electron microscopy indicates that at least some of the synthesis is taking place in mitochondria with a normal inner membrane structure. In *Xenopus* the labelling is almost all confined to the cloud (Fig. 2*b*) and a measure of the intensity of the DNA synthesis can be obtained by quantitative autoradiography (Al-Mukhtar, 1971). This indicates that the mitochondrial DNA synthesis is more intense in the smaller oocytes than the larger ones and leads to the idea that following a phase of DNA synthesis the mitochondrial mass might differentiate as a whole. The distribution of grains over the cloud appears to be random and no centres of replication can be detected.

The more rigorous methods of biochemistry (Webb & Smith, 1977) give a much clearer insight into the pattern of replication of DNA during the pre-vitellogenic phase in *Xenopus* and suggest that with the dispersion of the cloud the synthesis of DNA drops to a much lower rate – a maintenance synthesis to hold the level of mitochondrial DNA around 4 ng. It seems clear that by the time the oocyte is about 400 μm in diameter the maternal store of mitochondrial DNA at around 4 ng. It seems clear that by the cation of mitochondrial DNA is associated with displacement (D) loops (Robberson, Kasamatsu & Vinograd, 1972) and it is not surprising to find that isolated mitochondrial DNA from *Xenopus* oocytes has a high percentage of these loops (Hallberg, 1974), but unexpectedly they are a feature of the unfertilized egg. Accumulation of D loops occurs when larger oocytes of *Xenopus* are induced to mature *in vitro*; suggesting that the oocyte mitochondrial DNA may be stored in a partially replicated state (Barat *et al.*, 1977). In contrast the number of D loops in the mitochondrial DNA of mature sea urchin oocytes is said to be negligible (Matsumoto, Kasamatsu, Piko & Vinograd, 1974).

It is of interest to compare data for mitochondrial number (Marinos, 1978*a*) with those for DNA (Webb & Smith, 1977). There is a significant overlap for the two measurements in the region of the 250 μm diameter oocytes. In *Xenopus* oocytes at this stage approximately 1 ng of mitochondrial DNA corresponds to about 400000 mitochondria. The amount of DNA corresponds to 5×10^7 mitochondrial genomes or over 100 genomes for each mitochondrion. Leaving aside the tempting explanation that the calculation is wrong several interesting questions arise. Are the mitochondria highly polyploid? Are the mitochondrial units much smaller than $0.2 \ \mu m \times 2.0 \ \mu m$? Is it wrong to assume that DNA is only associated with profiles which are obviously mitochondrial? Do the small vesicles associated with the mitochondria in the cloud have any significance? Whatever the explanation the figures suggest that it may be more appropriate to regard the mitochondrial DNA as an index of the mitochondrial potential of the oocyte rather than to equate it with a store of mitochondria as generally understood.

A mixture of mitochondria?

Measurements of mitochondrial increase, either by calculating number or by following the minimum requirement for replication, an increase in mitochondrial DNA, give no indication of the functional status of the newly formed organelles. Mitochondriogenesis can only be considered complete when it results in the formation of a mitochondrion possessing a full complement of enzymes. An accepted morphological sign of differentiation is the presence of an inner membrane folded to form cristae. If this criterion is applied to oocytes then they all seem to possess at least some functional mitochondria throughout oogenesis. However, the presence of cristae, even the detection on them of enzyme assemblies by negative staining techniques, does not mean that all mitochondrial enzymes are present. A more positive way of detecting differentiation is to locate specific mitochondrial enzymes cytochemically and to assay their activity by biochemical means. Cytological techniques should reveal whether all or only some of the mitochondria are active and biochemical assay enable a comparison to be made between enzyme activity and other mitochondrial constituents. There are relatively few recent observations in this field. They are concerned mostly with cytochrome oxidase and indicate a general rise in the activity of this enzyme as the oocyte increases in size. This occurs in Amphibia (Petrucci, 1960) and in the worm *Urechis caupo* (Miller & Epel, 1973).

Cytochrome oxidase activity can be detected both at the light and electron microscope level by the diaminobenzidine (DAB) reaction (Seligman, Karnofsky, Wasserkrug & Hanker, 1968). Many oocytes, e.g. those of *Locusta* and *Xenopus*, show a strong DAB reaction and it is possible to quantify this reaction by scanning the stained areas with a microdensitometer. Using this method Marinos (1978b) has shown that in *Xenopus* the overall activity of the cloud increases until the oocyte reaches a diameter of about 200 μm and then declines before the cloud disperses. At the ultrastructural level the DAB reaction indicates that the cytochrome oxidase activity is distributed evenly both in the mitochondria themselves and throughout the cloud. Biochemical assay (Marinos, 1978a), using isolated mitochondrial fractions from the pre-vitellogenic oocytes, shows a peak of cytochrome oxidase activity in the 100 μm oocytes followed by a fall before the cloud disperses. Overall these results indicate that the amount of cytochrome oxidase per mitochondrial equivalent may vary as much as tenfold in the pre-vitellogenic oocyte.

When the cloud disperses the specific activity of the cytochrome oxidase in the mitochondrial pellet, isolated from the oocyte at this stage, is approximately 10 μmol/mg protein/min (Marinos, 1975a). In the mature oocyte the cytochrome oxidase activity is approximately five times this value: 0.31 μatom oxygen/min/egg (Chase & Dawid, 1972). This difference could be made up either by a further and substantial increase in the number of mitochondria during the vitellogenic phase, without a corresponding increase in the amount of mitochondrial DNA, or by increasing the amount of cytochrome oxidase per mitochondrion. In this connection it is of interest to note that in *Misgurnis fossilis* the number of cristae per mitochondrion increases tenfold as the oocyte increases in size (Ozernyuk & Pelmback, 1975). Both the fluctuation in the amount of cytochrome oxidase per mitochondrion during the early phase of oogenesis and the indication that the cytochrome oxidase activity continues to increase after DNA replication effectively comes to an end support the idea that the composition of oocyte mitochondria may vary greatly during oogenesis and that the embryo may inherit a somewhat mixed population.

Activity and survival of oocyte mitochondria

Although most of the mitochondrial material formed during oogenesis will be passed to the embryo the maternal store must be depleted to some extent by the needs of the oocyte itself. The growth of the oocyte is accompanied not only by the formation of carbohydrate, lipid and protein yolk, but also by the manufacture of much of the machinery for protein synthesis and of other cytoplasmic components for future use by the embryo. The enzyme and respiratory activity of the oocyte and the ultrastructural appearance of the mitochondria themselves point to the involvement of many of the mitochondria in these biosynthetic processes.

The best-known involvement of oocyte mitochondria relates to their apparent connection with the formation of protein yolk. In vertebrates such as amphibians and birds, the yolk precursor, vitellogenin, is formed in the liver, transported to the oocyte–follicular interface by the blood-stream, and taken into the oocyte by pinocytosis. A similar process, involving the fat body, occurs in insects such as *Locusta*. The peripheral location of the mitochondria suggests a role in active transport across the oocyte surface. Vitellogenin is converted into yolk by the activity of a phosphokinase of mitochondrial origin. This is the biochemical basis for the conversion of yolk precursor bodies into yolk platelets 'under the influence of mitochondria' (Gopal & Inoue, 1975).

The direct transformation of mitochondria into yolk has been reported frequently and is based on the presence of intramitochondrial granules which bear a close similarity to yolk (Lanzavecchia & Le Coultre, 1958; Ward, 1962; Massover, 1971; Hsu & Li, 1975). However, although this evidence is strong it is not conclusive and it has been suggested that in Amphibia the resemblance of the paracrystalline inclusion bodies to yolk is misleading as the composition of the two bodies is different (Kress & Spornitz, 1974). Whether or not these inclusions are yolk their large size and the way in which they distort the mitochondrial structure make it seem unlikely that the mitochondria involved would be of any further use to the embryo.

The possible derivation of yolk platelets from mitochondria is related to the problem of yolk DNA. In sea urchin eggs, for instance, about 30 % of the cytoplasmic DNA is associated with yolk granules (Piko, Tyler & Vinograd, 1967). In Amphibia there is a similar association (Hanocq-Quertier, Balthus, Ficq & Brachet, 1968). DNA has also been isolated from intracellular yolk granules derived from the chick blastoderm (Bruce & Emmanuelson, 1975), but this DNA was found to be linear with a buoyant density identical to that of nuclear DNA. In spider oocytes DNA bodies, associated with yolk formation, have been described (Verma, Patra & Das, 1975). There is, in fact, no really firm evidence that the yolk DNA is derived from oocyte mitochondrial DNA. The view that it is an artifact of preparative procedures (Dawid, 1972) is the simplest and most acceptable explanation.

The active participation of mitochondria in many of the synthetic activities of the oocytes means that at least some of the mitochondria will be somewhat secondhand, if not at the end of their useful life, by the time they reach the fertilized egg. The half-life of a mitochondrion can only be guessed at and will be dependent on the stability of its constituent parts; mitochondrial DNA, for instance, is said to have a half-life of only one week (Gross & Rabinowitz, 1969). The maternal store of mitochondria must be regarded not only as a mixed but also as a labile population. Its value to the embryo will depend on its composition and to assess this we need to know how many active mitochondria are present at the beginning of development, at what stage either inactive or incomplete mitochondria become functional, and when mitochondriogenesis based on maternally derived mitochondrial DNA commences.

Many biochemical and cytological observations suggest that mitochondrial biogenesis occurs before mitochondrial DNA synthesis can be detected in the embryo. In mouse embryos, for instance, mitochondria increase in size, in inner membrane complexity and in ribosomal content from the 2-cell to the blastocyst stage (Piko & Chase, 1973); these events precede mitochondrial DNA synthesis (Piko, 1970). Similarly in sea urchins mitochondrial RNA synthesis commences at the beginning of cleavage in the presence of large amounts of mitochondrial DNA (Chamberlain

& Metz, 1972) and in the mollusc *Nassarius* an actual increase in mitochondrial number occurs at the trefoil stage (Schmekel & Fiorini, 1975). Among crustaceans mitochondriogenesis commences at the gastrula stage and here biochemical changes clearly precede morphological ones (Schmitt, Grossfield & Littauer, 1973). Curiously in this species mitochondria are said to be derived from yolk platelets (Marco & Vallejo, 1976). In *Xenopus*, however, neither cytochrome oxidase activity nor an increase in mitochondrial protein generally can be detected until about the hatching stage and here the onset of mitochondriogenesis appears to be coincident with the renewal of mitochondrial DNA synthesis. Interestingly mitochondrial RNA synthesis can be detected at a much earlier stage (Chase & Dawid, 1972).

The effect of specific inhibitors of mitochondrial-based synthesis on the early development of embryos also indicates that mitochondriogenesis, facilitated by the presence of the maternally derived mitochondrial DNA, occurs early in development. For instance, ethidium bromide (which inhibits mitochondrial DNA transcription) prevents the large increase of mitochondrial ribosomes which occurs in mouse embryos from late cleavage to the blastula stage. In sea urchins chloramphenicol inhibits mitochondrial protein synthesis in fertilized eggs (Joel, 1975) and causes vegetalization when applied to 8–32-cell stages (Fujiwara & Yasamasu, 1974). Some earlier observations (Billett, Collini & Hamilton, 1965) showed that D-threo-choramphenicol inhibits the development of chick embryos explanted at pre-somite stages. These results can be interpreted as being due to an inhibition of mitochondrial protein synthesis and show that, in agreement with Schjeide, McCandless & Munn (1964), mitochondriogenesis must commence fairly early in avian embryos despite the massive amount of cytoplasmic, presumably mitochondrial, DNA contained in the yolk (Solomon, 1957).

In general both direct observation and the use of inhibitors suggest that in many embryos mitochondriogenesis is completed on existing structure, including maternally derived mitochondrial DNA, before new mitochondrial DNA synthesis is initiated. Such embryos receive a maternal legacy consisting of large numbers of mitochondria whose functional status may initially be limited and, in many cases, a large amount of mitochondrial DNA whose potential is immense and which may not be strictly equivalent to the number of mitochondria present at fertilization. This maternal contribution is best regarded as a source which facilitates the rapid production of active mitochondria during early development. In most cases it seems unlikely that the embryo receives more than a modest share of fully functional mitochondria derived directly from the oocyte.

The photographs were taken by Elizabeth Adam (Fig. 1*b*, 2*a* and 2*c*), D. Kanobdee (Fig. 1*a*) and K. A. Al-Mukhtar (Fig. 2*a*). The paper includes work supported by a grant from the Science Research Council.

References

AL-MUKHTAR, K. A. K. (1971). Oogenesis in Amphibia with special reference to the formation, replication and derivatives of mitochondria. PhD thesis, University of Southampton.

AL-MUKHTAR, K. A. K. & WEBB, A. C. (1971). An ultrastructural study of primordial germ cells, oogonia and early oocytes in *Xenopus laevis*. *Journal of Embryology and Experimental Morphology*, **26**, 195–217.

BALINSKY, B. I. & DEVIS, R. J. (1963). Origin and differentiation of cytoplasmic structures in the oocytes of *Xenopus laevis*. *Acta Embryologiae et Morphologiae Experimentalis*, **6**, 55–108.

BEALE, G. & KNOWLES, J. (1978). *Extranuclear Genetics*. Edward Arnold: London.

BARAT, M., DUFRESNE, C., PINON, H., TOURTE, M. & MOUNOLOU, J. (1977). Mitochondrial DNA synthesis in large and mature oocytes of *Xenopus laevis*. *Developmental Biology*, **55**, 59–67.

BILLETT, F. S. & ADAM, E. (1976). The structure of the mitochondrial cloud of *Xenopus laevis* oocytes. *Journal of Embryology and Experimental Morphology*, **33**, 697–710.

BILLETT, F. S., COLLINI, R. & HAMILTON, L. (1965). The effects of D- and L-threo-chloramphenicol on the early development of the chick embryo. *Journal of Embryology and Experimental Morphology*, **13**, 341–56.

BORST, P. (1970). Mitochondrial DNA: structure, information content, replication and transcription. In *Control of Organelle Development*, ed. P. L. Miller, *Symposium of the Society for Experimental Biology 24*, pp. 201–25. Cambridge University Press.

BORST, P. (1972). Mitochondrial nucleic acids. *Annual Review of Biochemistry*, **41**, 333–76.

BORST, P. & KROON, A. M. (1969). Mitochondrial DNA: physico-chemical properties, replication and genetic function. *International Review of Cytology*, **26**, 107–90.

BRANDT, J. T., MARTIN, A. P., LUCAS, F. V. & VORBECK, M. L. (1971). The structure of rat liver mitochondria: a re-evaluation. *Biochemical and Biophysical Research Communications*, **59**, 1097–103.

BRISTOW, D. A. & DEUCHAR, E. M. (1964). Changes in nucleic acid concentration during the development of *Xenopus laevis* embryos. *Experimental Cell Research*, **35**, 580–9.

BRUCE, L. & EMMANUELSON, H. (1975). Analysis of DNA isolated from intracellular yolk granules in the early chick blastoderm. *Experimental Cell Research*, **92**, 462–6.

CHAMBERLAIN, J. P. & METZ, C. B. (1972). Mitochondrial RNA synthesis in sea urchin embryos. *Journal of Molecular Biology*, **64**, 593–607.

CHASE, J. W. & DAWID, I. B. (1972). Biogenesis of mitochondria during *Xenopus laevis* development. *Developmental Biology*, **27**, 504–18.

CHÈVREMONT, M., BAECKLAND, E. & CHÈVREMONT-COMHAIRE, S. (1960). Contribution cytochemique et histoautoradiographique à l'étude du métabolisme et de la synthèse des ADN dans des cellules animales cultivées *in vitro*. II. Etude des acides desoxyribonucléiques dans des cellules animales soumises vivantes à l'action de desoxyribonucleases neutre ou acide synthèse et accumulation cytoplasmic d'ADN. *Biochemical Pharmacology*, **4**, 67–78.

COOPER, S. & HELMSTETTER, C. E. (1968). Chromosome replication and the division cycle of *E. coli* B/r. *Journal of Molecular Biology*, **31**, 519–40.

DAWID, I. B. (1965). Desoxyribonucleic acid in amphibian eggs. *Journal of Molecular Biology*, **12**, 581–99.

DAWID, I. B. (1966). Evidence for the mitochondrial origin of frog egg DNA. *Proceedings of the National Academy of Sciences of the USA*, **56**, 269–73.

DAWID, I. B. (1972). Cytoplasmic DNA. In *Oogenesis*, ed. J. D. Biggers & A. W. Schuetz. Baltimore: University Park Press.

DAWID, I. B. & BROWN, D. D. (1970). The mitochondrial and ribosomal DNA components of oocytes of *Urechis caupo*. *Developmental Biology*, **22**, 1–14.

FUJIWARA, A. & YASAMASU, I. (1974). Some observation of abnormal embryos induced by short period treatment with chloramphenicol during the early development of sea urchin. *Development, Growth and Differentiation*, **16**, 83–92.

GOPAL, D. N. & INOUE, S. (1975). Mechanism of yolk formation in the oocytes of *Triturus pyrrhogaster*. *Proceedings of the Indian Academy of Science, Series B*, **82**, 41–5.

GRANT, P. (1978). *Biology of Developing Systems*. New York & London: Holt, Reinhard & Winston.

GROSS, N. J. & RABINOWITZ, M. (1969). Synthesis of new strands of mitochondrial and nuclear DNA by semiconservative replication. *Journal of Biological Chemistry*, **244**, 1563–6.

HALLBERG, R. L. (1974). Mitochondrial DNA in *Xenopus laevis* oocytes. I. Displacement loop occurrence. *Developmental Biology*, **38**, 346–55.

HANOCQ-QUERTIER, J., BALTUS, E., FICQ, A. & BRACHET, J. (1968). Studies on the DNA of *Xenopus laevis* oocytes. *Journal of Embryology and Experimental Morphology*, **19**, 273–82.

HOFF-JORGENSEN, E. & ZEUTHEN, E. (1952). Evidence of cytoplasmic desoxyribosides in the frog's egg. *Nature, London*, **169**, 245.

HSU, C. Y. & LI, K. L. (1975). Presence of intramitochondrial yolk crystals in oocytes of hypophysectomized tadpoles. *Anatomical Record*, **181**, 534A.

JAYARAMAN, J., COTMAN, C., MAHLER, H. R. & SHARP, C. W. (1966). Biochemical correlates of respiratory deficiency. VII. Glucose repression. *Archives of Biochemistry and Biophysics*, **116**, 224–51.

JOEL, DE LA N. (1975). Phosphate transport appearance in sea urchin eggs. I. Effects of protein synthesis inhibitors on fertilized eggs and embryos. *Canadian Journal of Physiology and Pharmacology*, **52**, 1178–85.

KANOBDEE, D. K. (1975). Oogenesis in the gold fish (*Carassius auratus*) with particular reference to mitochondriogenesis. PhD thesis. University of Southampton.

KRESS, A. & SPORNITA, U. M. (1974). Paracrystalline inclusions in mitochondria of frog oocytes. *Exprientia*, **30**, 786–8.

LANZAVECCHIA, G. & LE COULTRE, A. (1958). Origine di mitocondri durante lo sviluppo embrionale di *Rana esculenta*. Studio al microscopio elettrono. *Archivio Italiano di Anatomia e Embriologia*, **63**, 445–58.

LEISTER, D. E. & DAWID, I. B. (1974). Physical properties and protein constituents of cytoplasmic and mitochondrial ribosomes of *Xenopus laevis*. *Journal of Biological Chemistry*, **249**, 5108–18.

MAHAWOLD, A. P. (1972). Oogenesis. In *Developmental Systems: Insects*, ed. S. J. Counce & C. H. Waddington. New York & London: Academic Press.

MARCO, R. & VALLEJO, C. G. (1976). Mitochondrial biogenesis during *Artemia salina* development; storage of precursors in yolk platelets. *Journal of Cell Biology*, **70**, 321A.

MARINOS, E. (1978a). Cytochemical and biochemical studies of mitochondria in *Xenopus laevis* oocytes. PhD thesis, University of Southampton.

MARINOS, E. (1978b). DAB reactivity in *Xenopus laevis* oocytes. *Journal of Histochemistry and Cytochemistry*, in press.

MASSOVER, W. H. (1971). Intramitochondrial crystals of frog oocytes. I. Formation of yolk crystal inclusions by mitochondria during bullfrog oogenesis. *Journal of Cell Biology*, **48**, 206–79.

MATSUMOTO, L., KASAMATSU, H., PIKO, L. & VINOGRAD, J. (1974). Mitochondrial DNA in sea urchin oocytes. *Journal of Cell Biology*, **63**, 146–59.

MERCKER, H. J., HERBST, R. & KLOSS, K. (1968). Electronenmikroskopische Untersuchungen an den Mitochondrion des menschlichen Uterusepithels wahrend der Sekretionsphase. *Zeitschrift für Zellforschung und Mikroskopische Anatomie*, **86**, 139–52.

MILLER, J. H. & EPEL, D. (1973). Studies of oogenesis in *Urechis caupo*. II. Accumulation during oogenesis of carbohydrate, RNA, microtubule protein and soluble mitochondrial lysosomal enzymes. *Developmental Biology*, **32**, 331–44.

NASS, N. K., NASS, S. & AFZELIUS, B. A. (1965). The general occurrence of mitochondrial DNA. *Experimental Cell Research*, **37**, 516–39.

NØRREVANG, A. (1968). Electron microscope morphology of oogenesis. *International Review of Cytology*, **23**, 114–76.

OZERNYUK, N. D. & PALMBACK, L. R. (1975). Growth and reproduction in loach oocytes. *Ontogenez*, **6**, 442–9.

PETRUCCI, D. (1960). La citocromo-c-oxidase nell'oogenesi degli Amfibi. *Acta Embryologiae et Morphologiae Experimentalis*, **3**, 237–59.

PIKO, L. (1970). Synthesis of macromolecules in early mouse embryos cultured *in vitro*: RNA, DNA and a polysaccharide component. *Developmental Biology*, **21**, 257–79.

PIKO, L. & CHASE, D. G. (1973). Role of the mitochondrial genome during the early development in mice. *Journal of Cell Biology*, **58**, 357–78.

PIKO, L., TYLER, A. & VINOGRAD, J. (1967). Amount, location, priming capacity, circularity and other properties of cytoplasmic DNA in sea urchin eggs. *Biological Bulletin*, **132**, 68–90.

RAFF, R. D. & MAHLER, H. R. (1975). The symbiont that never was: an inquiry into the evolutionary origin of the mitochondrion. in *Symbiosis*, ed. D. H. Jennings & D. L. Lee, *Symposium of the Society for Experimental Biology 29*, pp. 41–92. Cambridge University Press.

RANCOURT, M. W., McKEE, A. P. & POLLACK, W. (1975). Mitochondrial profile of a mammalian lymphocyte. *Journal of Ultrastructure Research*, **51**, 418–24.

RAVEN, CHR. P. (1961). *Oogenesis: The Storage of Developmental Information*. Oxford: Pergamon Press.

ROBBERSON, D. L., KASAMATSU, H. & VINOGRAD, J. (1972). Replication of mitochondrial DNA. Circular replicative intermediates in mouse L cells. *Proceedings of the National Academy of Sciences of the USA*, **69**, 737–41.

SHJEIDE, O. A., McCANDLESS, R. C. & MUNN, R. J. (1964). *De novo* synthesis of mitochondria in 1 day old chick embryos. *Nature, London*, **203**, 158–60.

SCHMEKEL, L. & FIORINI, P. (1975). Cell differentiation during the early development of *Nassarius reticulatus* L. *Cell and Tissue Research*, **159**, 503–22.

SCHMITT, H., GROSSFIELD, H. & LITTAUER, U. Z. (1973). Mitochondrial biogenesis during the differentiation of *Artemia salina*. *Journal of Cell Biology*, **58**, 643–9.

SELIGMAN, A. M., KARNOFSKY, M. J., WASSERKRUG, H. L. & HANKER, J. S. (1968). Non-droplet ultrastructural demonstration of cytochrome oxidase activity with a polymerizing osmiophilic reagent, diaminobenzidine (DAB). *Journal of Cell Biology*, **38**, 1–14.

SINGHIH, S. (1977). Development of follicular cell–oocyte interface in *Aphyosemion scheeli*. PhD thesis, University of Southampton.

SOLOMON, J. B. (1957). Nucleic acid content of early chick embryos and the hen's egg. *Biochimica et Biophysica Acta*, **24**, 584–91.

SZE, L. C. (1953). Changes in the amount of desoxyribonucleic acid in the development of *Rana pipiens*. *Journal of Experimental Zoology*, **122**, 577–601.

VERMA, P. G., PATRA, K. C. & DAS, C. C. (1975). Cytoplasmic DNA activity during vitellogenesis in the spider *Araneus nauticus*. *Experientia*, **31**, 976–7.

WARD, R. T. (1962). The origin of protein and fatty yolk in *Rana pipiens*. II. Electron microscopical and cytochemical observations of young and mature oocytes. *Journal of Cell Biology*, **14**, 309–41.

WEAKLEY, B. S. (1976). Variations in mitochondrial size and ultrastructure during germ cell development. *Cell and Tissue Research*, **169**, 531–50.

WEBB, A. C., LaMARCA, M. J. & SMITH, L. D. (1975). Synthesis of mitochondrial RNA by full grown and maturing oocytes of *Rana pipiens* and *Xenopus laevis*. *Developmental Biology*, **45**, 44–55.

WEBB, A. C. & SMITH, L. D. (1977). Accumulation of mitochondrial DNA during oogenesis in *Xenopus laevis*. *Developmental Biology*, **56**, 219–25.

WEIBEL, E. R. (1969). Stereological principles for morphometry in electron microscope cytology. *International Review of Cytology*, **26**, 235–302.

Germinal plasm and germ cell determinants in anuran amphibians

L. DENNIS SMITH AND MARILYN WILLIAMS

Dept of Biological Sciences, Purdue University, West Lafayette,
Indiana 47907, USA

It is well established that the functional gametes of many organisms
are derived solely from primordial germ cells whose formation is
specified through the action of cytoplasmic determinants present
at the time of fertilization. Direct evidence to support this
statement has existed for some time in those invertebrate organisms displaying chromosome diminution or elimination (reviews
by Wilson, 1925; Beams & Kessel, 1974; Davidson, 1976).
Equally compelling evidence exists in other invertebrates such as
Drosophila (Illmensee & Mahowald, 1974). In this case, the
electron dense polar granules first identified at the ultrastructural
level by Mahowald (1962) are suspected, but not proven, of being
the germ cell determinants (review by Mahowald *et al.*, 1979).

In vertebrates, Bounoure (1934) first reported the existence of
a cytoplasmic substance, recognized in the vegetal cortical region
of fertilized frog (*Rana temporaria*) eggs, which eventually was
confined to endodermal cells that migrated into the genital ridges.
The endodermal cells were designated as primordial germ cells
and the specially straining cortical cytoplasm was designated the
'germinal plasm'. This report has since been substantiated and
extended to include a number of anuran amphibians (reviews by
Blackler, 1966, 1970; Czolowska, 1969; Whitington & Dixon,
1975). That a structure similar to polar granules might also exist
within the germinal plasm of amphibian eggs was first indicated
by Balinsky (1966). He suggested that granular electron dense
bodies observed in association with mitochondria at the vegetal
hemisphere of the eggs of a South African frog might represent
the basophilic areas of cytoplasm known as germinal plasm. These
electron dense bodies, designated 'germinal granules', were

rediscovered in the vegetal hemisphere of *Rana pipiens* eggs five years later (Kessel, 1971; Mahowald & Hennen, 1971; Williams & Smith, 1971) and have since been identified in *Xenopus laevis* eggs (Czolowska, 1972; Kalt, 1973; Ikenishi, Kotani & Tanabe, 1974) and developing embryos (Ikenishi & Kotani, 1975) as well as in *Engystomops pustulosus* eggs (Smith & Williams, 1975).

Considerable experimental evidence has implicated the germ plasm, identified in anuran eggs, in the formation of primordial germ cells (reviews by Blackler, 1966, 1970; Beams & Kessel, 1974; Smith & Williams, 1975; Eddy, 1975). In contrast, a germinal plasm, recognizable by its affinity for specific stains, has not been identified in urodele amphibians (Blackler, 1966). The only indication that such an entity exists is the electron micrograph of Williams & Smith (1971) showing electron dense bodies, analogous to germinal granules, near the equator of a fertilized axolotl egg. This observation was considered significant in that, in urodeles, the presumptive germ cells are known to appear, at least by the gastrula stage, in lateral plate mesoderm (Nieuwkoop, 1947; Smith, 1964). Recently, Sutasurya & Nieuwkoop (1974) have reported that primordial germ cells, along with other mesodermal structures, can be induced in the blastula ectoderm when xenoplastic recombinants of animal hemisphere ectodermal caps are made with the ventral yolk mass of several urodeles. The designation of cells as primordial germ cells was based solely on cytological characteristics, and not on a demonstrated capacity of the 'induced' germ cells to form functional gametes. Also, in the absence of a specific marker such as germ plasm, the distribution of any presumptive germ cells in urodele blastulae is not known. Thus, predetermined presumptive germ cells might not yet be localized specifically to presumptive lateral plate mesoderm in blastulae. Nevertheless, Sutasurya & Nieuwkoop conclude that, in urodeles, primordial germ cells do not arise from predetermined elements ˙(such as germ plasm) but from totipotent animal hemisphere ectodermal cells. Eyal-Giladi, Kochau & Menashi (1976) have raised the possibility that in avian embryos also there is no specific germ plasm. Rather, germ cells could appear relatively late in development as a result of inductive processes.

In mammals (mouse), the earliest embryonic stage at which primordial germ cells have been identified is 8–8.5 days of gestation (Clark & Eddy, 1975). The primordial germ cells were detectable because of a faint accumulation of alkaline phosphatase reaction products. Other characteristics such as the germinal granules have not been seen at this time or in earlier stages of development, although they have been identified in presumptive primordial germ cells later in development (Spiegelman & Bennett, 1973). On the other hand, Mintz & Illmensee (1975) have shown that teratocarcinoma stem cells, transferred into mouse blasto-cysts, can in some cases give rise to functional sperm. These stem cells were derived originally from undetermined cells of a 6-day male embryo, and not from presumptive primordial germ cells (Stevens, 1970). Similarly, germ line chimaeras have been obtained after injection of embryonic ectoderm cells into host blastocysts (see Gardner & Rossant, 1976). These results also imply that germ cells can arise from totipotent embryonic cells as a result of inductive processes. Thus, it would appear that the only verte-brates in which the germ plasm theory has received substantial experimental support are the anuran amphibians.

Classically, two kinds of experimental approaches have been used to verify the role of germ plasm in the formation of primordial germ cells: removal of germ plasm surgically or destruction of germ plasm by UV irradiation (reviews by Blackler, 1970; Smith & Williams, 1975; Eddy, 1975). In the former case the most complete experiments have been performed by Buehr & Blackler (1970) using eggs of *Xenopus laevis*. Incisions were made into the vegetal pole of 2-cell and 4-cell embryos, resulting in the formation of exudates which were shown histologically, in some cases, to contain germ plasm. While the number of survivors from such an operation was low (average 25 %), about one-third of the blastulae which developed lacked cells containing germ plasm and, correspondingly, about one-third of the tadpoles which developed from operated embryos were totally sterile. Similar results were obtained by Gipouloux (1971) using embryos of *Rana dalmatia* and *Bufo bufo*.

The major experimental approach implicating a germ plasm in the control of germ cell formation has been the irradiation of the region of the egg suspected of containing germ cell determinants with UV light. Such experiments first were reported by Bounoure (1934) using *Rana temporaria* eggs. UV irradiation of the vegetal pole of fertilized eggs caused partial or total sterility in a proportion of the frogs which developed from such eggs. Subsequently a body of evidence, using UV irradiation, has accumulated which confirms and extends Bounoure's results on several anuran species (Bounoure, Aubry & Huck, 1954; Padoa, 1963; Blackler, 1966, 1970; Smith, 1966; Tanabe & Kotani, 1974; Ikenishi *et al.*, 1974; Ijiri & Egami, 1975; Ijiri, 1976, 1977; Züst & Dixon, 1977). In the best cases, primordial germ cells could be completely eliminated in all tadpoles which developed from eggs irradiated at the vegetal hemisphere at about the time of the first cleavage division or earlier.

Since UV-irradiated embryos usually developed into sterile tadpoles with no other observable defects, it was assumed that UV light had destroyed the germ cell determinants, thereby preventing the formation of primordial germ cells (Smith, 1966; Smith & Williams, 1975). However, based on recent evidence (Züst & Dixon, 1977; Williams, 1978; Williams & Smith, 1979), it appears that germ cell formation is not prevented, at least not entirely, by UV irradiation. This has raised questions concerning the nature of the effects of UV irradiation on germ cell formation and, hence, the role of a germ plasm in directing the formation of primordial germ cells. The purpose of the current review is to discuss these questions. In particular, we discuss the nature of the evidence that UV irradiation of the germ plasm region inactivates specific germ cell determinants.

Correlation between UV effects and germ plasm location

According to Blackler (1958) germinal plasm first appears after fertilization, and islets of germ plasm are concentrated close to the vegetal pole surface immediately prior to the first cleavage division. In subsequent cleavage stages the germ plasm is found closely applied to the vertical cleavage cell membranes in the vegetal macromeres. Thus, it has been shifted away from the egg surface and into the interior of the cleaving embryo. An example of these observations is shown in Fig. 1, which represents Epon sections through the germinal plasm region in *Rana pipiens* eggs and early embryos. Smith (1966) originally reported that, in *Rana pipiens*, UV irradiation at the time of the first cleavage division had the greatest effect on germ cell formation. After this, increasingly few embryos which had been irradiated at the vegetal hemisphere

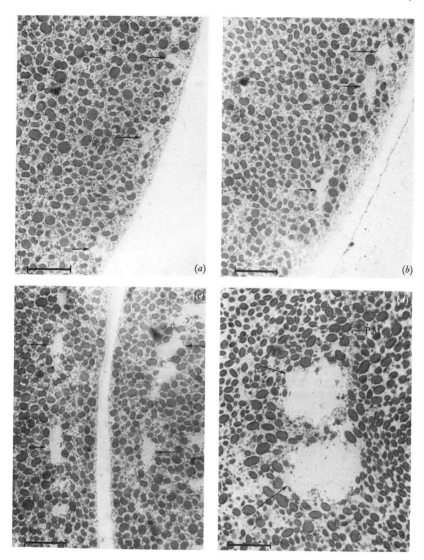

Fig. 1. Sections (1 μm thick) in the vegetal hemisphere of fertilized *Rana pipiens* eggs showing germinal plasm areas (arrows): (*a*) 1.5 hours after fertilization; (*b*) 2-cell stage; (*c*) 4-cell stage, showing primary cleavage furrow; (*d*) 8-cell stage, note plasma membrane (PM) between the two opposing blastomeres. Scale bars represent 50 μm. (From Williams & Smith, 1971.)

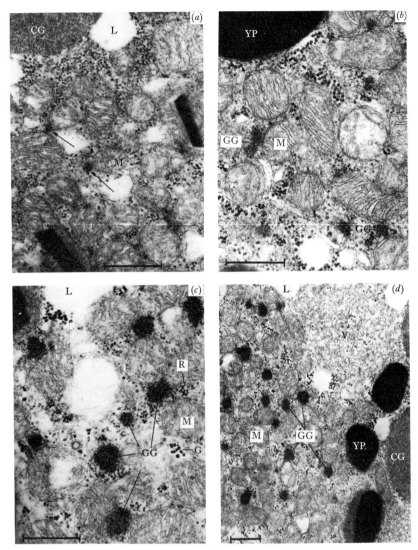

Fig. 2. Vegetal polar regions of *Rana pipiens* oocytes. (*a*) Full-grown oocyte. Small electron dense areas (arrows) are present within the aggregations of mitochondria which lie below the cortical granules. (*b*) Oocyte undergoing maturation, 11 hours after progesterone treatment. Germinal granules are present within the subcortical aggregations of mitochondria. (*c*) Enucleated oocyte 23 hours after progesterone treatment. Germinal granules are present within an aggregation of mitochondria. Yolk platelets and lipid droplets surround the mitochondria. Ribosomes and glycogen are interspersed between the mitochondria. (*d*) Unfertilized egg. Numerous germinal granules are present within

developed into larvae without germ cells. When 8-cell stage embryos were irradiated, complete elimination of germ cells was not observed. These results were interpreted to mean that, as cleavage progressed, the movement of germ plasm into the interior rendered it less accessible to the relatively low penetration power of UV and thus the germ plasm was less affected by UV irradiation (Smith, 1966).

In *Xenopus laevis* the effect of UV irradition at different stages of development has been correlated directly with the subcortical depth of stainable germ plasm. Tanabe & Kotani (1974) showed that as the distance of germ plasm from the vegetal hemisphere surface increased, the sterilizing effect of UV irradiation decreased. Likewise, displacement of germ plasm into the interior by centrifugation, followed by UV irradiation, resulted in a substantially decreased effect on germ cell formation.

In contrast to Blackler's original claim, it now is clear that germ plasm, recognizable in the light microscope, exists in unfertilized eggs (Czolowska, 1969). Similarly, electron dense germinal granules have been identified in unfertilized eggs of both *Xenopus laevis* (Czolowska, 1972; Kalt, 1973; Smith & Williams, 1975) and *Rana pipiens* (Williams & Smith, 1971; Mahowald & Hennen, 1971). As shown in Fig. 2, definitive germinal granules in *Rana pipiens* are not present in full-grown oocytes (Fig. 2*a*) but appear during the course of oocyte maturation (Fig. 2*b*), irrespective of whether or not the oocyte nucleus is present (Fig. 2*c*). Thus, germinal granules become localized just beneath the vegetal pole cortex prior to fertilization (Fig. 2*d*). Similar observations have been made on *Xenopus laevis* oocytes (Smith & Williams, 1975).

Recently, Ijiri (1977) has demonstrated that UV irradiation of the vegetal hemisphere of unfertilized *Xenopus laevis* eggs also results in the absence of germ cells in feeding tadpoles. Moreover, unfertilized eggs were equally as sensitive to UV as fertilized eggs while, as cleavage proceeded, the effects of UV decreased. By the 16–32-cell stages, irradiation had no observable effect on germ cell formation. Thus, in all cases, the detrimental effects of UV irradiation on the subsequent appearance of germ cells correlated

---

a mitochondrial cluster lying below the cortical granules. Yolk platelets and vesicles surround the mitochondrial cluster. CG, cortical granule; GG, germinal granule; G, glycogen; L, lipid; R, ribosomes; V, vesicles; YP, yolk platelet. Scale bars represent 0.5 μm. (From Williams & Smith, 1971.)

with the presence of germ plasm near the vegetal hemisphere surface and with no other obvious parameter.

Transfer of 'germ cell determinants'

Smith (1966) first reported that the effects of UV irradiation on germ cell formation could be reversed by injecting subcortical vegetal pole cytoplasm (unirradiated), but not animal pole cytoplasm, into the vegetal hemisphere of irradiated embryos. Significant numbers of the embryos developed into tadpoles which contained germ cells, albeit few in number, in the genital ridges. This so-called 'rescue' experiment was first repeated on *Drosophila* by Okada, Kleinman & Schneiderman (1974) and Warn (1975). The posterior pole plasm of *Drosophila* eggs was irradiated with UV light and normal pole plasm was injected back into the same region of the irradiated eggs. A high percentage of the irradiated injected eggs failed to survive, but among the survivors sterility was prevented in a significant proportion of the cases. Similar studies have now been repeated in frog eggs by Wakahara (1977, 1978). Initially Wakahara (1977) injected a crude homogenate of control, vegetal hemisphere cytoplasm into UV-irradiated *R. chensinensis* eggs. Tadpoles which developed from such injected eggs contained appreciable numbers of germ cells, while animal hemisphere cytoplasm had no effect on germ cell numbers. Further experiments showed that the active material could be pelleted by centrifugation for 30 minutes at $15\,000g$. Preliminary electron microscopic observations of the fractionated cytoplasm (Wakahara, 1977) revealed the presence of membranous structures, mitochondria, and aggregates of electron dense granules resembling components of the germinal granules. However, since the amount of such material required to 'rescue' irradiated eggs might be expected to be important, it would seem necessary to show that dilutions of the fraction have progressively less effect; such experiments were not performed. Furthermore, none of the UV controls (irradiated but not injected) were completely sterile. In the second report Wakahara (1978) injected non-irradiated *Xenopus* eggs with material from a $20\,000g$ pellet of vegetal pole cytoplasm. The number of germ cells in larvae developing from the injected eggs was only about 30% greater than in controls.

In spite of suggestions by Wakahara (1977, 1978) that the component which 'induces' germ cell formation when injected into eggs could be RNA, the putative active factor has not been positively identified. More importantly, neither the identity of the UV target nor the nature of the UV effect upon it are known. Thus, while the data discussed above are intriguing, and supports the hypothesis that a UV-sensitive component, present within the germ plasm region, is necessary for germ cell formation, none of the data is yet unequivocal.

Identity of the UV target

The staining characteristics of germinal plasm observed with the light microscope remain unaltered during early development of UV irradiated *Xenopus* embryos (Blackler, 1966; Züst & Dixon, 1975), while in post neurula embryos the number of germ cells containing germ plasm is reported to decline rapidly (Züst & Dixon, 1975). This observation has been interpreted to mean that germinal plasm loses its affinity for specific stains as a result of UV irradiation (Züst & Dixon, 1975), although other interpretations remain possible. Ikenishi & Kotani (1978) indicate that the amount of germ plasm in presumptive primordial germ cells decreases at post-cleavage stages after irradiation of early 2-cell embryos. Blackler (1970) has stated that 'the later behavior of sex plasm (particularly at the time of the intracellular movement) is plainly abnormal' as a result of irradiation. In the absence of corresponding data, presumably this means that movement of germ plasm from the cell periphery to a juxtanuclear position between late blastula and the gastrula stage is prevented.

As a result of electron microscopic observations, Ikenishi *et al.* (1974) reported that UV irradiation of the vegetal pole of *Xenopus* embryos (early 2-cell stage) caused damage both to the germinal granules and to the mitochondria present within germinal plasm. In the former case, the granules were reported to fragment as well as to undergo a decrease in diameter. These changes were evident shortly after irradiation. Thus, Ikenishi *et al.* (1974) interpret their data as suggestive of a direct UV effect on the cytoplasmic 'organelle' suspected of being the 'germ cell determinant'. Unfortunately we do not confirm the view that UV irradiation alters the structural integrity of the electron dense germinal granules (Smith & Williams, 1975).

In our studies (Smith & Williams, 1975; Williams, 1978) germ plasm distribution and ultrastructural characteristics were examined during early development (4-cell to gastrula) in order to detect morphological changes which could be ascribed to the consequences of irradiation. Electron microscopic examination of the germinal plasm of 4-cell *Rana* embryos revealed few ultrastructural alterations in the irradiated animals. Both control and irradiated embryos (dose: 18000 erg/mm^2) appeared to contain granules of comparable number and size. Electron microscopic examination of this germinal plasm of blastula and gastrula stages of *Xenopus* embryos also revealed very few ultrastructural alterations in the irradiated animals. No consistent effect of UV irradiation on germinal granule morphology or distribution was seen. For example, Fig. 3 shows that both control gastrulae and gastrulae which developed from irradiated 2-cell embryos appear to contain granules in comparable numbers and of comparable size, composed in both cases of electron dense foci about 17 nm in diameter. Likewise, UV irradiation does not appear to prevent mitochondrial aggregates, containing the germinal granules, from undergoing the intracellular shift to a juxtanuclear position (Fig. 3*a*). Thus, if UV irradiation actually produces molecular alterations in the structural components of germinal plasm (germinal granules) these were not manifest as prominent or numerous structural alterations, either during the 4-cell stage (*Rana*) or during the blastula and gastrula stages (*Xenopus*). The diminution of germinal granule size which Ikenishi *et al.* (1974) observed in irradiated embryos could be due to fixation artifacts (see Smith & Williams, 1975). In summary, while the idea that UV irradiation affects components of the germ plasm directly is very appealing, the data supporting such an idea are scant and by no means compelling.

The nature of the UV effect on germ cell formation

A number of years ago, Nieuwkoop (1947) advanced the argument that UV irradiation could prevent germ cell formation in amphibians by creating conditions unfavourable for germ cell differentiation. In this regard, Okada *et al.* (1974) observed that after UV irradiation of the pole plasm in *Drosophila* migration of nuclei into the pole plasm was delayed and constriction of pole cells was

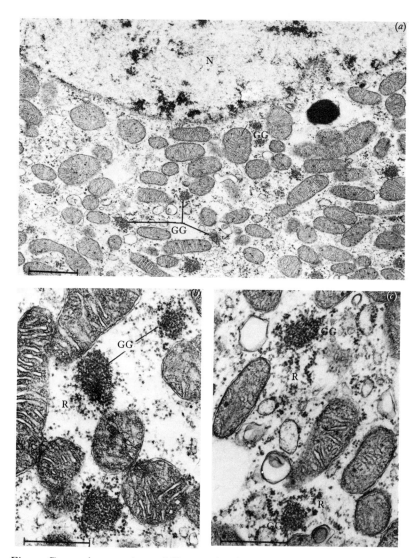

Fig. 3. Germ plasm regions of *Xenopus laevis* gastrulae (stage 12) derived from control and UV-irradiated eggs. (*a*) Numerous germinal granules (GG) are present in the germ plasm adjacent of the nucleus (N) from an irradiated embryo. Scale bar represents 1 μm. (*b*) and (*c*) Germ plasm from control (*b*) and irradiated (*c*) embryos exhibit germinal granules (GG) surrounded by clusters of ribosomes (R). (*c*) represents a higher magnification of a part of the juxtanuclear area shown in (*a*). Fixation as per Kalt & Tandler (1971). Scale bars represent 0.5 μm.

prevented. As a result of these observations Okada *et al.* (1974) questioned whether UV acts indirectly to block pole cell (germ cell) formation rather than specifically destroying germ cell determinants. Similar questions have been raised by Dixon and colleagues in recent work on *Xenopus laevis* (Züst & Dixon, 1975; Beal & Dixon, 1975). In these experiments UV irradiation of the vegetal hemisphere (2–4-cell stage) was reported to delay cleavage, but not karyokinesis, such that a syncytium formed in the vegetal hemisphere. Presumptive germ cells (cells containing germ plasm) eventually formed, but only when the syncytium broke down into individual cells, between morula and late blastula stages. Thus, it is apparent that the orderly processes which normally result in the segregation of presumptive germ cells (Whitington & Dixon, 1975) were deranged by the UV treatment.

In normal *Xenopus* embryos the germ plasm is partitioned between the first four blastomeres by the first two cleavages. During subsequent cleavage the number of presumptive germ cells remains fairly constant due to the eccentric intracellular location of the plasm, which restricts its post-mitotic distribution to only one daughter cell during this time (Whitington & Dixon, 1975). The number of cells containing germ plasm increases between gastrula stage and the end of their endodermal residence (stages 38–41; Nieuwkoop & Faber, 1956); after germ cells enter the median genital ridge (stages 43/44) they no longer divide. At about this time (stages 39–44) germ plasm loses its affinity for specific stains. Finally, mitosis recommences between stages 48 and 52, several days after the last previous division (Whitington & Dixon, 1975). This quantitative study was corroborated by an autoradiographic analysis of DNA synthesis in presumptive primordial germ cells (Dziadek & Dixon, 1975, 1977). Taken together, these studies indicate that proliferation of germ cells is separable into two distinct phases. The first phase, between stages 10 and 44, involves two to three divisions of each presumptive primordial germ cell (see also Kamimura *et al.*, 1976). These divisions have been referred to as 'cloning divisions' (Whitington & Dixon, 1975; Dziadek & Dixon, 1977), implying that the number of germ cells increases but the determined state of the cells is not altered. These findings are significant, since they invalidate a previous suggestion that presumptive primordial germ cells become mitotically inhibited after germ plasm assumes a juxta-nuclear position (Blackler, 1958). At the end of the first phase germ cells migrate out of the endoderm and, after a period of residence in the genital ridges, the second proliferative phase ensues. This results in the formation of a generation of gonial cells followed by sexual differentiation (Züst & Dixon, 1977; Dziadek & Dixon, 1977).

The presumptive germ cells in UV-irradiated embryos also synthesized DNA and divided, at least between gastrula and neurula stages (Beal & Dixon, 1975). Beyond the neurula stage, however, the numbers of presumptive germs cells which could be recognized declined rapidly. At stages 43 to 46, when germ cells were located in the genital ridges of control embryos, no germ cells were found in the genital ridges of tadpoles which developed from irradiated embryos (Züst & Dixon, 1975). Two possible interpretations were proposed to account for the inability to detect presumptive germ cells in post-neurula embryos (Züst & Dixon, 1975): either they were lost or the germ plasm changed so that it was no longer detectable with the usual staining procedures.

Recently, Züst & Dixon (1977) reported that when *Xenopus* tadpoles which developed from UV-irradiated embryos were assayed between stages 41 and 46, no germ cells were found, confirming the earlier study by the same authors (Züst & Dixon, 1975) as well as work by others referred to earlier. However, in tadpoles from the same population examined at stage 48, the genital ridges did contain germ cells. Tadpoles were not examined at this late a stage in development in the other UV studies referred to on *Xenopus*. Furthermore, these germ cells differed morphologically from those in controls of a similar stage in that they were much smaller, lacked cytoplasmic yolk, and had a more highly lobed nucleus, characteristics attributable to germ cells in older control embryos. Züst & Dixon (1975, 1977) concluded from these observations that UV caused a delayed segregation of germ plasm, the consequences of which are manifest in the delayed migration of germ cells to the genital ridges. The smaller size of these late-arriving germ cells was further attributed to a disrupted division pattern during their residence in the endoderm, i.e. presumptive germ cells while still in the endoderm enter into the second phase of proliferation which in normal animals begins after they have entered the genital ridges. By analogy, the germ plasm in such cells would also precociously lose its affinity for the usual staining procedure.

In the studies reported by Smith (1966) on *Rana pipiens* eggs,

UV irradiation at a wavelength of 254 nm, at doses resulting in sterility, caused no obvious retardation in developmental events. In contrast irradiation at a wavelength of 230 nm resulted in severe developmental abnormalities with little effect on germ cell formation. However, a careful study of potential effects of UV on cleavage in the vegetal hemisphere, particularly the early cleavage divisions, was not performed. In order to determine whether such a response had been overlooked, the vegetal hemispheres of irradiated and control *Rana* embryos were examined in the dissecting microscope for cessation of cleavage (Williams, 1978; Williams & Smith, 1979). Upon reaching stage 25 (Shumway, 1940) germinal ridges were assayed for primordial germ cells. Five groups of embryos were irradiated, with approximately one-third of each group receiving 12000 erg/mm², and another third receiving 18000 erg/mm², while the remaining third were not irradiated. Cleavage of the vegetal hemisphere was delayed in various proportions in all embryos in the irradiated groups (587 embryos/high dose, 578 embryos/low dose). The 718 unirradiated embryos cleaved normally. The embryos of one female, examined at the 4-cell stage, exhibited 100 % and 92 % delayed cleavage at the high and low dose respectively. Apparently this response was very transitory, for by the 16-cell stage the proportion dropped to 28 % in those embryos which received the high dose. Embryos from four other females, examined at the 16- to 32-cell stage, displayed an average of only 11 % delayed cleavage (range of 4–17 %) for the high dose and an average of only 4 % delayed cleavage (range of 3–7 %) for the low dose. Thus, while the fraction of *Rana* embryos exhibiting delayed cleavage was quite variable, it was considerably smaller, with the exception of one group, than the 100 % cleavage delay or cessation observed in *Xenopus* embryos after similar UV doses of 12000 and 18000 erg/mm² (Beal & Dixon, 1975; Züst & Dixon, 1975). In those *Rana* embryos in which cleavage was restricted, spontaneous resumption of cleavage occurred prior to stages 8 to 10. Consequently, the percentage of *Rana* embryos arrested at the gastrula stage was very low and 96 to 99 % developed to stage 25. In contrast, more than 50 % of *Xenopus* embryos were unable to complete gastrulation due to the persistence of large vegetal syncytia (Beal & Dixon, 1975; Züst & Dixon, 1975). These results demonstrate that factors influencing cleavage are considerably less sensitive to UV irradiation in *Rana* embryos than in *Xenopus* embryos.

The effect of UV irradiation on germ cell formation in *Rana* was not as variable as the delayed cleavage response. All embryos which received a UV dose of 18 000 erg/mm² developed into sterile stage 25 tadpoles, and most of those which received the lower UV dose also developed as sterile tadpoles. Thus, there is no obvious relationship between delayed cleavage and the subsequent absence of germ cells in the genital ridges of stage 25 tadpoles. We conclude that the delayed cleavage response, when present, and sterility, are independent consequences of UV irradiation of the vegetal hemisphere, at least in *Rana pipiens*.

The UV doses which resulted in large vegetal syncytia and 50 % inhibited gastrulation in *Xenopus* were two to three times greater than the UV doses reported necessary for larval (stages 43–46) sterility (Blackler, 1970; Tanabe & Kotani, 1974; Ikenishi *et al.*, 1974; Züst & Dixon, 1975). We have noted (unpublished data) that a lower UV dose, 8400 erg/mm², applied to *Xenopus* embryos appeared to have no effect on vegetal cleavage that was discernible in the dissecting microscope during stage 6 (32-cell). Subsequent development appeared normal in all but one embryo, an exogastrula, and 93 % reached stage 44. Serial sections of the posterior half of the 18 stage 44 larvae revealed 66 % of the animals totally sterile. Thus, even in *Xenopus*, sterility can be an independent consequence of irradiation.

In our initial studies designed to determine whether UV affects the ultrastructural integrity of germinal granules (Smith & Williams, 1975; Williams, 1978), *Xenopus* embryos at the 2-cell stage were irradiated at a dose of 7200 erg/mm². From the evidence of previous reports (Tanabe & Kotani, 1974; Ikenishi *et al.*, 1974), such a dose should have been more than sufficient to cause sterility. Nevertheless, in one group of tadpoles examined at stages 50–52, serial sections of the genital ridges showed that only 75 % of the tadpoles were sterile. Surprisingly when tadpoles from the same population were examined a few weeks later (stage 63), only 25 % were sterile. In a group of 12 additional animals raised for 11 months, 8 males contained testes which produced motile sperm and 4 females contained growing oocytes. These observations in themselves are not necessarily revealing since tadpoles from the same population had not been examined at younger stages.

Contrary to published reports, circumstances could be conceived in which UV irradiation at the dose used is not 100 % effective. Conceivably, these observations reflect the extreme of sampling errors from a population of tadpoles which originally (stages 44–46) would have been only partially sterile. Nevertheless, the observations provide a tentative confirmation of the work of Züst & Dixon (1977) in suggesting that tadpoles originally designated as sterile subsequently reacquire germ cells.

As mentioned earlier, stage 25 *Rana* tadpoles derived from irradiated 2-cell embryos are completely devoid of detectable germ cells. Animals from these same populations were also raised, and paraffin sections of the gonads were examined at early (stage II) and late (stages XXII–XXIV) metamorphic stages (Taylor & Kollros, 1946). As early as stage II (about 3 days beyond larval stage 25) the proportion of sterile animals had declined. Of the gonads examined in 12 tadpoles which had received 18000 erg/mm^2, only 33 % still were sterile. Total germ cell number in the remainder ranged from 2 to 13 (Williams, 1978; Williams & Smith, 1979). These cells appeared to be typical gonia with large, pale nuclei and two distinct nucleoli. None of 8 irradiated animals examined at stages XXII to XXIV (70–80 days beyond larval stage 25) was sterile. At these late metamorphic stages differentiated germ cells were found in both irradiated and control groups, although the testes and ovaries in controls were more than double the size of the gonads in the irradiated group. The medulla of the testes contained clusters of germ cells: pale spermatogonia with lobulated nuclei and distinct nucleoli; spermatocytes with nuclei covered with a fine reticulum of condensed chromatin; young spermatids whose oval nuclei were darkly stained due to the onset of chromatin condensation. Primary oocytes with maximal diameters of about 100 μm were present in the ovaries of both irradiated and control groups. Nucleolar amplification (Kalt & Gall, 1974) was apparent with as many as 15 nucleoli per section evident within the peripheral nucleoplasm. Thus, the UV-induced sterility observed in *Rana pipiens* at stage 25 (Shumway, 1940) appears to be a transient response, readily reversed by germ cells appearing in the gonads during later metamorphic stages. Similar results have recently been obtained by Subtelny (personal communication), although not all metamorphosing frogs were observed to contain germ cells.

Gonads of stage V (Taylor & Kollros, 1940) *Rana* tadpoles also were examined in the electron microscope to discern the ultrastructural characteristics of germ cells which ultimately reached the gonads of animals derived from irradiated eggs. In control animals the numbers of germ cells observed during stage V was greatly increased relative to larval stage 25, presumably due to their mitotic proliferation on reaching the genital ridges (Züst & Dixon, 1977). In irradiated animals germ cells were relatively scarce. The germ cell shown in Fig. 4(*a*) is readily distinguishable from the surrounding gonadal epithelial cells due to its large size, large oval nucleus with diffuse chromatin and, in particular, by the presence of electron dense cytoplasmic inclusions associated with mitochondria. The latter material appears to be identical to nuage as previously described in germ cells and gonial stages of gamete differentiation (Fig. 4*b*; see Smith & Williams, 1975).

In addition to structures common to germ cells in control and irradiated animals, germ cells of irradiated animals contained ultrastructural modifications not found in their non-irradiated counterparts. Most prominently, mitochondria which appear to be damaged to varying degrees were present throughout the cytoplasm. Mitochondria vacuolation, membrane distortions, and apparent disintegration into membranous whorls were observed (Fig. 4*c*). These alterations are not considered to be fixation artifacts since normal-looking mitochondria were found adjacent to the damaged ones and distorted mitochondria were not seen in the enveloping epithelial cells. Autophagosomes with their characteristic trilamellar boundaries were also present in the irradiated cytoplasm, as were structures resembling primary and secondary lyzosomes. These structural abnormalities did not appear to prevent germ cell mitosis and subsequent differentiation, since germ cell numbers increased and differentiation was apparent in later metamorphic stages.

Ikenishi *et al.* (1974) observed vacuolated and swollen mitochondria in *Xenopus* embryos shortly after UV irradiation (early 2-cell stage), and degradation of mitochondria within the germ plasm region appeared to advance progressively during development to the late blastula stage. However, similar changes were often observed in control embryos. Thus, the possibility that these changes resulted from fixation artifacts cannot be excluded (see Czolowska, 1972; Smith & Williams, 1975). We have not usually

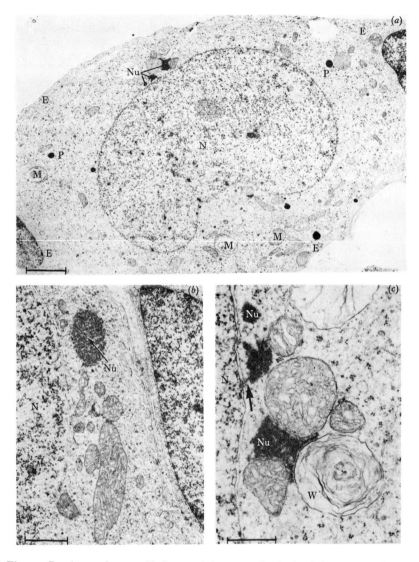

Fig. 4. Portions of stage V *Rana pipiens* gonads derived from control and UV-irradiated embryos. (*a*) Germ cell from UV-irradiated embryo. A single layer of germinal epithelial cells (E) encompasses the germ cell, which is distinguished from the epithelial cells by its large size, large nucleus (N) with diffuse chromatin, and electron dense bodies resembling nuage (Nu). Some germ cell mitochondria (M) appear damaged. Pigment granules (P) are distributed with the cytoplasm. Scale bar represents 2 μm. (*b*) and (*c*) Portions of germ cells from control and irradiated embryos respectively. (*b*) Electron dense body of nuage (Nu) and mitochondria located near the nucleus (N). Scale bar represents

observed such damage within the germ plasm mitochondria in either irradiated *Rana pipiens* embryos (4-cell stage) or *Xenopus* embryos (blastula and gastrula stages) (Smith & Williams, 1975; Williams, 1978). On the other hand, primary UV-induced changes in the mitochondrial population might not be immediately expressed as gross morphological changes. Since presumptive germ cells are not readily identified in irradiated embryos between neurula stages and the time they finally enter the genital ridges (Züst & Dixon, 1977), the appearance of mitochondrial abnormalities during these times obviously would not be detected.

Dawid & Blackler (1972) have shown that the mitochondria in germ cells are inherited from the mitochondrial population present in the fertilized egg. Thus, we suggest that the damaged mitochondria present in 'late-appearing' germ cells are derived from mitochondria present within the germ plasm at the time of UV irradiation. By the same token, we conclude that UV irradiation of the vegetal pole of anuran amphibian embryos delays, but does not always prevent, the appearance of germ cells in the developing gonads. Our results confirm and extend the observations of Züst & Dixon (1977), who were the first to note this phenomenon in *Xenopus*.

Deciding whether UV inactivates germ cell determinants directly, or affects other processes necessary for germ cell differentiation, depends on interpretations regarding how and when germ cell determinants act. Kerr & Dixon (1974) have suggested that germ plasm is 'inactive' during cleavage but that it becomes active soon after, perhaps at gastrula when germ plasm becomes associated with the nucleus (see also Blackler, 1970). Thus, germ plasm would become active some time between gastrulation and the time germ cells localize in the genital ridges. In this context, one might not expect to see dramatic differences between presumptive germ cells (cells containing germ plasm) prior to gastrula when comparing irradiated and control embryos. This expectation is fulfilled (Smith & Williams, 1975; Williams, 1978). However, in post-gastrula embryos, cells containing germ plasm progres-

---

1 μm. (*c*) Enlargement of a part of the germ cell shown in (*a*). Electron dense bodies of nuage (Nu) are located near the nucleus (N) and at one point contact the nuclear envelope (arrow). Mitochondria and a membranous whorl (W) are apposed to one of the dense bodies of nuage. Scale bar represents 0.5 μm.

sively disappear (cannot be detected) in UV-irradiated embryos, while germ cells appear in the genital ridges later than usual.

We have considered two alternative explanations for these observations. First, it remains possible that UV irradiation is simply less efficient in destruction of the 'germ cell determinants' than previously assumed. Thus, in post-gastrula embryos which develop from irradiated eggs, most presumptive germ cells die, and the few surviving germ cells migrate later than usual into the developing gonads. In this case, mitochondrial damage (see above) resulting from irradiation could contribute to a reduced ability to migrate. Alternatively, in irradiated post-gastrula embryos the number of presumptive germ cells remains constant but undetectable and they undergo a precocious differentiation as suggested by Züst & Dixon (1977), i.e. the presumptive germ cells precociously enter the 'second proliferative phase' characteristic of normal germ cells in the genital ridges, and the smaller cells produced are, for some reason, less efficient at migration. In a sense, this implies that germ plasm prevents germ cell differentiation as postulated by Blackler (1958) years ago.

Very recently, Ikenishi & Kotani (1979) have followed the fate of presumptive primordial germ cells, using 0.5 μm Epon sections, in irradiated *Xenopus* embryos. In control embryos the number of presumptive germ cells was reported to increase 3–4-fold between stages 12 and 46, confirming earlier studies (Whitington & Dixon, 1975; Kamimura *et al.*, 1976; Dziadek & Dixon, 1977). In UV-irradiated embryos the number of germ cells increased at most by 2-fold up to the mid-tail-bud stage (stage 28) then decreased and disappeared by the feeding tadpole stage (stage 46). Thus, in contrast to the suggestion of Züst & Dixon (1977), presumptive primordial germ cells divided more slowly in post-gastrula embryos derived from irradiated eggs than in controls. Perhaps more significantly, at stages 33/34, when presumptive germ cells have begun to migrate from the deep endoderm in controls and are found in the lateral or dorsal part of the endoderm (Kamimura *et al.*, 1976), the presumptive germ cells in irradiated embryos were still found in the central part of the endoderm. This again suggests a defect in the ability to migrate, which correlates with the subsequent disappearance of presumptive primordial germ cells.

Kamimura *et al.* (1976) suggest that the displacement of primordial germ cells from a deep endodermal position at the neurula stage to the dorsal endodermal crest in tadpoles is due to an active migration. This remains to be proven directly. However, a series of observations by Wylie and colleagues on *Xenopus* tadpoles indicates that germ cells are motile at stages 44–49 and probably localize in the presumptive gonads, at least in part, by active locomotion (Wylie & Heasman, 1976; Wylie, Bancroft & Heasman, 1976; Wylie & Roos, 1976).

Since a major effect of UV irradiation appears to be the prevention or retardation of germ cell migration, it becomes important to determine whether this reflects an intrinsic property of cells containing irradiated germ plasm or whether it results indirectly from UV-induced disruption of other processes. In this regard, recent experiments by Subtelny and colleagues (unpublished data) are very instructive. They made parabionts by joining the endoderms (at the early tail-bud stage) of UV-irradiated and control embryos and analysed the embryos subsequently (at stage 25) for the presence or absence of germ cells. They observed in many cases that the UV-parabiont acquired germ cells of normal size and at the same time as did controls. By using isozyme variants these germ cells were demonstrated to have originated from the control embryos. Similar results were obtained when control endoderm, containing germ cells, was grafted (Blackler, 1961) into the germ cell containing region of irradiated embryos. Since normal presumptive germ cells migrated through the endoderm of irradiated embryos, we conclude from their data that delayed or inhibited migration of 'irradiated' presumptive germ cells is a property intrinsic to these cells and not the overall endodermal environment in irradiated embryos.

In summary, we have discussed several lines of evidence which correlate the deleterious effects of UV irradiation on the subsequent appearance of germ cells with the presence and location of a stainable germ plasm. We suggest that the results, taken together, support the view that UV irradiation modifies a component(s) contained within the germ plasm that somehow specifies that cells containing germ plasm, and only those cells, eventually differentiate into germ cells. This view is further supported by the fact that, in *Drosophila*, the definitive experiments of Illmensee & Mahowald (1974) have shown that the pole plasm, essentially identical in appearance to germ plasm in anuran eggs, directs the formation of germ cells. Thus, we conclude that the germ plasm theory remains valid in vertebrates, at least in anuran amphibians. The nature and mode of action of germ plasm remain problematical. However, based on observations concerning germ cell migration, we suggest that at least one function of germ plasm relates to the ability of germ cells to move to and localize within the genital ridges. Whether germ plasm also functions at other levels of regulation in germ cell differentiation remains to be answered.

Nature of the germ cell determinants

Having arrived again at the conclusion that cells containing germ plasm, and only those cells, eventually differentiate into germ cells, it becomes critical to establish which component(s) contained within the germ plasm functions as a germ cell determinant.

Cytochemical studies have demonstrated that the germinal plasm recognized in the light microscope stains positive for RNA, both in amphibian eggs (Blackler, 1958; Czolowska, 1969) and insect eggs (Mahowald, 1971; Mahowald et al., 1979). In both cases the RNA staining disappears relatively early in development, during the gastrula stage in amphibians (Blackler, 1958) and the blastoderm stage in Drosophila (Mahowald, 1971). Thus, loss of specific affinity for basophilic stains correlates with the intracellular shift of germ plasm to a juxtanuclear position, at least in anurans. Unfortunately, the germ plasm region contains, in addition to the electron dense germinal granules, large aggregates of mitochondria and considerable numbers of ribosomes as well as polysomes. Cytoplasm adjacent to germinal plasm regions is sparsely populated by these components but contains large quantities of yolk. Thus, it is not unlikely that in amphibians, cytochemical procedures identify largely ribosomal RNA contained and perhaps concentrated within the yolk-free germinal plasm. Perhaps the strongest indication for involvement of RNA in germ cell formation is the UV action spectrum reported by Smith (1966), which suggests that the UV-sensitive material contains nucleic acid. However, this study involved a limited number of different wavelengths and, as already pointed out, identity of the UV target is not known.

Mahowald (1971) has reported that polar granules in Drosophila stain positive for RNA, and suggested that they represent the site of localization of maternal messenger RNA which codes for the synthesis of proteins necessary for germ cell formation. In this hypothesis, only the proteinaceous moiety of polar granules would be present continuously throughout the life cycle, with RNA added in oogenesis and disappearing once it had been translated during early development.

Webb (1976) has reported that when genital ridges from stages 54–57 *Xenopus* tadpoles are cultured with [³H]uridine, followed by high-resolution autoradiography, the density of silver grains associated with cytoplasmic 'nuage' is about 4-fold greater than that in the surrounding cytoplasm. Since some investigators feel that nuage and germinal granules may represent the same material (review by Eddy, 1975), such results would imply that germinal granules contain RNA synthesized very early during gametogenesis.

Mahowald & Hennen (1971) have reported moderate staining of *R. pipiens* germinal granules with the indium trichloride procedure, reported by Watson & Aldridge (1961) to be fairly specific for RNA. After cold perchloric acid extraction of the ribonucleic acids from the egg, the germinal granules lost most of their electron density and the intense indium staining of ribosomes was obliterated. Therefore, they concluded that ribonucleic acid was a constituent of germinal granules. In contrast, Kalt (1973) interpreted the faint indium staining of *Xenopus* germinal granules to be the result of background staining not attributable to nucleic acids.

Irrespective of these opposite interpretations, observations that polysomes are closely associated with amphibian germinal granules, from their initial appearance during oocyte maturation through early embryogenesis (Mahowald & Hennen, 1971; Williams & Smith, 1971; Kalt, 1973), suggest that germinal granules are small foci of protein synthesis. In this regard, the germinal plasm of *Xenopus* embryos (8-cell to gastrula) became labelled 1 hour after the injection of [³H]leucine (Hogarth & Dixon, 1976). However, since this study was based entirely on light microscopic evidence, it could not reveal whether the germinal granules themselves actually account for the label.

We have recently completed a preliminary high-resolution autoradiographic study in which germ plasm regions of *R. pipiens*

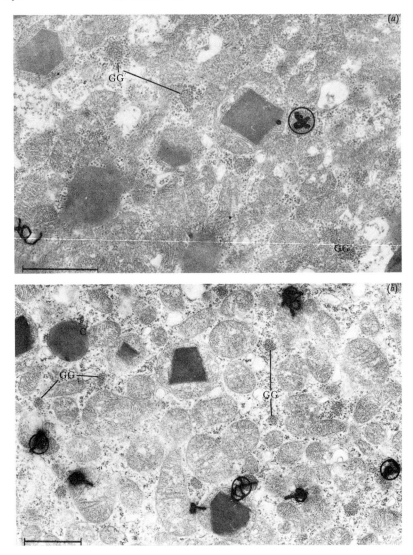

Fig. 5. Autoradiograms of germinal plasm regions of 8-cell *Rana pipiens* embryos injected with [³H]leucine at the beginning of the 2-cell state. Sections were exposed to photographic emulsion for 19 weeks (*a*) and 8 months (*b*). Thirty per cent 'probability circles' (Salpeter & McHenry, 1973) superimposed over the grains associated with germinal granules (GG) also encompass portions of adjacent mitochondria. Scale bar represents 1 μm.

embryos were examined during the 8-cell stage after the injection of [³H]leucine into the vegetal hemisphere at the time of initial cleavage (Williams, 1978). Therefore, the autoradiograms exhibit proteins synthesized during a select period of early development. As shown in Fig. 5, grains can be observed in association with germinal granules. Unfortunately, since the granules are fre- quently, if not always, associated intimately with mitochondria, it is difficult to exclude the possibility that such grains do not have a mitochondrial origin. In order to approach this problem, the distribution of grains in the germinal plasm region was analysed quantitatively according to the 'probability circle' analysis de- scribed by Salpeter & McHenry (1973). The autoradiograms revealed the presence of relatively few silver grains, with 120 grains distributed over a total germ plasm area of 1465 μm^2. However, the grain density of the germ plasm (1 grain per 12.8 μm^2) was about three times greater than that of the yolk-rich cytoplasm adjacent to the germ plasm regions. Within the germ plasm less than 10 % of the total germinal granules observed (433) were associated with silver grains, i.e. the 30 % probability circle encompassing the grain also covered a portion of a germinal granule. Nevertheless, estimates of the germinal granule grain density (1 grain per 2.2 μm^2) were about 6 times higher than the grain density in the total germ plasm area. Alternatively, autoradiographic labelling can be expressed as per cent density, which is simply the ratio of per cent total grains associated with a specific organelle to the per cent of total area occupied by that organelle. If the distribution of grains over an organelle is random, as might be expected for germinal granules if the associated grains are derived from adjacent mitochondria or ground cytoplasm, the per cent density will be 1.0 (Salpeter & McHenry, 1973). While 30 % of the total number of grains observed were associated with germinal granules, the granules occupied only 5.5 % of the total area examined. Thus, the per cent density associated with germinal granules was calculated to be 5.7.

The value of this analysis may be questioned because of the low number of grains actually observed; 100 silver grains has been accepted as the absolute minimum for quantitative analysis (Williams, 1969). Nevertheless these results indicate that the grain distribution associated with germinal granules is highly non- random. Thus, the data support the view that germinal granules are either sites of protein synthesis or sites of protein accumulation. The autoradiograms may, in fact, be recording both processes. The former possibility would support the premise that germinal granules contain maternal messenger RNA, while interpretation of the latter possibility is more equivocal.

Because the electron dense bodies are associated with the germ plasm in animals as diverse as frogs and flics, they have long been considered as visible manifestations of germ cell determinants (see Mahowald, 1975). Thus, the suggestion that they contain maternal messenger RNA is very appealing. It is, of course, also conceivable that the electron dense bodies play no role in germ cell determination. Electron dense bodies very similar in structure to germinal (polar) granules have been identified in cell types other than germ cells (see Smith & Williams, 1975; Dohmen & Lok, 1975), although, in at least one case (Dohmen & Lok, 1975), their presence has also been correlated with the existence of morphogenetic determinants (see also Dohmen & Verdonk, 1979). Perhaps, structures such as germinal granules, acting in concert with other components of the germinal plasm (such as mitochondria) ultimately are required for normal germ cell differentiation. Clearly, what needs to be accomplished is a direct demonstration that a specific component of germ plasm, such as the electron dense bodies, can or cannot direct germ cell formation. By isolating these structures and testing their activity, as already done for pole plasm (Illmensee & Mahowald, 1974), it should be possible to identify their biological function(s) as well as their chemical identity. The demonstration by Mahowald and colleagues that a partially purified polar granule fraction has been isolated from *Drosophila* pole cells (Mahowald *et al.*, 1979) indicates that answers to these questions may be forthcoming.

The original work described here was supported by a research grant from the National Institute of Health (HDO4229). We thank Drs W. Wasserman and I. R. Konigsberg for helpful comments in the preparation of this paper.

References

BALINSKY, B. I. (1966). Changes in the ultrastructure of amphibian eggs following fertilization. *Acta Embryologiae et Morphologiae Experimentalis*, **9**, 132–54.

BEAL, C., M. & DIXON, K. E. (1975). Effect of UV on cleavage of *Xenopus laevis*. *Journal of Experimental Zoology*, **192**, 277–83.

BEAMS, H. W. & KESSEL, R. G. (1974). The problem of germ cell determinants. *International Review of Cytology*, **39**, 413–79.

BLACKLER, A. W. (1958). Contribution to the study of germ cells in the Anura. *Journal of Embryology and Experimental Morphology*, **6**, 491–503.

BLACKLER, A. W. (1961). Transfer of primordial germ-cells in *Xenopus laevis*. *Journal of Embryology and Experimental Morphology*, **9**, 634–41.

BLACKLER, A. W. (1966). Embryonic sex cells of Amphibia. *Advances in Reproductive Physiology*, **1**, 9–28.

BLACKLER, A. W. (1970). The integrity of the reproductive cell line in the Amphibia. *Current Topics in Developmental Biology*, **5**, 71–87.

BOUNOURE, L. (1934). Recherches sur la lignée germinale chez la Grenouille rousse aux premiers stades du développement. *Annales des Sciences Naturelles*, *10th Series*, **17**, 67–248.

BOUNOURE, L., AUBRY, R. & HUCK, M. L. (1954). Nouvelles recherches expérimentales sur les origines de la lignée reproductrice chez la Grenouille rousse. *Journal of Embryology and Experimental Morphology*, **2**, 245–63.

BUEHR, M. L. & BLACKLER, A. W. (1970). Sterility and partial sterility in the South African clawed toad following the pricking of the egg. *Journal of Embryology and Experimental Morphology*, **23**, 375–484.

CLARK, J. M. & EDDY, E. M. (1975). Fine structural observations on the origin and associations of primordial germ cells of the mouse. *Developmental Biology*, **47**, 136–55.

CZOLOWSKA, R. (1969). Observations on the origin of the germinal cytoplasm in *Xenopus laevis*. *Journal of Embryology and Experimental Morphology*, **22**, 229–51.

CZOLOWSKA, R. (1972). The fine structure of the 'germinal cytoplasm' in the egg of *Xenopus laevis*. *Wilhelm Roux' Archiv für Entwicklungsmechanik der Organismen*, **169**, 335–44.

DAVIDSON, E. H. (1976). *Gene Activity in Early Development*, 2nd ed, chapter 7. New York & London: Academic Press.

DAWID, I. B. & BLACKLER, A. W. (1972). Maternal and cytoplasmic inheritance of mitochondrial DNA in *Xenopus*. *Developmental Biology*, **29**, 152–61.

DOHMEN, M. R. & LOK, D. (1975). The ultrastructure of the polar lobe of *Crepidula fornicata*. *Journal of Embryology and Experimental Morphology*, **34**, 419–38.

DOHMEN, M. R. & VERDONK, N. H. (1979). The ultrastructure and role of the polar lobe in development of molluscs. In *Determinants of Spatial Organization*, ed. I. R. Konigsberg & S. Subtelny. New York & London: Academic Press (in press).

DZIADEK, M. & DIXON, K. E. (1975). Mitosis in presumptive primordial germ cells in post-blastula embryos of *Xenopus laevis. Journal of Experimental Zoology*, **192**, 285–91.

DZIADEK, M. & DIXON, K. E. (1977). An autoradiographic analysis of nucleic acid synthesis in the presumptive primordial germ cells of *Xenopus laevis. Journal of Embryology and Experimental Morphology*, **37**, 13–31.

EDDY, E. M. (1975). Germ plasm and the differentiation of the germ cell line. *International Review of Cytology*, **43**, 229–80.

EYAL-GILADI, H., KOCHAU, S. & MENASHI, M. K. (1976). On the origin of primordial germ cells in the chick embryo. *Differentiation*, **6**, 13–16.

GARDNER, R. L. & ROSSANT, J. (1976). Determination during embryogenesis. In *Embryogenesis in Mammals, Ciba Foundation Symposium 40*, pp. 5–18. Amsterdam: Elsevier.

GIPOULOUX, J.-D. (1971). Effets de l'extrusion totale ou partielle du cytoplasme germinal au cours des premiers stades de la segmentation sur la fertilité des larves d'amphibiens anoures. *Comptes Rendus de l'Academie des Sciences, Paris, Series D*, **273**, 2627–9.

HOGARTH, K. & DIXON, K. E. (1976). Protein synthesis and germ plasm in cleavage embryos of *Xenopus laevis. Journal of Experimental Zoology*, **198**, 429–35.

IJIRI, K. (1976). Stage-sensitivity and dose-response curve of UV effect on germ cell formation in embryos of *Xenopus laevis. Journal of Embryology and Experimental Morphology*, **35**, 617–23.

IJIRI, K. (1977). Existence of ultraviolet-labile germ cell determinant in unfertilized eggs of *Xenopus laevis* and its sensitivity. *Developmental Biology*, **55**, 206–11.

IJIRI, K. & EGAMI, N. (1975). Mitotic activity of germ cells during normal development of *Xenopus laevis* tadpoles. *Journal of Embryology and Experimental Morphology*, **34**, 687–94.

IKENISHI, K. & KOTANI, M. (1975). Ultrastructure of the 'germinal plasm' in *Xenopus* embryos after cleavage. *Developmental Growth and Differentiation*, **17**, 101–10.

IKENISHI, K. & KOTANI, M. (1979). UV effects on presumptive primordial germ cells (pPGCs) in *Xenopus laevis* after cleavage stage. *Developmental Biology*, in press.

IKENISHI, K., KOTANI, M. & TANABE, K. (1974). Ultrastructural changes associated with UV irradiation in the 'germinal plasm' of *Xenopus laevis. Developmental biology*, **36**, 155–68.

ILLMENSEE, K. & MAHOWALD, A. P. (1974). Transplantation of posterior pole plasm in *Drosophila*. Induction of germ cells at the anterior pole of the egg. *Proceedings of the National Academy of Sciences of the USA*, **71**, 1016–20.

KALT, M. R. (1973). Ultrastructural observations on the germ line of *Xenopus laevis. Zeitschrift für Zellforschung*, **138**, 41–62.

KALT, M. R. & GALL, J. G. (1974). Observations on early germ cell development and premeiotic ribosomal DNA amplification in *Xenopus laevis. Journal of Cell Biology*, **62**, 460–72.

KALT, M. R. & TANDLER, B. (1971). A study of fixation of early amphibian embryos for electron microscopy. *Journal of Ultrastructure Research*, **36**, 635-45.

KAMIMURA, M., IKENISHI, K., KOTANI, M. & MATSUNO, T. (1976). Observations on the migration and proliferation of gonocytes in *Xenopus laevis*. *Journal of Embryology and Experimental Morphology*, **36**, 197-207.

KERR, J. B. & DIXON, K. E. (1974). An ultrastructural study of germ plasm in spermatogenesis of *Xenopus laevis*. *Journal of Embryology and Experimental Morphology*, **32**, 573-92.

KESSEL, R. G. (1971). Cytodifferentiation in the *Rana pipiens* oocyte. II. Intramitochondrial yolk. *Zeitschrift für Zellforschung*, **112**, 313-22.

MAHOWALD, A. P. (1962). Fine structure of pole cells and polar granules in *Drosophila melanogaster*. *Journal of Experimental Zoology*, **151**, 201-15.

MAHOWALD, A. P. (1971). Origin and continuity of polar granules. In *Origin and Continuity of Cell Organelles*, ed. J. Reinert & H. Ursprung, pp. 159-69. New York: Springer-Verlag.

MAHOWALD, A. P. (1975). Ultrastructural changes in the germ plasm during the life cycle of *Miastor*. *Wilhelm Roux' Archiv für Entwicklungsmechanik der Organismen*, **176**, 223-40.

MAHOWALD, A. P., ALLIS, C. D., KARRER, K. M., UNDERWOOD, E. M. & WARING, G. L. (1979). Germ plasm and pole cells of *Drosophila*. In *Determinants of Spatial Organization*, ed. I. R. Konigsberg & S. Subtelny. New York & London: Academic Press (in press).

MAHOWALD, A. P. & HENNEN, S. (1971). Ultrastructure of the 'germ plasm' in eggs and embryos of *Rana pipiens*. *Developmental Biology*, **24**, 37-53.

MINTZ, B. & ILLMENSEE, K. (1975). Normal genetically mosaic mice produced from malignant teratocarcinoma cells. *Proceedings of the National Academy of Sciences of the USA*, **72**, 3585-9.

NIEUWKOOP, P. D. (1947). Experimental investigations on the origin and determination of the germ cells and on the development of the lateral plate and germ ridges in Urodeles. *Archives Néerlandaises de Zoologie*, **8**, 1-205.

NIEUWKOOP, P. D. & FABER, J. (1956). A normal table of *Xenopus laevis*. Amsterdam: North-Holland.

OKADA, M., KLEINMAN, A. & SCHNEIDERMAN, H. A. (1974). Restoration of fertility in sterilized *Drosophila* eggs by transplantation of polar cytoplasm. *Developmental Biology*, **37**, 43-54.

PADOA, E. (1963). Le gonadi di girini de *Rana esculenta* de nova irradiate con ultraviolette. *Monitore Zoologico Italiano*, **71**, 238-49.

SALPETER, M. M. & MCHENRY, F. A. (1973). Electron microscope autoradiography. In *Advanced Techniques in Biological Electron Microscopy*, ed. J. K. Koehler, pp. 113-52. New York: Springer-Verlag.

SHUMWAY, W. (1940). Stages in the normal development of *Rana pipiens*. I. External form. *Anatomical Record*, **78**, 139-47.

SMITH, L. D. (1964). A test of the capacity of presumptive somatic cells to transform into primordial germ cells in the Mexican axolotl. *Journal of Experimental Zoology*, **156**, 229-42.

SMITH, L. D. (1966). The role of a 'germinal plasm' in the formation of primordial germ cells in *Rana pipiens*. *Developmental Biology*, **14**, 330-47.

SMITH, L. D. & WILLIAMS, M. A. (1975). Germinal plasm and determination of the primordial germ cells. In *The Developmental Biology of Reproduction*, ed. C. L. Markert & J. Papaconstantinou, pp. 3–24. New York & London: Academic Press.

SPIEGELMAN, M. & BENNETT, D. (1973). A light and electron-microscopic study of primordial germ cells in the early mouse embryo. *Journal of Embryology and Experimental Morphology*, **30**, 97–118.

STEVENS, L. (1970). The development of transplantable teratocarcinomas from intratesticular grafts of pre- and postimplantation mouse embryos. *Developmental Biology*, **21**, 364–82.

SUTASURYA, L. A. & NIEUWKOOP, P. D. (1974). The induction of the primordial germ cells in the urodeles. *Wilhelm Roux' Archiv für Entwicklungsmechanik der Organismen*, **175**, 199–220.

TANABE, K. & KOTANI, M. (1974). Relationship between the amount of the 'germinal plasm' and the number of primordial germ cells in *Xenopus laevis*. *Journal of Embryology and Experimental Morphology*, **31**, 89–98.

TAYLOR, A. C. & KOLLROS, J. J. (1946). Stages in the normal development of *Rana pipiens* larvae. *Anatomical Record*, **94**, 7–23.

WAKAHARA, M. (1977). Partial characterization of 'primordial germ cell-forming activity' localized in vegetal pole cytoplasm in anuran eggs. *Journal of Embryology and Experimental Morphology*, **39**, 221–33.

WAKAHARA, M. (1978). Induction of supernumerary primordial germ cells by injecting vegetal pole cytoplasm into *Xenopus* eggs. *Journal of Experimental Zoology*, **203**, 159–64.

WARN, R. (1975). Restoration of the capacity to form pole cells in UV-irradiated *Drosophila* embryos. *Journal of Embryology and Experimental Morphology*, **33**, 1003–12.

WATSON, M. L. & ALDRIDGE, W. G. (1961). Methods for the use of indium as an electron stain for nucleic acids. *Journal of Biophysical and Biochemical Cytology*, **11**, 257–72.

WEBB, A. C. (1976). An autoradiographic study of tritiated uridine incorporation into the larval ovary of *Xenopus laevis*. *Anatomical Record*, **184**, 285–300.

WHITINGTON, P. McD. & DIXON, K. E. (1975). Quantitative studies of germ plasm and germ cells during early embryogenesis of *Xenopus laevis*. *Journal of Embryology and Experimental Morphology*, **33**, 57–74.

WILLIAMS, M. A. (1969). The assessment of electron microscopic autoradiography. *Advances in Optical and Electron Microscopy*, **3**, 219–72.

WILLIAMS, M. A. (1978). Ultrastructural and experimental studies on anuran germinal plasm. PhD thesis, Purdue University.

WILLIAMS, M. A. & SMITH, L. D. (1971). Ultrastructure of the 'germinal plasm' during maturation and early cleavage in *Rana pipiens*. *Developmental Biology*, **25**, 568–80.

WILLIAMS, M. A. & SMITH, L. D. (1979). Effects of UV irradiation on germinal plasm, cleavage and sterility. *Developmental Biology*, in preparation.

WILSON, E. B. (1925). *The Cell in Development and Heredity*, pp. 310–28. New York: Macmillan.

WYLIE, C. C., BANCROFT, M. & HEASMAN, J. (1976). The formation of the gonadal ridge in *Xenopus laevis*. II. A scanning electron microscope study. *Journal of Embryology and Experimental Morphology*, **35**, 139–48.

WYLIE, C. C. & HEASMAN, J. (1976). The formation of the gonadal ridge in *Xenopus laevis*. I. A light and transmission electron microscope study. *Journal of Embryology and Experimental Morphology*, **35**, 125–38.

WYLIE, C. C. & ROOS, T. B. (1976). The formation of the gonadal ridge in *Xenopus laevis*. III. The behavior of isolated primordial germ cells *in vitro*. *Journal of Embryology and Experimental Morphology*, **35**, 149–57.

ZÜST, B. & DIXON, K. E. (1975). The effect of UV irradiation of the vegetal pole of *Xenopus laevis* eggs on the presumptive primordial germ cells. *Journal of Embryology and Experimental Morphology*, **34**, 209–20.

ZÜST, B. & DIXON, K. E. (1977). Events in the germ cell lineage after entry of the primordial germ cells into the genital ridges in normal and UV-irradiated *Xenopus laevis*. *Journal of Embryology and Experimental Morphology*, **41**, 33–46.

The pole plasm of *Drosophila*

RICHARD M. WARN

School of Biological Sciences, University of East Anglia,
Norwich NR4 7TJ, UK

Structure and function of pole plasm

The pole plasm of *Drosophila* is probably the best described example of an area of egg cytoplasm endowed with special properties. In this case alone is there conclusive evidence that nuclei which enter it become channelled along a particular path of development to form mainly the germ cell line. The fate of these nuclei depends on an interaction with certain elements of the cytoplasm classically known as 'determinants' (Hegner, 1914). Little is known at present about the nature of cytoplasmic determinants but they may operate at several levels. First, the molecular nature of the pole plasm directly affects the future activity of the nuclei which enter it, programming them in a particular pathway of development. But in addition the pole plasm forms the structure of the pole cell. Thus special characteristics of the cytoplasm are required to make this kind of cell. This process is quite separate from the programming of the nuclei.

The cortex of the *Drosophila* egg is a region about 5 μm in thickness where yolk granules and other inclusions are absent. At the posterior pole the cortical cytoplasm thickens to a maximum of 10–15 μm. The most posterior 5 μm or so of this cytoplasm is the pole plasm proper. It is characterized by the presence of polar granules. These structures are identified in the light microscope as being densely stained by haematoxylin. They contain protein and RNA (Mahowald, 1971a). Between different species of *Drosophila* there is considerable variation in the size and structure of the polar granules and in details of their behaviour at different developmental stages (Counce, 1963; Mahowald, 1968). However, it is possible to draw up a generalized scheme of the life history of polar granules (see Mahowald, 1971*b*, for further details). The polar granule in every species looked at consists of a densely staining fibril 10–15 nm in thickness which forms an interwoven mesh. During the latter part of oogenesis polar granules are found in close association with mitochondria. Shortly after fertilization, though, the granules become free from the mitochondria and ribosomes are found attached to their periphery. Then, by the completion of the blastoderm, the granules aggregate to form large masses. Sometimes these are paracrystalline in structure. The large fused masses remain until the formation of the embryonic gonad. At this time the fused masses of granules come into a close association with the nucleus and become changed into fibrous structures known as 'nuage' which are very common in germ line cells. During the earliest stages of oogenesis it is impossible to see any structures similar to polar granules and the 'nuage' has disappeared. Some fibrous structures are visible in the oocyte when it becomes established in the egg chamber, but there is no clear evidence that they are the precursors of polar granules. Only at stage 10, during mid-vitellogenesis, does the typical polar granule reappear.

A second special feature of the posterior pole is the presence of large numbers of microvilli in the plasmalemma. These are visible in the light microscope without staining after mild fixation and a gentle wax embedding procedure (Warn, unpublished data). Microvilli are present throughout the egg but are especially numerous at the posterior pole. The structure of these microvilli is somewhat variable, ranging from thread-like with a fine tip to a conspicuously club-like form. The microvilli have a length of about 1 μm. The head of the club-like form has a maximum diameter of about 0.5 μm. They are present throughout the nuclear cleavage stages and do not seem to change during the cycle of nuclear division. By the 64-nuclei stage, after 1 hour of development, larger club-like microvilli, with a length of 0.5–1.5 μm and a maximum diameter of 1 μm, have become evident at the posterior pole. Club-like microvilli appear to play a major role in the formation of the pole cells. As the nuclei approach the pole plasm there are many microvilli present. As the pole cells begin to bud off, the microvilli in the plasmalemma immediately above the nuclei become smaller and less obviously club-like in

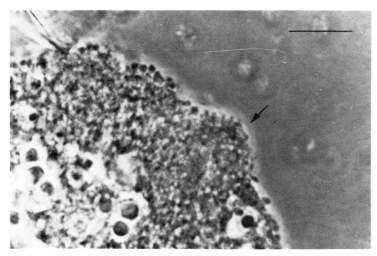

Fig. 1. Pole cell (arrowed) beginning to bulge out from the posterior pole. Above it is a region with club-like microvilli. Scale bar represents 10 μm.

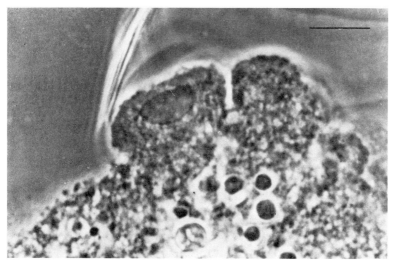

Fig. 2. Second stage in pole cell formation where the nascent cell has a flattened form. Scale bar represents 10 μm.

shape (Fig. 1). However, areas between pole cells retain the club-like microvilli. Whilst the pole plasm is forming itself into a cell it is initially oblong in shape (Fig. 2). However, it rapidly rounds up (Fig. 3) and is then pinched off from the egg (Fig. 4). As the cell forms, its surface becomes progressively smoother and

R. Warn

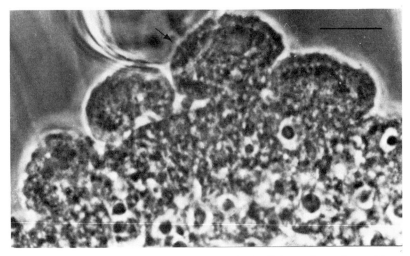

Fig. 3. The pole cell arrowed is being pinched off from the embryo. Note a conspicuous club-like microvillus present on its lower left margin. Scale bar represents 10 μm.

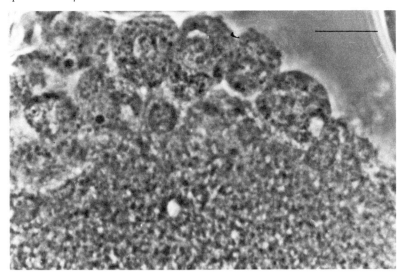

Fig. 4. Fully formed pole cells. Scale bar represents 10 μm.

the microvilli disappear. This process is quite different from what occurs during the formation of the somatic blastoderm cells. Furrow canals do not seem to be present and the whole organization of the process is quite different. The time course of events is much faster, taking only a few minutes as compared with half an hour for the formation of the blastoderm cells. Club-shape microvilli remain in the areas between the pole cells. They also extend into areas beyond the immediate region of the pole cells. Thus they do not appear to be special characteristics of pole plasm but rather of the whole posterior pole region. What might the functions of the club-like microvilli present immediately before pole cell formation be? The simplest and most reasonable explanation is that they represent stored membrane material for the formation of the pole cells. As the cells bubble up, the microvilli in that area gradually disappear. In the regions between cells they remain. The continuous presence from fertilization onwards of large numbers of smaller microvilli in the posterior pole region suggests other functions as well. Possibly they are involved in the expansions and contractions of the embryo during nuclear cleavage (Imaizumi, 1954).

The presence of a concentration of microvilli at the posterior pole is a second example of a morphological specialization of one region of the *Drosophila* egg. Some of this variation is quite possibly concerned with the formation of a special cell type and probably reflects differences in the morphogenetic potential of this region. Variations in the distribution of the microvilli in different parts of the newly fertilized egg of *Xenopus* have also been found (Monroy & Baccetti, 1975). The dimensions of the club-shape microvilli they described are broadly similar to those present at the posterior pole of *Drosophila melanogaster*. A dense accumulation of microvilli occurs at the site of cleavage in the animal pole immediately prior to the first cell division in *Xenopus* (Denis-Donini, Baccetti & Monroy, 1976). The authors conclude that in this case microvilli might represent the site of new membrane insertion.

It is reasonable to suggest that the ultrastructural specializations of the posterior pole will be reflected in the presence of proteins not found in other parts of the egg. However, until now there have been few data on the cytoplasmic location of specific macromolecules. Graziosi & Roberts (1975) made antibodies against various parts of *Drosophila* eggs and demonstrated at least two antigens present only in the posterior pole region. Sections of eggs incubated with anti-posterior pole antiserum showed specific labelling in this region.

Experimental manipulation of pole plasm

The pole plasm of *Drosophila* has proved itself to be eminently suitable for manipulative studies. The most important result to come from these studies is the definite conclusion that one or more specific components (the so-called 'determinants') are present which cause the formation of a specific cell line, the pole cells. These cells have a very restricted fate to form only two types of cell. Many of them later come to form the germ cells. Others are probably incorporated into the gut as cuprophilic cells.

Centrifugation

These experiments were carried out by Jazdowska-Zagrodzinska (1966). Eggs were centrifuged with the centrifugal force acting either anteriorly (away from the pole plasm), posteriorly (towards the pole plasm), or laterally (perpendicular to the long axis of the egg). Eggs showing a distinct stratification were either allowed to develop until blastoderm formation or were immediately analysed. In the case of both anterior and posterior centrifugation polar granules were no longer visible and it was not possible to find out where they had migrated to. In anteriorly centrifuged eggs the amount of clear cytoplasm at the posterior tip had increased, whereas in posteriorly centrifuged eggs the posterior ends were filled with large yolk granules. In both these cases no pole cells were formed. Laterally centrifuged eggs had polar granules which had been displaced onto the centrifugal side. Pole cells were found to develop on this side only. Eggs centrifuged anteriorly did however manage to produce normal-looking blastoderm cells at the region of the posterior pole, as did some of the posteriorly centrifuged cells. In other cases of posterior centrifugation so much yolk was present that the blastoderm closed above the yolk, thus cutting this region off from the embryo. Two conclusions can be drawn here. First, the presence or absence of polar granules is a clear guide to whether or not pole cells will form. This does not mean that polar granules are essential for this process, they could be just markers, but it is consistent with the idea that they are normally involved in pole cell formation. Secondly, after centrifugation, regions of the egg which would usually form pole cells form instead blastoderm cells. This is a surprising finding.

Abnormal cells of any description were not reported. Thus, that area of the egg which normally formed pole cells was now divided up into blastoderm cells. However, in this case it is not known whether constituents of the original pole plasm were included in these cells.

UV irradiation

Historically, Geigy (1931) was the first to find that UV irradiation of the posterior pole of a newly fertilized *Drosophila* egg is followed by a loss of the ability to form pole cells and consequent sterility of the adults. This has since been confirmed by a number of workers (Poulson & Waterhouse, 1960; Hathaway & Selman, 1961; Warn, 1972; Graziosi & Micali, 1974; Okada, Kleinman & Schneiderman, 1974a). When a suitable dose is used no pole cells are formed, but a normal layer of blastoderm cells forms in the region of the pole plasm, as well as around the rest of the egg (compare Figs. 5 and 6). The layer of blastoderm cells which normally underlies the pole cells is now immediately beneath the vitelline membrane. In effect the pole plasm has been respecified to form blastoderm cells. At a higher dosage level more drastic effects occur (Warn, unpublished results; Okada *et al.*, 1974a). One effect is that the nuclei are much reduced in size and do not expand as the cell membranes of the blastoderm form. At still higher levels of irradiation another effect is common. The nuclei are abnormally distributed over the whole of the posterior pole region instead of being in a single line. The cytoplasm is broken up and disorganized, whilst the nuclei are irregular in size, sometimes being enormously swollen, and may or may not contain nucleoli. At blastoderm formation fragments rather than whole cells occur and the nuclei rapidly become pycnotic and degenerate in this area. Thus at higher doses UV irradiation causes a failure of further development to occur in the pole plasm.

Okada *et al.* (1974a) have investigated further what happens when irradiation of the pole plasm results in the formation of blastoderm cells where pole cells would usually form. They observed living eggs in KEL-F polymer oil. In unirradiated embryos protrusions were formed by the pole plasm as the nuclei entered. Approximately two minutes later protrusions appeared on all other parts of the cortex. Those at the posterior pole budded off

R. Warn

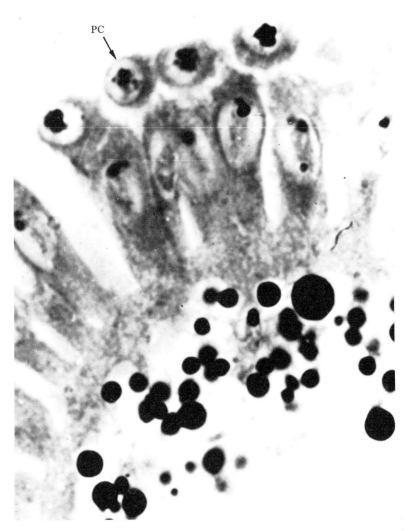

Fig. 5. Normal pole cells (PC) and posterior blastoderm cells stained with Heidenhain's haematoxylin. Scale bar represents 10 μm.

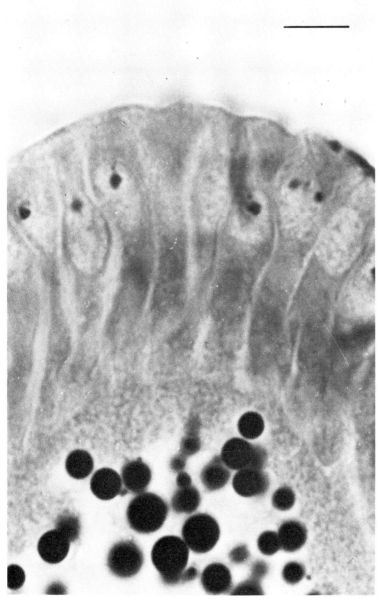

Fig. 6. Posterior tip of a slightly later stage than Fig. 5 which was irradiated with UV shortly after fertilization. Pole cells are completely absent. Scale bar represents 10 μm.

to become pole cells. In irradiated eggs, on the other hand, protrusions appeared at approximately the same time on all surfaces of the periplasm including that of the posterior pole. The protrusions at the posterior pole of irradiated eggs grew further out from the surface of the embryo than protrusions in other parts of the periplasm. They resembled those protrusions which form during pole cell formation but they never detached from the surface. Subsequently these large protrusions became incorporated into the posterior part of the surface of the blastoderm.

It is not clear how UV light affects pole cell formation. One possibility is that alteration of RNA, perhaps that in the polar granules, occurs by the formation of pyrimidine dimers. Thus loss of the ability to form pole cells would occur as the result of a loss of stored informational molecules. However, until an action spectrum is obtained for different wavelengths and photoreversion with visible light demonstrated, caution must be exercised in such an interpretation. In fact there are a number of other possible molecular targets. These include breakage of S–S bonds, destruction of tryptophan, and changes in ionic permeability or in the intracellular pH. It is known that appropriate doses of UV will cause a delay in the cytokinesis of *Xenopus* embryos (Beal & Dixon, 1975; Züst & Dixon, 1975). Higher doses will permanently damage this process resulting in a syncytium at the irradiated pole, with seemingly normal development at the other pole. The target of the UV irradiation in this case is again not known. However, it suggests an alternative interpretation for the results of UV irradiation of *Drosophila* pole plasm: that UV inhibits the mechanism by which pole cells are budded off.

In eggs which have been centrifuged or irradiated with UV light, areas where pole cells would normally form can be switched into a second developmental pathway of blastoderm cell formation. A similar result is seen in mutants of the *grandchildless* class. These are maternal sterile mutants where defective offspring are produced as a result of an abnormality of oogenesis. The progeny appear normal in every respect except that they are sterile. In both of two different studies of *grandchildless* mutants the abnormality produces progeny embryos lacking pole cells. In the *grandchildless* of *Drosophila subobscura*, it was found that there was a generalized delay in the migration of nuclei into the posterior regions of the egg (Fielding, 1967). It was found that the pole plasm vacuolated and degenerated during this time and was cut off from the blastoderm. However, in this study only aged females could be used. This was because homozygous *grandchildless* mothers could only be identified by the appearance of their progeny. Mahowald & Gehring (quoted in Gehring, 1973) were able to create a stock carrying a marker on the same chromosome as *grandchildless* and to balance it against an inversion. Using this, eggs could be obtained from much younger mothers. They were able to confirm the delay in nuclear migration at the posterior end of the embryo. However, they found that nuclei could secondarily migrate into the posterior pole from the adjacent lateral periplasm and form blastoderm cells there. Polar granules with a very similar structure to normal were present and were included in these blastoderm cells. The subsequent fate of these polar granules is not known.

The second *grandchildless* mutation is found in *D. melanogaster* and is known as *grandchildless* 87 (Thierry-Mieg, 1976). In this mutant adult sterility is also due to the absence of pole cells in the blastema-stage embryo. However, the polar granules are apparently normal in morphology when examined in the electron microscope. In about 20 % of the embryos there is defective nuclear migration. These embryos contain abnormal multiple arrangements of nuclei rather than the usual single layer covering the blastema. Up to 60 % of the blastema-stage embryos contain no pole cells. In these cases the nuclei are distributed normally in all areas except the pole plasm. Here the nuclei form a continuous single layer along the cortex in an identical way to other regions of the blastema and the pole plasm has become incorporated in this layer. They exactly resemble embryos where the pole plasm has been previously irradiated with a suitable dose of UV light to prevent pole cell formation.

Described above are two and perhaps three situations where the cytoplasmic constituents of pole plasm have become incorporated into a completely different kind of cell rather than either forming abnormal cells or being excluded altogether from the embryo and degenerating. In the case of the two *grandchildless* mutants it seems very possible that all the constituents of the pole plasm, including apparently normal polar granules, have been incorporated into blastoderm cells and that the defect is due to the time of arrival of the nuclei in the posterior tip. This could also be so in the case of the embryos irradiated with UV light. This finding that pole plasm can form part of blastoderm cells is surprising, particularly as a pole cell is so different from a blastoderm cell. Major differences must exist between the composition of pole plasm and that of the cortex of the rest of the egg, particularly in whatever structures determine cell shape and form, e.g. microfilament bundles and microtubule organization. The presence of polar granules and the organization of the microvilli at the posterior pole are only a part of this special organization.

There are several possible explanations of what might be occurring. One simple possibility is that any region of the pole plasm is programmed to become part of the underlying blastoderm if it does not become incorporated into a pole cell at a particular time. As the precise area of a pole cell is presumably not determined in advance then the whole posterior cortex may well have this property. Furthermore, at the periphery of the pole plasm polar granules have been recorded as being incorporated into blastoderm cells (Illmensee, Mahowald & Loomis, 1976) in normally developing eggs. Thus constituents of pole plasm can be incorporated into normal blastoderm cells during the course of development.

An alternative explanation is that some kind of cytoplasmic 'respecification' has occurred. A spectacular example is the result of UV irradiation of the anterior pole of *Smittia* eggs (Kalthoff, 1971). As a result, all the cells which would normally have formed as progenitors of the head, thorax, and anterior abdomen structures now form as progenitors of a second posterior abdomen in mirror image symmetry to the first. The UV probably acts upon RNA laid down during oogenesis. This effect is mimicked by the *Drosophila* mutant *bicaudal* (Bull, 1966). Thus instead of forming cells determined for anterior structures, the irradiated anterior part of the *Smittia* egg forms cells determined for posterior structures. This can be defined as a 'respecification' of the information contained in the cytoplasm. It is possible that a similar effect occurs after irradiation of pole plasm or as a result of the *grandchildless* mutations in *Drosophila*.

Pole plasm transfers

Two different approaches have been undertaken to show that 'germ cell determinants' are indeed present in polar plasm. The first has been to eliminate the capacity of newly fertilized eggs to make pole cells by irradiating them with a suitable dose of UV light. Fresh pole plasm was then injected into the pole plasm region of the irradiated eggs. Then either the embryos were histologically analysed for the formation of pole cells (Warn, 1975) or flies developing from the embryos were examined for the presence of eggs or spermatozoa (Okada *et al.*, *1974a*). The basis of the experiment is shown in Fig. 7. Controls, where cytoplasm from the cortex of the anterior tip was injected after irradiation, did not have the capacity to form pole cells restored. From these experiments it can be concluded that the pole plasm, where injected into irradiated posterior poles, has the capacity either to restore the destroyed component(s) (e.g. polar granule material) or alternatively to form pole cells *de novo*.

Illmensee & Mahowald (1974, 1976) have been able to demon-strate that pole plasm can function to produce 'fertile' germ cells in regions other than the posterior pole. They removed pole plasm from genetically marked donor eggs and transferred it to the anterior tip or mid-ventral region of recipient eggs (Fig. 8). The recipients were then allowed to develop to the blastoderm stage. Normal-appearing pole cells were visible at the site of injection in addition to the usual complement of pole cells at the posterior pole. Electron microscopy showed that these pole cells, produced at sites far distant from the posterior pole, contained both polar granules and the characteristic nuclear bodies present in the nuclei of pole cells. To show whether these experimentally induced pole cells can produce functional germ cells Illmensee & Mahowald transplanted them to the posterior tips of second recipient embryos of the same age. These recipients were genetically different from both the previous donor genotypes. In a small number of cases (3 %) germ line mosaics were formed with a proportion of the offspring carrying genetic markers from the anteriorly produced pole cells. As in half the cases pole cells of one sex were inserted in recipient embryos of the other sex and do not develop (van Deusen, 1977), the theoretical maximum is only 50 %. As about 20 % of transplanted posterior pole cells produced germ cell

R. Warn

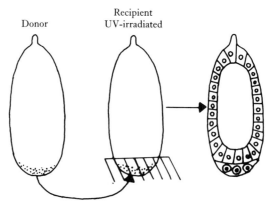

Fig. 7. Schematic diagram of pole plasm transfer into the posterior pole of a UV-irradiated embryo. (After Warn, 1975.)

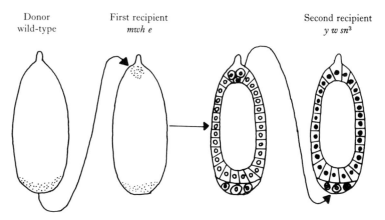

Fig. 8. Schematic diagram of pole plasm transfer into the anterior tip. (After Illmensee & Mahowald, 1974.)

mosaics in other experiments (Illmensee, 1973), it was possible to calculate that about one-quarter of the pole cells formed anteriorly led to 'fertile' germ cells.

The transferred pole cells did not contribute only to the ovaries and testes. In addition, they participated in the formation of larval and adult midgut tissue. This is in line with the view that pole cells are also progenitors of the midgut cuprophilic cells (Poulson & Waterhouse, 1960).

These experiments demonstrate that pole plasm does contain 'determinants' in the classical sense of the word (Hegner, 1914). Such determinants programme a precise developmental commit-

ment in the cells which contain them regardless of the position
of the pole plasm. The characteristic of 'determinants' is that they
are cell-type specific. In contrast, other types of cytoplasmic
transfer experiments (e.g. Garen & Gehring, 1972; Okada *et al.*,
1974*b*) have restored factors which are not cell-type specific and
are presumably involved in general metabolism.

At what stage of oogenesis does the ability to form fully
functional pole cells occur? Illmensee *et al.* (1976) answered this
by transplanting pole plasm from unfertilized eggs and from
oocytes at stages 10–14 of *D. melanogaster* into the anterior region
of cleaving embryos. They found that pole plasm from unfertilized
eggs and from oocytes of stages 13 and 14 behaved just as pole
plasm from fertilized eggs. Pole cells were formed after injection
into an ectopic site, and upon re-implanting these into the
posterior tip of another recipient embryo 'fertile' germ cells could
be produced in a small percentage of cases. Polar plasm from stage
10 (the earliest stage at which polar granules are visible) to stage
12 was not capable of forming pole cells at the site of injection.
This was in spite of the fact that polar granules were clearly
observable near the site of the injection and close to nuclei which
showed the structure of blastoderm nuclei. In many cases the polar
plasm became closed off within cytoplasmic vesicles and cut off
from the cytoplasmic constituents of the rest of the embryo. What
occurs between stages 12 and 13 that causes the maturation of the
pole plasm is not at all clear. The most likely explanation is that
some component is absent or not fully formed in the posterior pole
region in spite of the presence of polar granules.

Pole plasm can be transferred from the early embryo of one
species into that of another species and still produce pole cells
capable of forming a germ line (Mahowald, Illmensee & Turner,
1976). The experiment was carried out by injecting pole plasm
from *D. immigrans* early cleavage embryos into the anterior tips
of similar stages of *D. melanogaster*. 'Hybrid' pole cells were
formed with *immigrans* cytoplasm and *melanogaster* nuclei. These
had characteristics of both the species. When these 'hybrid' cells
were transferred into the posterior poles of a second set of
melanogaster recipients of a different genotype from the first, a
small proportion of them gave rise to germ cells (Fig. 9). These
recipients had previously been irradiated with UV light to eliminate their own pole cells. When crossed, offspring were produced

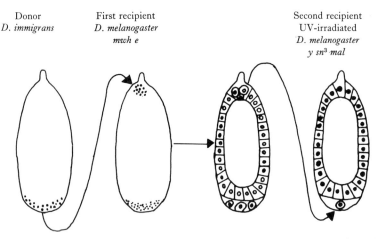

Fig. 9. Schematic diagram of interspecific pole plasm transfer. (After Mahowald, Illmensee & Turner, 1976.)

in a small percentage of cases whose genotype was of the anteriorly produced pole cells. This experiment shows that there is no barrier between the two species of *Drosophila* in the ability of their pole plasm to produce fertile germ cells. Presumably the molecular structure of the germ cell determinants has been conserved during the evolution of the two groups. In a very small number of cases female imagines were found which contained only germ cells derived from anteriorly induced pole cells. These could be used to test the hypothesis that polar granules have a structural continuity throughout the life cycle of the organism (Mahowald, 1971*b*, *c*). Eggs from the female progeny of these flies were analysed to determine the kind of polar granule formed. In all cases they were of *melanogaster* type rather than the *immigrans* type. There was no hint of any hybridity in any structures formed. Thus it is possible to conclude that, although polar granules do retain their structure for a long period of development, they are not autonomous cell organelles but must be re-formed under nuclear control.

Are polar granules a unique characteristic of pole plasm?

Polar granules are the major characteristic of the pole plasm of *Drosophila* and the most obvious candidate for the location of 'germ cell determinants'. Because they contain RNA (Mahowald,

1971*a*) it is possible that they act as 'informosomes' (Spirin, 1966) and release their information into the cytoplasm for translation sometime after fertilization. Their fragmentation and the attachment of ribosomes to their periphery after fertilization (Mahowald, 1971*b*) is consistent with this idea although significant protein synthesis does not occur in this region until blastoderm formation, when it is much stronger for the pole cells than for other regions (Zalokar, 1976). It is therefore an important question as to whether structures with similar properties are found in similar developmental situations. Such structures are characteristic of the germ plasm of many different organisms (extensively reviewed in Beams & Kessel, 1975; Eddy, 1975). A whole variety of insect orders have clearly distinguishable granules in their germ plasm as do Amphibia and Chaetognatha. In addition there are many groups (Eddy, 1975, lists 80 species in 8 phyla) where 'nuage' has been reported. It has been visualized in sperm as well as oocytes and eggs. Thus special cytoplasmic granular structures are a characteristic of germ cells. Do they occur in other situations as well? Mahowald (1971*c*) mentions that fibrous bodies similar to polar granules are found throughout the mature egg of *Drosophila*. Therefore there are likely to be parallels. Verdonk describes several examples among the Mollusca, elsewhere in this volume.

Furthermore Raven (1974) has shown the existence of haematoxylin-staining granules (subcortical accumulations or SCAs) in the cytoplasm of the *Lymnaea* embryo. This darkly staining cytoplasm is initially present as six patches just below the cortex of the equatorial region of the egg. The SCAs appear as a result of cytoplasmic streaming within the egg as it passes through the female genital tract after being fertilized. Raven finds that, in the uncleaved egg, numbers of small granules occur in the SCA cytoplasmic matrix. The SCAs fuse into a single band which, just before the first cleavage, is distributed among the two, then four blastomeres. At the third cleavage this material passes into the lower blastomeres (macromeres) and during successive divisions becomes concentrated at the lower ends of these cells. During these divisions larger granules form apparently by fusion of smaller haematoxylin-staining bodies.

There are several parallels between the SCA plasm of *Lymnaea* and the pole plasm of *Drosophila*. In both, the cytoplasm shows a differential staining and contains granules. Both sorts of granules are rich in RNA (RNAse treatment destroys the SCA granules).

Both show at least one phase where the granules are compacted and another where they disperse or disintegrate. As yet there is no direct evidence for a function of the SCA plasm. However, the fact that after fertilization it comes to lie in a position which coincides with the median plane of the future embryo could suggest an important role in development, perhaps in 'mapping out' the major axes of the organism.

A second example comes from the oocytes of the lizard *Lacerta sicula* obtained whilst the animal is in hibernation (Taddei, 1972). They contain arrays of paracrystalline bodies which have similarities with the structure of polar granules, particularly the paracrystalline granules of *Drosophila immigrans* (Mahowald, 1968). Electron microscopy shows that they are in fact stacks of ribosomal tetramers embedded in a matrix. Early in the spring these bodies slowly disintegrate and the component ribosomes appear to disperse in the cytoplasm. Protein synthesis is much lower in oocytes from overwintering lizards compared with oocytes dissected out during summer (Taddei, Gambino, Metaforo & Monroy, 1973) and it may well be that the accumulations represent a storage system. Whether or not mRNA is included with the ribosomes is not known. Polar granules are dissimilar from these paracrystalline bodies in that they probably do not contain ribosomes (Mahowald, 1971*a*). However, the similarity of the paracrystalline structures, particularly in the case of *Drosophila immigrans* polar granules, suggests an analogy of structure perhaps related to a storage function.

Polar granules are remarkable structures. Why they should be such conspicuous markers of pole plasm is not clear. Clearly structures of this kind are characteristic of germ cells but it seems unlikely that they are unique to this cell type. Quite possibly they are one end of a spectrum of ribonucleoprotein complexes, containing 'masked' messengers. The granules could be exceptionally large because of the huge number of informational molecules which may be needed for the formation of the next generation of germ cells.

References

BEAL, C. M. & DIXON, K. E. (1975). Effect of UV on cleavage of *Xenopus laevis*. *Journal of Experimental Zoology*, **192**, 277–83.

BEAMS, H. W. & KESSEL, R. G. (1975). The problem of germ cell determinants. *International Review of Cytology*, **39**, 413–74.

BULL, A. L. (1966). *Bicaudal*, a genetic factor which affects the polarity of the embryo in *Drosophila melanogaster*. *Journal of Experimental Zoology*, **161**, 221–42.

COUNCE, S. J. (1963). Developmental morphology of polar granules in *Drosophila* including observations on pole cell behaviour and distribution during embryogenesis. *Journal of Morphology*, **112**, 129–45.

DENIS-DONINI, S., BACCETTI, B. & MONROY, A. (1976). Morphological changes of the surface of the eggs of *Xenopus laevis* in the course of development. II. Cytokinesis and early cleavage. *Journal of Ultrastructure Research*, **57**, 104–12.

EDDY, E. M. (1975). Germ plasm and the differentiation of the germ cell line. *International Review of Cytology*, **43**, 229–75.

FIELDING, C. J. (1967). Developmental genetics of the mutant *grandchildless* of *Drosophila subobscura*. *Journal of Embryology and Experimental Morphology*, **17**, 375–84.

GAREN, A. & GEHRING, W. J. (1972). Repair of the lethal developmental defect in *deep orange* embryos of *Drosophila* by injection of normal egg cytoplasm. *Proceedings of the National Academy of Sciences of the USA*, **69**, 2982–5.

GEHRING, W. J. (1973). Genetic control of determination in the *Drosophila* embryo. In *Genetic Mechanisms of Development*, ed. F. H. Ruddle, pp. 103–28. New York & London: Academic Press.

GEIGY, R. (1931). Action de l'ultraviolet sur le pôle germinal dans l'œuf de *Drosophila melanogaster* (castration et mutabilité). *Revue Suisse de Zoologie*, **38**, 187–288.

GRAZIOSI, G. & MICALI, F. (1974). Differential responses to ultraviolet irradiation of the polar cytoplasm of *Drosophila* eggs. *Wilhelm Roux' Archives of Developmental Biology*, **175**,. 1–11.

GRAZIOSI, G. & ROBERTS, D. (1975). Molecular anisotropy of the early *Drosophila* embryo. *Nature, London*, **258**, 157–9.

HATHAWAY, D. S. & SELMAN, G. G. (1961). Certain aspects of cell lineage and morphogenesis studied in embryos of *Drosophila melanogaster* with an ultraviolet micro-beam. *Journal of Embryology and Experimental Morphology*, **9**, 310–25.

HEGNER, R. W. (1914). *The Germ-Cell Cycle in Animals*. New York: Macmillan.

ILLMENSEE, K. (1973). The potentialities of transplanted early gastrula nuclei of *Drosophila melanogaster*. Production of their imago descendants by germ-line transplantation. *Wilhelm Roux' Archives of Developmental Biology*, **171**, 331–43.

ILLMENSEE, K. & MAHOWALD, A. P. (1974). Transplantation of posterior pole plasm in *Drosophila*. Induction of germ cells at the anterior pole of the egg. *Proceedings of the National Academy of Sciences of the USA*, **71**, 1016–20.

ILLMENSEE, K. & MAHOWALD, A. P. (1976). The autonomous function of germ plasm in a somatic region of the *Drosophila* egg. *Experimental Cell Research*, **97**, 127–40.

ILLMENSEE, K., MAHOWALD, A. P. & LOOMIS, M. R. (1976). The ontogeny of germ plasm during oogenesis in *Drosophila*. *Developmental Biology*, **49**, 40–65.

IMAIZUMI, T. (1954). Recherches sur l'expression des facteurs létaux héréditaires chez l'embryon de la drosophile. I. La variation du volume de l'embryon pendant la première periode du développement. *Protoplasma*, **44**, 1–10.

JAZDOWSKA-ZAGRODZINSKA, B. (1966). Experimental studies on the role of polar granules in the segregation of pole cells in *Drosophila melanogaster*. *Journal of Embryology and Experimental Morphology*, **16**, 391–9.

KALTHOFF, K. (1971). Position of targets and period of competence for UV-induction of the malformation 'double abdomen' in the egg of *Smittia*. *Wilhelm Roux' Archives of Developmental Biology*, **168**, 63–84.

MAHOWALD, A. P. (1968). Polar granules of *Drosophila*. II. Ultrastructural changes during early embryogenesis. *Journal of Experimental Zoology*, **167**, 237–44.

MAHOWALD, A. P. (1971*a*). Polar granules of *Drosophila*. IV. Cytochemical studies showing loss of RNA from polar granules during early stages of embryogenesis. *Journal of Experimental Zoology*, **176**, 345–52.

MAHOWALD, A. P. (1971*b*). Origin and continuity of polar granules. In *Origin and Continuity of Cell Organelles*, ed. J. Reinhert & H. Ursprung, pp. 158–69. New York: Springer-Verlag.

MAHOWALD, A. P. (1971*c*). Polar granules of *Drosophila*. III. The continuity of polar granules during the life cycle of *Drosophila*. *Journal of Experimental Zoology*, **176**, 329–44.

MAHOWALD, A. P., ILLMENSEE, K. & TURNER, F. R. (1976). Interspecific transplantation of polar plasm between *Drosophila* embryos. *journal of Cell Biology*, **70**, 358–73.

MONROY, A. & BACCETTI, B. (1975). Morphological changes of the surface of the egg of *Xenopus laevis* in the course of development. *Journal of Ultrastructure Research*, **50**, 131–42.

OKADA, M., KLEINMAN, I. A. & SCHNEIDERMAN, H. A. (1974*a*). Restoration of fertility in sterilized *Drosophila* eggs by transplantation of polar cytoplasm. *Developmental Biology*, **37**, 43–54.

OKADA, M., KLEINMAN, I. A. & SCHNEIDERMAN, H. A. (1974*b*). Repair of a genetically caused defect in oogenesis in *Drosophila melanogaster* by transplantation of cytoplasm from wild type eggs and by injection of pyrimidine nucleosides. *Developmental Biology*, **37**, 55–62.

POULSON, D. F. & WATERHOUSE, D. F. (1960). Experimental studies on pole cells and midgut differentiation in Diptera. *Australian Journal of Biological Sciences*, **13**, 541–67.

RAVEN, CHR. P. (1974). Further observations on the distribution of cytoplasmic substances among the cleavage cells in *Lymnaea stagnalis*. *Journal of Embryology and Experimental Morphology*, **31**, 37–59.

SPIRIN, A. S. (1966). On 'masked' forms of messenger RNA in early embryogenesis and in other differentiating systems. *Current Topics in Developmental Biology*, **1**, 2–36.

TADDEI, C. (1972). Ribosome arrangement during oogenesis of *Lacerta sicula*. *Experimental Cell Research*, **70**, 285–92.

TADDEI, C., GAMBINO, R., METAFORA, S. & MONROY, A. (1973). Possible role of ribosomal bodies in the control of protein synthesis in pre-vitellogenic oocytes of the lizard *Lacerta sicula*. *Experimental Cell Research*, **78**, 159–67.

THIERRY-MIEG, D. (1976). Study of a temperature sensitive mutant *grandchildless*-like in *Drosophila melanogaster*. *Journal de Microscopie et de Biologie Cellulaire*, **25**, 1–6.

VAN DEUSEN, E. B. (1977). Sex determination in germ line chimeras in *Drosophila melanogaster*. *Journal of Embryology and Experimental Morphology*, **37**, 173–85.

WARN, R. (1972). Manipulation of the pole plasm of *Drosophila melanogaster*. *Acta Embryologiae et Morphologiae Experimentalis, Supplement*, 415–27.

WARN, R. (1975). Restoration of the capacity to form pole cells in UV-irradiated *Drosophila* embryos. *Journal of Embryology and Experimental Morphology*, **33**, 1003–11.

ZALOKAR, M. (1976). Autoradiographic study of protein and RNA formation during early development of *Drosophila* eggs. *Developmental Biology*, **49**, 425–37.

ZÜST, B. & DIXON, K. E. (1975). The effect of UV irradiation of the vegetal pole of *Xenopus laevis* eggs on the presumptive primordial germ cells. *Journal of Embryology and Experimental Morphology*, **34**, 209–20.

The maternal environment and the control of morphogenesis in insects

A. D. LEES

Agricultural Research Council Insect Physiology Group, Dept of Zoology and Applied Entomology, Imperial College, London, UK

The life strategies of insects demand rapid adjustments to environmental change, whether this involves the disappearance of a temporary habitat, the onset of winter or, in a social insect, a lack of balance in the internal morph structure of the colony. These exigencies may be met by radical changes in behaviour (e.g. migration) or by adopting a modified pattern of morphogenesis. The latter strategy includes the controlled development of alternate morphs (polymorphism) and the entry of the insect into a state of diapause. Since sudden environmental changes are rarely permanent and may be strictly seasonal, a system embodying a developmental switch which is finely tuned to the available environmental signals will clearly permit a more rapid and flexible response than would be possible through irreversible genotype selection. Indeed, by this means the insect can often anticipate environmental change.

Since the senses of adult insects are usually better developed than those of larvae, it would not be surprising if the life cycle stages responsive to the environment were centred on the adult with the morphogenetic effects appearing in the progeny. Although this emphasis on the maternal role is justified, it should be noted that important 'token' stimuli such as photoperiod are not perceived by the conventional sense organs; indeed, light-operated control mechanisms have been evolved in pupae, larvae and even in eggs. The interval between the critical stage and the response can therefore span any part of the life cycle.

The maternal effects which I now wish to describe form a somewhat miscellaneous collection. But they seem to fall into two main categories, namely maternal switching mechanisms which operate within one generation, and a variety of phenotypically imposed morphological and physiological effects which show progressive changes over several generations.

The maternal control of diapause

A maternally controlled diapause is not uncommon in insects and mites. There are, however, many species variations. The stage of arrest may occur early or late in the progeny generation. The mother may herself be environmentally sensitive or she may transmit the environmental 'message' received earlier in her development.

In the fruit tree red spider mite *Panonychus ulmi* the egg diapause and the oviposition behaviour of the mother are controlled by a variety of factors including the day length, temperature and the food quality of the leaf epidermal cells. These act on the last nymphal stage and adult female mite. Under long day conditions rapidly developing eggs are laid on the leaves of the apple, while in short days the mites deposit diapause eggs in crevices in the bark. The latter remain dormant for over 200 days (Lees, 1953). Besides their rate of development the two egg types differ in size, in the amount of carotenoid pigment they contain and in the time of deposition of an internal wax layer (Beament, 1951).

A conspicuous external feature of tetranychid mite eggs is the secondary outer covering of hard wax which is secreted in a plastic state after the egg has been laid. This material is drawn into a thin tapering 'whisker' as the female withdraws her everted 'ovipositing pouch' from the substratum. In another species, *Petrobia latens*, which lays eggs on stones lying in grass, the non-diapause egg carries the usual spine while the diapause egg is pork pie shaped (Fig. 1). The abundant outer wax layer, glistening white in colour, is fashioned by the female mite, who allows her abdomen to remain in contact with the egg until the wax is solidified. The finished product is an elaborate wax structure with a cavitated, air-containing inner layer which presumably permits an adequate uptake of oxygen while assisting also in water retention.

The stage of arrested development is more remote from the maternal generation in the tiny wasp *Nasonia vitripennis*, a parasite of the blowfly *Sarcophaga*. Diapause supervenes in the last larval

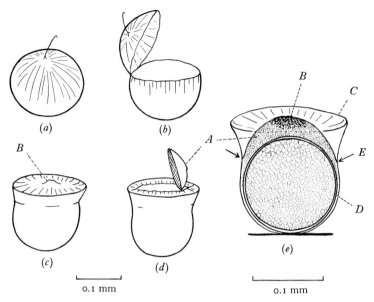

Fig. 1. Eggs of the mite *Petrobia latens* Müller before and after hatching. (*a*), (*b*) non-diapause eggs; (*c*), (*d*) diapause eggs; (*e*) the same viewed in optical section to show respiratory air spaces. Note that the structure of outer wax covering is controlled maternally. In the more elaborate diapause egg it consists of a thin external skin of non-porous wax (*C*) and a thicker, domed layer of highly cavitated wax (*A*) which invests the 'shell layer' (*D*) of the egg. The aeroscopic layer slightly protrudes from the centre of the cap at *B* and fractures at *E* on hatching. (From Lees, 1961.)

instar if the mother has experienced a day length less than $15\frac{1}{2}$ hours at a suitable temperature (Saunders, 1966). The silkworm *Bombyx mori* also provides a particularly striking example of a delayed response to photoperiod. The committal to diapause takes place when eggs at a late stage of embryonic development or young larvae are exposed to long days and high temperatures, but diapause does not supervene until eggs of the next generation are laid (Kogure, 1933). The latter part of the sequence of events is under endocrine control, the production of diapause eggs being induced by a neurohormone secreted by two prominent neuro-secretory cells located in the suboesophageal ganglion (Hasegawa, 1957; Fukuda & Takeuchi, 1967). The precise mode of action of the diapause hormone is uncertain. The embryonic growth arrest has been attributed to the increasing impermeability of the egg shell (chorion) which produces anoxic conditions within the egg.

This could lead to the development of the observed deficiencies in the electron transport system and to the conversion of glycogen into the polyhydric alcohols sorbitol and glycerol (Chino, 1963; Okada, 1971).

Returning to the maternal control system: the output of diapause hormone from the suboesophageal ganglion is itself controlled by nerves which arise in the protocerebrum. Transplantation experiments in moths have shown that the brain either promotes the release of hormone in those individuals determined for diapause or restrains its production in non-diapause egg producers (Fukuda, 1952). But little is known of the way the brain is programmed by photoperiod many instars earlier. Since the central nervous system may be both receptor and effector it is possible that the nervous or neuroendocrine systems are the repositories of this information. However, this cannot be so in *Nasonia* since the latent message is transmitted through the egg before the nervous system has differentiated.

An alternative and at present entirely speculative suggestion is that delayed effects on morphogenesis may be controlled maternally by the storage of other insect hormones in the egg. In this connection, it is known that the topical application of juvenile hormone (JH) to post-blastokinesis embryos of the heteropteran *Pyrrhocoris* causes late disturbances in larval development: after a series of normal larval ecdyses the corpus allatum of the recipient fails to 'shut down' prior to metamorphosis. As a result of this 'reprogramming' supernumerary moults and metathetelic effects are observed (Truman & Riddiford, 1972). In this context, it is also known that in some species larval diapause is associated with a high and stable level of JH in the blood (Fukaya & Mitsuhashi, 1961). Although the JH content of the eggs does not yet seem to have been assayed in any insect with a larval diapause, recent work has shown that hormones (in this instance ecdysone) can be sequestered in large amounts in the eggs (Ohtaki & Tanaka, 1978; Hsiao & Hsiao, 1978).

An intrinsic maternal effect, additional to the environmental switching mechanism, has been described in a few mites and insects. This is revealed by the increasing propensity of individuals in successive post-diapause generations to enter diapause under constant photoperiodic conditions. Table 1 shows that the red spider mite *Tetrachychus urticae* can hardly be induced to enter

Table 1. *Progressive changes in the response of the red spider mite*
Tetranychus urticae *to a short inductive day length (10L 14D,
18 °C) in the first six post-diapause generations*

Generation	1	2	3	4	5	6
% diapause	0	2	36	18	89	95

From Dubynina (1965).

diapause in the first two post-diapause generations but that the
response to short and long photoperiods becomes fully established
by the sixth generation (Dubynina, 1965). In *Panonychus ulmi* a
generation effect is much less in evidence (Lees, 1953; Razumova,
1967). But in *Petrobia latens* entry into diapause is wholly dependent on intrinsic changes, the mites laying dormant eggs after one
or at most two generations, even in long-day conditions (Lees,
1961).

Cumulative morphogenetic effects in locusts

Polymorphism in locusts is said to be continuous, in contrast to
the discontinuous or alternative type seen in such insects as aphids
(Kennedy, 1961). Typically, the two extreme forms, the *gregaria*
and *solitaria* phases, differ considerably in form, size, colour and
most of all in behaviour. Yet all grades of intermediates between
the two extreme phases are found, both in the laboratory and in
the field. These are transient forms which may be proceeding
either towards the *solitaria* phase (*dissocians*) or towards the
gregaria phase (*congregans*) in response to the density of the locust
population. This gradual two-way process, which requires at least
three generations to reach completion, is constantly moulded by
environmental influences except during the egg stage. It therefore
lacks the characteristics of a switching mechanism in which both
environmental sensitivity and the point of divergence of the
developmental pathways is stage-specific.

Uvarov's (1921) phase theory of locust transformations em-
bodied the concept that there was a cumulative transmission of
phase characters from one generation to the next. Observations on
newly hatched larvae have supported this view. Hatchlings of
Locusta migratoria migratorioides derived from egg pods deposited

by crowded mothers are darker and march more vigorously than those from *solitaria* mothers (Hunter-Jones, 1958; Ellis, 1959). Although these trends can be either intensified or reduced by the rearing conditions prevailing during the larval period, the total efect produced by crowding or isolation during one period of larval development is limited (Ellis, 1962). In *Schistocerca gregaria*, hatchlings from crowded parents are also larger, have more lipid reserves in the fat body and pass through one larval instar fewer than the hatchlings from *solitaria* mothers (Albrecht, 1955). It is probable that these features serve to pre-adapt the insects to the type of habitat experienced by their parents and grandparents (Kennedy, 1961). Thus the heavier hatchlings from crowded parents also survive longer in a dry habitat whereas the smaller hatchlings from isolated mothers survive better in wet ones (Albrecht, 1965).

Visual and tactile responses to other locusts are particularly important in promoting the phase change from *solitaria* to *gregaria* although a gregarizing pheromone also plays a role in *Schistocerca* (Ellis, 1962; Gillett & Phillips, 1977). These stimuli evoke imme-diate changes in behaviour but the more delayed morphometric and pigmentary changes are no doubt controlled by the endocrine system. It is known that the implantation of corpora allata into *gregaria* larvae causes them to assume the green coloration char-acteristic of *solitaria* reared in moist conditions (Joly, 1954; Staal, 1961). But applied JH has much less effect on the morphometric characters, merely causing supernumerary moulting and metath-etely. The endocrinology of phase transformation is clearly more complex than was first supposed (Pener, 1976).

In order to account for the accumulation of phase characters during successive generations some form of non-genic inheritance has been proposed (Ellis, 1959). It is worth noting, however, that in an early experiment by Albrecht, Verdier & Blackith (1958) differences in the colour, as well as the size of hatchlings were induced by applying ligatures which removed a portion of the egg yolk. This suggests that the storage of physiologically active substances, perhaps hormones, in the yolk may be of significance.

Maternally controlled morphogenetic effects in social insects

Caste regulation in lower termites is accomplished very largely by a series of inhibitory mechanisms, individuals of each caste producing pheromones which inhibit the competent larvae and nymphs from differentiating into the same morph (Lüscher, 1976). Removal of the queen from a colony of *Kalotermes flavicollis* results in the differentiation of replacement reproductives which have responded to the disappearance of the queen's inhibitory pheromone. The pheromone, which is distributed throughout the colony by trophallaxis, exerts its morphogenetic effects through the endocrine system of the recipient (probably via the corpus allatum). In the higher termites, where pheromonal communication is much less prominent, the behavioural link between maternal cause and morphogenetic effect is absent, for the castes are already predetermined in the egg. Assays have shown that the eggs may contain large concentrations of JH which vary seasonally and are lowest when nymphs and winged imagos are developing. It is suggested that hormones which have been dispensed by the queen in the egg play an important role in regulating caste development.

In the honey bee the differentiation of the two female castes is also controlled by a maternal pheromone, although only indirectly. 'Queen substance', a product of the mandibular glands, probably consists of two pheromones which are distributed by workers who lick and groom the queen. 'Queen substance' not only suppresses the growth of the ovaries but also modifies the behaviour and physiology of the adult workers, inhibiting them from building queen cells (Butler, 1969). If the queen dies, the workers begin constructing emergency queen cells and feed the larvae royal jelly which contains an allatotropic substance, probably a peptide. This results in the stimulation of corpora allata at a critical stage in larval growth and channels development towards the alternative morph – the queen (Wirtz, 1973).

Maternal control of polymorphism in aphids

The differentiation of many aphid morphs depends on maternal switch mechanisms which are responsive to an array of environmental variables including photoperiod, temperature, the presence or absence of other aphids and nutrition. These mechanisms are found in the viviparous morphs and they serve to direct the development of embryos along different developmental pathways which are usually strictly canalized.

In a typical life cycle a sequence of spring and summer generations consisting of alate or apterous virginoparae (parthenogenetic viviparous females) alternates with an autumn generation of males and sexual females (oviparae). The latter lay fertilized eggs which overwinter. The members of the first spring generation of virginoparae arising from the diapause eggs are usually morphologically distinctive and are termed fundatrices. They are usually apterous. In species without host alternation (monoecy) the oviparae and males are the progeny of apterous or alate virginoparae. The vetch aphid *Megoura viciae* is an example. In species which exhibit host alternation (heteroecy), such as the black bean aphid *Aphis fabae*, the oviparae are born on the primary host plant (the spindle tree *Euonymus*) and are the daughters of an additional alate morph, the gynopara, which migrates there from the secondary herbaceous host.

The virginoparae from which all morphs except the fundatrix are derived are paedogenetic, their 12–16 ovarioles containing chains of precociously developing embryos. Because the parthenogenetic eggs are ovulated sequentially, the embryos in the chain are of different ages. If the embryo sequence is traced back during the developmental period of the mother it is found that adults have about seven embryos in the longest chains, 3rd instar larvae have three or four, 1st instar larvae two. Ovulation therefore begins when the mother is herself an embryo within the abdomen of the grandmother (Fig. 2). To find the first ovulations it is necessary to dissect the largest embryos present in the 4th instar grandparent. The oogonial divisions of the germarium are to be found in the most advanced (but slightly smaller) embryos present in the 3rd instar grandparent.

A further note must be added on germarial structure. Germaria of virginoparae that are generating diploid parthenogenetic eggs are relatively small because the nurse cells are undeveloped and the eggs apparently yolkless. On the other hand, germaria of oviparae that are forming yolky haploid eggs by normal meiotic divisions are much larger and contain prominent nurse cells of the type described in other insects with telotrophic ovaries.

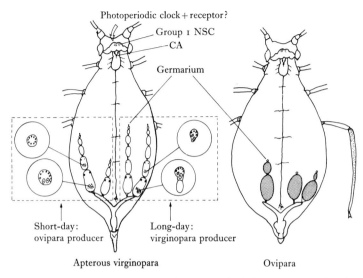

Fig. 2. Diagram illustrating certain anatomical features associated with the maternal control of morph determination in the aphid *Megoura viciae*. Embryos in the virginopara can develop either as oviparae (left rectangle) or alternatively as virginoparae (right rectangle), according to the photoperiod experienced by the mother. The maternal switching mechanism involves the Group I neurosecretory cells (NSC) and the corpus allatum (CA). Note that three generations are represented simultaneously in long-day mothers since the most advanced embryos already contain the germaria and first embryos of a third generation (single germaria shown enlarged in circles).

It might be expected that 'emboitement', as described above, where three generations overlap, might favour the evolution of maternal switching mechanisms. Certain of these will now be considered, together with some instances of through-generation cumulative effects not apparently involving any switching process.

The maternal control of sex determination

In many species, such as the pea aphid *Acyrthosiphon pisum*, the production of males by virginoparae can be induced by placing grandmother and mother in a short-day, medium-temperature regime. Nevertheless, despite this treatment the first oocytes formed by the germarium invariably give rise to XX parthenogenetic eggs and never become males. But later oocytes, which differ cytologically from the first (Blackman, 1978), may have either an XX or XO constitution, depending on the environmental conditions. The determination of the later oocytes as male or female begins just before the birth of the mother and continues to the 3rd instar.

Analysis of the photoperiodic response in *A. pisum* has shown that time measurement is related to night length rather than to day length (Lees, unpublished results). Although this is also true of the mechanism governing ovipara production in the same species, the two response curves are quite unlike one another, suggesting that the clocks are different. The location of the male-determining clock and its mode of action are unknown.

Maternal control of virginopara production

Photoperiod and temperature are also the principal factors controlling the determination of embryos as virginoparae or oviparae. Aphids transferred from a short to a long day length give birth first to oviparae and then to a series of virginoparae. The increasing number of embryos committed to develop as oviparae as the mother develops is indicated by the 'switch over' point in the progeny sequence, which becomes progressively later as the change in photoperiod is deferred. Determination occurs when the embryo is about three-quarters grown (Lees, 1959).

Switching experiments are also useful in deciding whether individuals are responding to photoperiod. For example, one may prove that the control mechanism is maternal by giving short-day mothers extended (long-day) illumination over restricted areas of the body. *Megoura* virginoparae switch from ovipara to virginopara production when microilluminators (light-transmitting fibres or capillaries) are attached to the head but fail to do so when the extra illumination is restricted to the embryo-filled abdomen (Lees, 1964).

When smaller areas of the head were examined it became apparent that the photoreceptors were extra-ocular and probably lay within the brain near the pars intercerebralis area – an anterior region which contains a prominent group of neurosecretory cells (the Group I NSC). When small brain lesions were made with a microcautery the photoperiodic switching mechanism failed if the Group I cells were eliminated and also if certain lateral areas of the protocerebrum were damaged, leaving the Group I cells intact (Steel & Lees, 1977). It seems likely that this part of the protocerebrum contains the neural photoreceptor and clock and that these neurones are connected synaptically with the Group I cells. The latter are regarded as primary effectors responsible for the long-day response. It is notable, however, that they are not required for the production of oviparae under short-day conditions, indicating that the virginopara production is the positive event.

The putative clock neurones do not show any obvious cytological peculiarities. Yet experiments in which light and dark are varied independently, show that the clock is quite complex and acts as a very accurate hour glass which measures night length (Lees, 1973). Nights shorter than the critical length of 9.5 hours promote virginopara production, presumably by stimulating the endocrine effectors. When the critical length is exceeded the endocrine system is not activated, with the result that oviparae are differentiated. In this system the duration of light in a long or short day is immaterial as this component is not 'measured'. It is also notable that several short night cycles are required to elicit a response, possibly because the NSC act as a 'counter' for accumulating the daily output from the clock.

Further insight into the maternal control mechanism has been gained by applying juvenile hormone (JH) or JH analogues to developing long- or short-day virginoparae. JH has no effect when administered to long-day virginopara producers. But ovipara producers treated from the 4th instar onwards subsequently give birth to groups of virginoparae, despite continued short-day exposure. Many grades of intermediate between unmodified oviparae and normal alatae are found among the progeny of treated mothers. Often these have mixed ovaries, some ovarioles forming yolky eggs from large germaria, while others contain embryos and have small germaria (Lees, 1978). The same effects have been described when other ovipara producers, such as the gynoparae of *Myzus persicae*, are treated with exogenous JHA (Mittler, Nassar & Staal, 1976). These observations suggest that JH may be directly responsible in normal long-day parents for parthenogenetic egg production by the germaria and also for transforming the external form of competent embryos. In other insects a JH-controlled switch operating in the last larval instar would itself disturb the development of the mother by preventing metamorphosis. However, it may be significant that last (4th) instar aphids are exceptionally insensitive to applied JH although highly sensitive in their 3rd instar (Lees, 1977).

It may well be that in *Megoura* the maternal switching mechanism involves both JH and the Group I cells. If the latter produce an allatotropin, this might serve to delay the response while the critical number of short night lengths is accumulated.

Maternal control of aptera and alata production

Depending on the species of aphid this morph-determining mechanism may be either prenatal (i.e. maternal) or postnatal. I shall consider some examples of the former type. In *Megoura* and *Acyrthosiphon* interactions between the aphids themselves provide a potent stimulus to alata production. Adult virginoparae isolated during their last larval instar produce only apterous daughters, whereas those from the crowded colonies produce many alatae. The mobility of the aphids, and therefore the frequency of their contacts, can be greatly increased by confining groups of two or more aphids for 24 hours in an empty tube. Alata production then starts almost immediately they are re-isolated on the plant. This maternal crowding response has proved to be entirely tactile, the visual and olfactory senses playing no part (Lees, 1967). The tactile stimulation is rather unspecific and can be imitated by allowing other small insects to jostle the aphid or, in a sensitive strain of *A. pisum*, by confining a single individual in a muslin bag which is sufficiently hairy in texture to provide abundant tactile stimulation (Sutherland, 1969). Mechanoreceptive bristles on the antennae, body and legs are believed to be implicated. Other stimuli which influence aptera–alate dimorphism include temperature, nutrition and even attendance of the aphid colony by ants (El-Ziady, 1960; Mittler & Sutherland, 1969).

The further steps in the maternal switching process are unknown. Several experiments suggest that some form of neural control is important. Thus previously crowded aphids immediately cease producing alatae if anaesthetized for 1 minute in carbon dioxide or ether. And the same result is observed if the nerve cord is sectioned between the suboesophageal and the fused thoracic–abdominal ganglion – a procedure which also releases the parturition reflex from inhibition. These time relations show that the determination of the virginoparous embryo as an alata does not take place until it is poised for release from the ovariole or is actually passing down the oviducts.

Transgeneration effects

Several morph characters or morph control systems in aphids show progressive changes which can be followed over two or more generations. One of the most striking concerns the inhibition which prevents fundatrices and their descendants in young clones from producing sexuales in response to the short photoperiods that would normally prevail in spring. This has been referred to as a 'facteur fondatrice' (Bonnemaison, 1951) or, since this function has the characteristics of a timing device, as an 'interval timer' (Lees, 1960, 1966). In *Megoura*, which has a strongly developed interval timer system, no oviparae or males are produced by young clones for some 90 days in a constant long-night regime (12 hours light, 12 hours darkness) at 15 °C (Fig. 3). The most interesting feature of this response is that it is independent of generation number. During this time a lineage selected from early-born mothers of each generation will have passed through some four post-fundatrix generations before ovipara producers appear, whereas only two generations are required in a parallel lineage derived from progeny born later in the reproductive sequence. Time dependence rather than generation dependence seems to imply that the timing mechanism is replicated and is not merely diluted out at each mitosis.

Using the sycamore aphid *Drepanosiphum platanoides*, Dixon (1971) has shown that the interval timer mechanism does not function independently of the environment while it is 'running out'. With the passage of time the ratio of oviparae to virginoparae produced by aphids from successive generations gradually increases but does so more rapidly in a 'strong' short day (8 hour photoperiod) than in a longer photoperiod which approaches critical length (15 or 16 hours) (Fig. 4). The progressive change in the intensity of inhibition recalls the post-diapause status of some red spider mites (p. 224). Although, in *Drepanosiphum*, even the fundatrix can be induced to give birth to a few oviparae under the appropriate photoperiodic conditions, some individuals still retain a trace of the inhibition in autumn. This permits a part of the population to continue reproducing parthenogenetically in

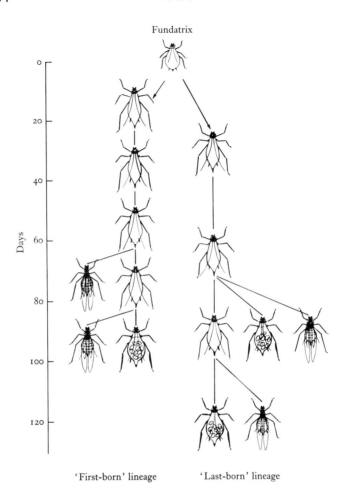

Fig. 3. A maternally inherited timing mechanism is present in young clones of the aphid *Megoura viciae* which delays the appearance of the sexual forms (oviparae and alate males). Comparison of the two lineages shows that the 'interval timer' or timers operate independently of generation number.

short days when the senescent sycamore foliage provides a rich source of food.

A second interval timer, insensitive to day length, controls the production of males in *Drepanosiphum*. Separate ovipara and male interval timers are also found in the lime aphid *Eucallipterus tiliae*. Both respond to photoperiod but seem less day-length-dependent since oviparae eventually appear even under long-day conditions (Dixon, 1972).

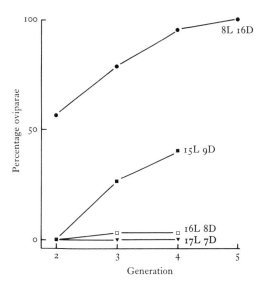

Fig. 4. Progressive changes in the ability of successive generations of the sycamore aphid *Drepanosiphum platanoides* to produce sexual females (oviparae) in response to short photoperiods. Generation 1 is the fundatrix. L, hours of light; D, hours of darkness. (After Dixon, 1971.)

Progressive anatomical changes are also seen in the individuals from young aphid clones. Apterous progeny of the fundatrix (the fundatrigeniae) have a narrower abdomen, and longer legs and antennae than the fundatrix (Fig. 3). Most of the transition takes place in one generation but smaller progressive changes may occur in later generations. It is noteworthy that the early-born progeny of the fundatrix is more fundatrix-like than later-born progeny (Lees, 1966; Dixon, 1974), suggesting that the transition is occurring continuously and may be related to time (Lees, 1966).

The facility with which aphids produce alate daughters after crowding is a further function that is subject to consistent inborn changes during successive generations. The capability is usually low or zero in alatae and in the early-born progeny of alatae but is higher in later-born progeny and higher still in the next generation (Lees, 1966). McKay (1977) has concluded that such trends are not controlled by an interval timer but by a response pattern that is characteristic of a particular generation. The mode of inheritance of this generation effect is nevertheless obscure.

A. D. Lees

Summary

Maternal effects which subsequently influence morphogenesis include switch mechanisms that direct development along alternative pathways. These operate within the maternal and progeny generations, are usually triggered by the environment and may be mediated by the maternal endocrine system. They occur commonly in diapausing and polymorphic insects. Other effects transmitted maternally are subject to progressive changes over two or more generations. They may be environmentally induced, as in the continuous polymorphism of locusts; or, after initiation at a certain developmental stage, they may proceed to 'run down' through several generations. This is seen in the diapausing stages of some mites and in several functions, sometimes time dependent, associated with the fundatrix stage of aphids.

References

ALBRECHT, F. O. (1955). La densité des populations et la croissance chez *Schistocerca gregaria* (Forsk). et *Nomadacris septemfasciata* (Serv.); la mue d'ajustement. *Journal d'Agriculture Tropicale et de Botanique Appliquée*, **2**, 109–92.

ALBRECHT, F. O. (1965). Influence du groupement, de l'état hygrométrique et de la photoperiode sur la résistance au jeûne de *Locusta migratoria migratorioides* R. et F. (Orthoptère acridien). *Bulletin Biologique de la France et de la Belgique*, **99**, 287–339.

ALBRECHT, F. O., VERDIER, M. & BLACKITH, R. E. (1958). Détermination de la fertilité par l'effet de groupe chez le criquet migrateurs (*Locusta migratoria migratorioides* (R. & F.)). *Bulletin Biologique de la France et de la Belgique*, **4**, 349–427.

BEAMENT, J. W. L. (1951). The structure and formation of the egg of the fruit tree red spider mite, *Metatetranychus ulmi* Koch. *Annals of Applied Biology*, **38**, 1–24.

BLACKMAN, R. L. (1978). Early development of the parthenogenetic egg in three species of aphids (Homoptera: Aphididae). *International Journal of Insect Morphology and Embryology*, **7**, 33–44.

BONNEMAISON, L. (1951). Contribution à l'étude des facteurs provoquant l'apparition des formes ailées et sexuées chez les Aphidinae. *Annales des Epiphyties, Séries C*, **2**, 1–380.

BUTLER, C. G. (1969). Some pheromones controlling honeybee behaviour. In *Proceedings of the VI Congress of the International Union for the Study of Social Insects*, Bern, pp. 19–32.

CHINO, H. (1963). Respiratory enzyme system of the *Bombyx* silkworm egg in relation to the mechanism of the formation of sugar alcohols. *Archives of Biochemistry and Biophysics*, **102**, 400–15.

DIXON, A. F. G. (1971). The 'interval timer' and photoperiod in the determination of parthenogenetic and sexual morphs in the aphid, *Drepanosiphum platanoides. Journal of Insect Physiology*, **17**, 251–60.

DIXON, A. F. G. (1972). The 'interval timer', photoperiod and temperature in the seasonal development of parthenogenetic and sexual morphs in the lime aphid, *Eucallipterus tiliae* L. *Oecologia, Berlin*, **9**, 301–10.

DIXON, A. F. G. (1974). Changes in the length of the appendages and the number of rhinaria in young clones of the sycamore aphid, *Drepanosiphum platanoides. Entomologia Experimentalis et Applicata*, **17**, 1–8.

DUBYNINA, T. S. (1965). Onset of diapause and reactivation in *Tetranychus urticae* Koch (Acarina, Tetranychidae). *Entomologicheskoe Obozrenie* (trans. and publ. by Scripta Technica for the Entomological Society of America), **44**, 159–61.

ELLIS, P. E. (1959). Some factors influencing phase characters in the nymphs of the locust, *Locusta migratoria migratorioides* (R & F). *Insectes Sociaux*, **6**, 21–39.

ELLIS, P. E. (1962). Physiologie, comportement et écologie des acridiens en rapport avec la phase. *Colloques Internationaux du Centre National de la Recherche Scientifique*, **114**, 1–16.

EL-ZIADY, S. (1960). Further effects of *Lasius niger* L. on *Aphis fabae* Scopoli. *Proceedings of the Royal Entomological Society of London, Series A*, **35**, 30–8.

FUKAYA, M. & MITSUHASHI, J. (1961). Larval diapause in the rice stemborer with special reference to its hormonal mechanism. *Bulletin of the National Institute of Agricultural Sciences, Japan, Series C*, **13**, 1–32.

FUKUDA, S. (1952). Function of the pupal brain and suboesophageal ganglion in the production of non-diapause and diapause eggs in the silkworm. *Annotationes Zoologicae Japonenses*, **25**, 149.

FUKUDA, S. & TAKEUCHI, S. (1967). Studies on the diapause factor-producing cells in the suboesophageal ganglion of the silkworm, *Bombyx mori* L. *Embryologia*, **9**, 333–53.

GILLETT, S. D. & PHILLIPS, M. L. (1977). Faeces as a source of a locust gregarisation stimulus. Effects on social aggregation and on cuticular colour of nymphs of the desert locust, *Schistocerca gregaria* (Forsk.). *Acrida*, **6**, 279–86.

HASEGAWA, K. (1957). The diapause hormone of the silkworm, *Bombyx mori. Nature, London*, **179**, 1300–1.

HSIAO, T. H. & HSIAO, C. (1978). Ovarian synthesis of ecdysteroids in the greater wax moth. In *Comparative Endocrinology*, ed. P. J. Gaillard & H. H. Boer. Amsterdam: Elsevier/North Holland Biomedical Press (in press).

HUNTER-JONES, P. (1958). Laboratory studies on the inheritance of phase characters in locusts. *Anti-Locust Bulletin, London*, **29**, 32pp.

JOLY, L. (1954). Résultats d'implantations systematiques de *corpora allata* à de jeunes larves de *Locusta migratoria* L. *Compte Rendus de la Société de Biologie*, **148**, 579–83.

KENNEDY, J. S. (1961). Continuous polymorphism in locusts. In *Royal Entomological Society Symposium 1*, ed. J. S. Kennedy, pp. 80–90. London: Royal Entomological Society.

KOGURE, M. (1933). The influence of light and temperature on certain characters in the silkworm, *Bombyx mori*. *Journal of the Department of Agriculture, Kyushu University*, **4**, 1.

LEES, A. D. (1953). Environmental factors controlling the evocation and termination of diapause in the fruit tree red spider mite *Metatetranychus ulmi* Koch (Acarina: Tetranychidae). *Annals of Applied Biology*, **40**, 449–86.

LEES, A. D. (1959). The role of photoperiod and temperature in the determination of parthenogenetic and sexual forms in the aphid *Megoura viciae* Buckton. I. The influence of these factors on apterous virginoparae and their progeny. *Journal of Insect Physiology*, **3**, 92–117.

LEES, A. D. (1960). The role of photoperiod and temperature in the determination of parthenogenetic and sexual forms in the aphid *Megoura viciae* Buckton. II. The operation of the 'interval timer' in young clones. *Journal of Insect Physiology*, **4**, 154–75.

LEES, A. D. (1961). On the structure of the egg shell in the mite *Petrobia latens* Müller (Acarina: Tetranychidae). *Journal of Insect Physiology*, **6**, 146–51.

LEES, A. D. (1964). The location of the photoperiodic receptors in the aphid *Megoura viciae* Buckton. *Journal of Experimental Biology*, **41**, 119–33.

LEES, A. D. (1966). The control of polymorphism in aphids. *Advances in Insect Physiology*, **3**, 207–77.

LEES, A. D. (1967). The production of the apterous and alate forms in the aphid *Megoura viciae* Buckton, with special reference to the role of crowding. *Journal of Insect Physiology*, **13**, 289–318.

LEES, A. D. (1973). Photoperiodic time measurement in the aphid *Megoura viciae*. *Journal of Insect Physiology*, **19**, 279–316.

LEES, A. D. (1977). Action of juvenile hormone mimics on the regulation of larval–adult and alary polymorphism in aphids. *Nature, London*, **267**, 46–8.

LEES, A. D. (1978). Endocrine aspects of photoperiodism in insects. In *Comparative Endocrinology*, ed. P. J. Gaillard & H. H. Boer. Amsterdam/North Holland Biomedical Press (in press).

LÜSCHER, M. (1976). Phase and caste determination in insects. Endocrine aspects. In *XV International Congress on Entomology*, ed. M. Lüscher, pp. 91–103. Oxford: Pergamon Press.

MACKAY, P. A. (1977). Alata-production by an aphid: the 'interval timer' concept and maternal age effects. *Journal of Insect Physiology*, **23**, 889–93.

MITTLER, T. E. & SUTHERLAND, O. R. W. (1969). Dietary influences on aphid polymorphism. *Entomologia Experimentalis et Applicata*, **12**, 703–13.

MITTLER, T. E., NASSAR, S. G. & STAAL, G. B. (1976). Wing development and parthenogenesis induced in progenies of kinoprene-treated gynoparae of *Aphis fabae* and *Myzus persicae*. *Journal of Insect Physiology*, **22**, 1717–25.

OHTAKI, T. & TANAKA, H. (1978). Ecdysones in the ovary and the embryo of arthropod species. In *Comparative Endocrinology*, ed. P. J. Gaillard & H. H. Boer. Elsevier/North Holland Biomedical Press (in press).

OKADA, M. (1971). Role of the chorion as a barrier to oxygen in the diapause of the silkworm, *Bombyx mori* L. *Experientia*, **27**, 658–60.

PENER, M. P. (1976). The differential effect of the *corpora allata* on yellow colouration in crowded and isolated *Locusta migratoria migratorioides* (R & F) males. *Acrida*, **5**, 267–85.

RAZUMOVA, A. P. (1967). Variability of the photoperiodic reaction in a number of successive generations of spider mites. *Entomologicheskoe Obozrenie* (trans. and publ. by Scripta Technica for the Entomological Society of America), **46**, 268–72.

SAUNDERS, D. S. (1966). Larval diapause of maternal origin. II. The effect of photoperiod and temperature on *Nasonia vitripennis*. *Journal of Insect Physiology* **12**, 569–81.

STAAL, G. B. (1961). Studies on the physiology of phase induction in *Locusta migratoria migratorioides* R & F. *Publikatie Fonds Landbouw Export Bureau* 1916–18, **40**, 1–125.

STEEL, C. G. H. & LEES, A. D. (1977). The role of neurosecretion in the photoperiodic control of polymorphism in the aphid *Megoura viciae*. *Journal of Experimental Biology*, **67**, 117–35.

SUTHERLAND, O. R. W. (1969). The role of crowding in the production of winged forms by two strains of the pea aphid, *Acyrthosiphon pisum*. *Journal of Insect Physiology*, **15**, 1385–410.

TRUMAN, J. W. & RIDDIFORD, L. M. (1972). Delayed effects of juvenile hormone on insect metamorphosis are mediated by the corpus allatum. *Nature, London*, **237**, 458.

UVAROV, B. P. (1921). A revision of the genus *Locusta*. L (= *Pachytylus*, Fieb.), with a new theory as to the periodicity and migrations of locusts. *Bulletin of Entomological Research*, **12**, 135–63.

WIRTZ, P. (1973). Differentiation in the honeybee larva. *Mededeelingen van de Landbouwhoogeschool te Wageningen*, **73–5**, 1–155.

Maternal effect genes in the Mexican axolotl (*Ambystoma mexicanum*)

GEORGE M. MALACINSKI AND JOHN SPIETH

Dept of Biology, Indiana University, Bloomington, Indiana 47401, USA

The pattern of early post-fertilization amphibian development has been well documented as being largely under the control of the maternal genome. Subsequent differentiation patterns of the embryo are presumably established through a series of nucleo-cytoplasmic interactions. The nuclei themselves are genetically identical, a fact which has been confirmed by a variety of nuclear transplantation experiments (reviewed by Briggs, 1977). It is, therefore, regional differences in the egg's cytoplasm which are thought to be responsible for the varied gene expression patterns of the embryo's population of nuclei.

That a general contribution of the maternal genome to the pattern of early amphibian post-fertilization is substantial was perhaps first demonstrated in interspecies hybrid embryos. Developmental arrest in most hybrid combinations occurs at the gastrula stage, while in a few combinations development persists much longer, even to the tail-bud stage. It is in large part the maternal characteristics which persist until the embryos reach the stage when development arrests. The rate of cleavage division in several hybrids of *Rana*, for example, is characteristic of the maternal species (Moore, 1941). Likewise, the rate of oxygen consumption in a cross of *Triton* × *Salamandra* remains maternal (Chen, 1953). Hybrids derived from very closely related species or races develop, needless to say, successfully through the entire course of embryogenesis, and in some cases yield sexually mature adults (Moore, 1955). In several of the combinations which provide hybrids that develop through gastrulation, it is at neurulation when deviations from the maternal pattern of morphogenesis are first detected. In a few cases, some features of neural morphogenesis are actually intermediate in character, suggesting that expression of both the maternal and paternal genomes is taking place (reviewed by Subtelny, 1974).

Evidence that specific macromolecules which are synthesized during oogenesis and stored in the egg cytoplasm participate in early embryogenesis is rapidly becoming available. A summary of the extent to which some maternally derived egg components persist in embryogenesis is contained in Fig. 1. A variety of proteins, including histones (Adamson & Woodland, 1974), and several enzymes, such as lactate dehydrogenase (Wright & Moyer, 1968), accumulate in the oocyte and are deployed during post-fertilization development. In the case of histones, sufficient amounts are present in the egg prior to fertilization to supply a major proportion of the requirements of cleavage division up to the blastula stage (Woodland & Adamson, 1977). For more than a dozen enzymes of intermediary metabolism, evidence is available indicating that catalytically active isozymes which are products of the maternal genome persist through the neural fold stage of development (Wright, 1975).

Populations of mRNA which are synthesized during oogenesis remain functionally active up to the blastula stage (Crippa & Gross, 1969). Ribosomal RNA synthesized during oogenesis persists even longer, up to the swimming tadpole stage (Brown & Gurdon, 1964). And the yolk proteins, and probably components of the germ plasm as well (Smith & Williams, this volume), persist all the way up through organogenesis.

In order to exploit the unique opportunity to study maternal control presented by amphibians, R. R. Humphrey set out to recognize mutant genes affecting embryogenesis (Humphrey, 1975). He chose to develop the Mexican axolotl, a neotenous salamander (Fig. 2) which can be conveniently maintained in the laboratory, as an experimental organism.

The large number of eggs obtained from single spawnings, the convenient observation of all stages of development, and the ability of axolotl embryos to withstand surgical manipulations and nuclear transplantation combine to make this organism a particularly useful one. Among the mutant genes he has recognized are several which are termed 'maternal effect' mutations. The altered phenotypes produced by these mutations are strictly determined by the female's genotype.

	Stage of development						
Approximate time (h) at 18 °C	Fertile egg	Early cleavage	Blastula	Gastrula	Neural folds	Tail bud	Organogenesis
Xenopus	o	2.5	5	12.5	17.5	28	48
Rana	o	6.5	16	42	62	84	140
Ambystoma	o	12.5	22	55	78	102	170

Maternal component	Histones ⟶
	Membrane components? ⟶
	Regulatory molecules (e.g. gene *o*) ⟶
	Various enzymes ⟶
	Ribosomal RNA ⟶
	Yolk proteins ⟶
	Germ plasm? ⟶

Fig. 1. Persistence of maternal contributions to early amphibian embryogenesis. Substantial experimental evidence derived from several species of Amphibia reveals that a variety of maternal egg components are employed during early stages of development. The developmental rates for the species most frequently used for analyses of maternal control are shown in top section of diagram.

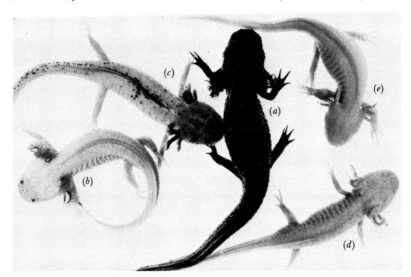

Fig. 2. Adult axolotls which display various pigment phenotypes. (*a*) Wild-type dark (*D/D*); (*b*) white – note presence of melanin in eyes and gills; (*c*) eyeless white (*dd/ee*) – note broad distribution of melanin; (*d*) golden albino (*Dd/aa*); (*e*) white albino (*dd/aa*). These pigment markers have proved to be very useful in the surgical manipulations employed in studies of axolotl developmental genetics.

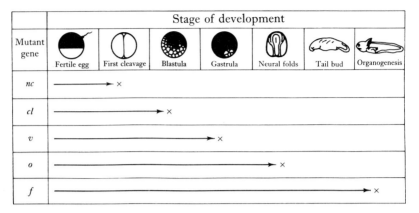

Fig. 3. Diagram of the developmental stage to which embryogenesis proceeds in eggs from female axolotls homozygous for a single maternal effect gene. The *nc* gene is now apparently extinct.

Among the maternal effect mutants which have been recognized in various organisms (see other articles in this volume), two general types are well known. The most prevalent type includes those mutations which lead to altered phenotypes that can be rescued by the introduction of the wild-type allele via the sperm at fertilization. This class of mutants apparently involves genes which act not only during oogenesis but at other stages of development as well. A second general type includes 'true' maternal effect mutations. In these cases the mutant phenotype is displayed by embryos which are spawned by mutant females, regardless of the paternal contribution to the zygote at fertilization.

Several 'true' maternal effect mutations have been recognized by Humphrey. They give rise to various effects on the pattern of early embryogenesis. A summary of the mutant phenotypes is included in Fig. 3. A detailed description of those maternal effect mutants in the axolotl is the main subject of this review.

nc: no cleavage mutant

The maternal effect mutation which acts earliest on post-fertilization development is the gene *nc* (no cleavage). It was first described by Raff, Brothers & Raff (1976) and further characterized by Raff & Raff (1978). Eggs spawned by females homozygous for *nc* appear to be morphologically normal. Fertilization and egg activation are accompanied by a normal cortical reaction, polar body formation, and pronuclear migration. The mitotic apparatus and cytasters, however, never form and cleavage division does not take place. This mutation appears to be a true maternal effect mutation in that fertilization by a wild-type sperm is incapable of rescuing the mutant phenotype. Interestingly, the *nc* can be phenocopied by electrically activating unfertilized eggs from normal females. Initially these activated eggs morphologically appear to resemble fertilized eggs but subsequently do not undergo cleavage.

The effect of the *nc* gene occurs within the ovary itself rather than indirectly through effects on tissues other than the ovary. Albino animals containing ovary grafts from an *nc/nc* animal produce darkly pigmented *nc* eggs in addition to normal albino eggs. The *nc* phenotype is not due to either a lack of, or an alteration in, egg tubulin protein. Mutant eggs contain a pool of soluble tubulin which is indistinguishable on SDS or urea acrylamide gels from that prepared from wild-type or albino eggs. The size of the tubulin pool does, however, appear to be only about 80 % as large as that of normal eggs. This amount should be sufficient, nonetheless, to permit cleavage to occur. Such an 'apparent' relatively lower pool size may be a reflection of a reduced metabolic stability of the tubulin, rather than an actual diminution of the absolute pool size (Raff, personal communication). A slightly less stable tubulin should not in itself prevent cleavage, since tubulin from normal axolotl embryos is less stable than tubulin from unfertilized eggs (Raff, 1977). Altered stability does not, therefore, appear to prevent cleavage.

The *nc* phenotype can be partially corrected by injection of microtubules obtained from fragments of outer-nine doublet tubules from the tails of sea urchin sperm (Raff *et al.*, 1976). The amount required for this rescue operation is actually less than 10 % of the total tubulin already present in *nc* eggs. The first cleavage divisions are apparently normal in the animal hemisphere but incomplete in the vegetal region. Later cleavages appear to be irregular and embryos finally arrest as partial blastulae. The correction is specific for microtubules and requires prior activation of the egg. Parallel arrays of microtubules are formed in corrected embryos and occasionally these arrays organize to resemble a spindle. Microtubules do not, however, assemble from or attach to kinetichores, and the chromosomes appear to be fragmented.

The defect in *nc* mutants appears to involve an early event of egg activation (Raff & Raff, 1978). Part of this defect is corrected by the injection of tubule fragments which function as mitotic microtubule organizing centres: *nc* tubulin is recruited into spindle-like arrays and cleavage is triggered. The lesion is, however, not simply due to an absence of mitotic organizing centres. After injection of microtubule fragments the resulting cleavages are not all normal and development arrests at a partial blastula stage. Likewise, some of the events leading up to mitosis are also defective. Fragmented chromosomes and improper attachment of microtubules to kinetichores can be observed. The mutant gene apparently arose spontaneously in the Indiana University axolotl colony. Before it was propagated it may have been lost. We are at present systematically searching our stocks for the *nc* gene, in the hope that we will recover it.

cl: cleavage defect mutant

Another maternal effect mutation which affects the cleavage process during early embryogenesis is the gene *cl* (cleavage defect). The first published description of this mutant phenotype was made by Carroll & Van Deusen (1973). The abnormality is observed in the eggs from *cl/cl* females. The eggs are quite fragile so that many break during spawning or flatten in the animal hemisphere when the egg's gelatinous capsule is removed. In addition, cytoplasm occasionally appears to leak from the crater-like areas which mark the points of entry of sperm (sperm pits). At the first cleavage division the furrow in mutant eggs is often seen as a line across the top of a flat egg rather than the usual crease formed on round normal eggs. Pigment moves out of the region around the sides of the grooves and collects in the crease, giving the appearance of a white band across the top of the egg. At this point the furrow may regress, leaving the egg uncleaved, or the egg may continue to divide to form several cells. Usually cleavage is normal only in the animal hemisphere, leaving the entire vegetal region uncleaved. A large proportion of the embryos may continue to divide only in the animal region to form a 'blastula cap' stage with little, if any, cleavage in the vegetal half. Fig. 4(*a*), (*b*) and

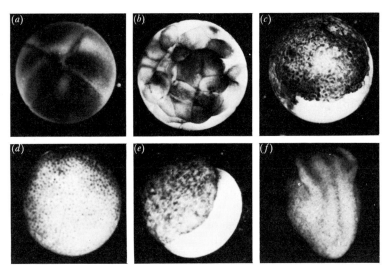

Fig 4. Eggs spawned by animals which are homozygous for either the *cl* (*a–c*) or *o* gene (*d–f*). (*a*) The cleavage furrows of an embryo from a *cl/cl* animal at the 4-cell stage do not penetrate the cortex deeply. (*b*) Same embryo at morula stage. (*c*) Continued development of the embryo to the blastula stage – cleavage continues only in the animal hemisphere. (*d*) Embryo from an *o/o* female starts to undergo developmental arrest at the late blastula/early gastrula stage. Note coalescence of pigment in centre of each blastomere. (*e*) Same embryo fails to develop beyond gastrulation. (*f*) Embryo from *o/o* animal, injected with normal egg cytoplasm before first cleavage division. Same as embryo in (*e*). (From Malacinski (1971). *Developmental Biology*, **26**, 442–51.)

(*c*) shows several typical stages in the development of eggs spawned from *cl/cl* mothers. Only eggs which are left in their natural jelly could ever develop to the blastula cap stage. Those which are manually dissected from the jelly flatten in the animal hemisphere, as mentioned above, and fail to continue any division. The surface of the egg appears to lack any rigidity, indicating the possibility of a defect in some aspect of the egg cortex. Even eggs left in their jelly capsules rarely complete gastrulation and development beyond gastrulation has been observed only in a single spawning from one female (Carroll & Van Deusen, 1973).

In eggs that arrest during the first few cleavages the positions of the succeeding cleavages are evident on the egg surface as a white line at the time when these cleavages would normally occur. For example, when a normal 12 hour embryo would have about 64 cells an uncleaved mutant egg might have the positions of the

64 cells traced on its surface by white lines. This important observation indicates that the planes of cleavage, at least for the first six cleavage divisions, are predetermined in the egg and do not rely on spatial orientation derived from preceding cleavages or the presence of blastomere cell membranes.

Cytological examination of embryos (Carroll & Van Deusen, 1973) reveals that nuclear division proceeds normally in the part of the animal hemisphere where cell cleavage occurs. Nuclear division also continues in the uncleaved regions of the egg. This is followed in the uncleaved regions by fusion of nuclei and a consequent polyploidization, as judged by nuclear size and number of nucleoli per nucleus. Many of the dividing nuclei are connected to multipolar spindles. Interphase fusion nuclei are often caught in an active mitotic spindle and pulled apart, which results in extensive chromosome breakage. In the vegetal hemisphere the cycle of centriole replication and spindle formation is apparently separated from the cycle of chromosome replication and condensation. It is not known whether this separation is an integral part of the *cl* lesion or simply a consequence of the absence of cell membranes which would normally separate asynchronously dividing cells.

In embryos in which the animal hemisphere cells undergo what appears to be normal cell divisions, the failure of gastrulation to occur is probably due to restrictions imposed on the embryo by the uncleaved vegetal half, which in normal development would cleave and undergo invagination movements. The ability of cells from the blastula cap to survive and differentiate neural structures when grafted into a homologous position in a normal recipient indicates that these cells are viabled and that the uncleaved cytoplasm is the most likely cause of the limited development of mutant embryos beyond gastrulation (Carroll & Van Deusen, 1973). Taken together, these results suggest the existence of an animal–vegetal gradient of the wild-type *cl* gene product. Such a component probably exists in a high concentration at the animal pole and a low concentration at the vegetal pole. A mild maternal effect might reduce the low concentration at the vegetal pole sufficiently to render it developmentally inactive. A similar reduction in concentration at the animal pole might still maintain the *cl* gene product above a threshold concentration. In eggs showing a severe maternal effect the *cl* gene product concentration would be reduced enough throughout the gradient to prevent any cleavage. An equally valid hypothesis would require that there be a uniform concentration throughout the egg of the *cl* gene product. More of it would, therefore, be needed in the vegetal region than in the animal region. As a consequence there would be a normal excess of the gene product in the animal region which would, in the case of a mild maternal effect, still be sufficient to allow cleavage to take place.

Cytochalasin B produces in normal eggs a phenocopy of the most severely affected *cl* mutants (Carroll & Van Deusen, 1973). In normal eggs treated with either 3 or 5 mg/ml of cytochalasin B the furrow begins to form normally but soon pigment collects in the crease and the sides of the groove whiten. The furrow always regresses but continued attempts at furrow formation result in the development of eggs with white lines marking the positions where cell boundaries would have been. There is, however, no extensive polyploidization in cytochalasin-treated embryos.

v: vasodilation mutant

The one maternal effect mutant in the axolotl which displays variable penetrance of the altered phenotype is the gene *v* (vaso-dilation). This mutation was first observed in a spawning from parents which were heterozygous for another of the maternal effect mutants, gene *f* (Humphrey, 1962), which will be discussed later. The *v* mutation yields a phenotype which is quite different from *f* and apparently arose by a relatively recent mutation in a single animal in the Indiana University axolotl colony. Embryos which are homozygous for gene *v* undergo normal embryogenesis up to the stage at which the gills develop and circulation is established. At this stage (pre-feeding) the mutants are first distinguished by reduced growth, most noticeable as a narrowing of the head (Fig. 5), as well as the presence of smaller than normal gills which contain a reduced number of filaments. Several days before their yolk supply is depleted a critical stage in the development of mutant larvae is reached. During this period there is an enlarge-ment of their pronephroi and a dilation of blood vessels, partic-ularly in the gills and pronephroi. Most of the *v/v* larvae become weak and die although a small, albeit variable, number recover in some spawnings and develop to adulthood. The growth rate of

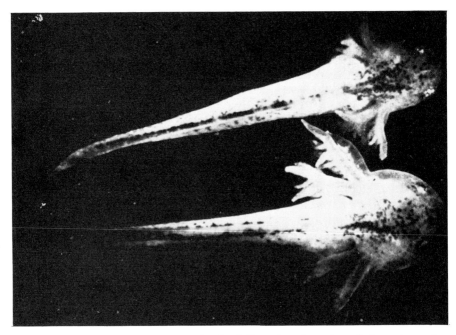

Fig. 5. Recently hatched normal animal (lower) and 12-day v/v larva (upper). In the v/v larva note the narrower head and smaller gills. An enlarged pronephroi is partially concealed by the gills. (From Humphrey (1962). *Developmental Biology*, **4**, 423–51.)

these few survivors is slower than normal and they are usually smaller than an average-sized adult.

The reduced growth and vasodilation of v/v embryos from heterozygous mothers might be related to a reduction in the ability of the mutant embryos to utilize various ions in the water. By raising v/v embryos in Steinberg's medium, which contains higher levels of sodium, potassium and calcium than the common tap water used for raising embryos, a greatly increased yield of surviving v/v larvae is obtained (Justus & Humphrey, 1964). It has been postulated that a deficiency in aldosterone, cortisone, and other steroids may be responsible for this apparent defect in either the uptake or utilization of ions (Tompkins, 1968).

A variable maternal effect is always seen in the offspring of the few surviving v/v females. Embryos from these homozygous females usually arrest between cleavage stages and the onset of gastrulation. A few embryos may survive longer, though only rarely is survival to the tail-bud stage observed. The extent of development is, in part, a function of the temperature at which the embryos are grown. Most embryos arrest in mid-blastula stages if grown at 25 °C. At 14–20 °C, however, 70% of the embryos develop to gastrula and neurula stages, with 30% proceeding to more advanced embryonic stages (Briggs & Humphrey, 1962). Likewise, the genotype of the fertilizing sperm can modify the expression of the maternal effect of *v*. Eggs from *v/v* females sometimes develop to advanced embryonic stages, but whether they survive beyond these stages depends on their zygotic genotype. Eggs fertilized by a wild-type (*V*) sperm often are viable, whereas eggs fertilized by a mutant (*v*) sperm invariably arrest (Humphrey, 1962).

In arrested blastulae and gastrulae from eggs of a *v/v* female there is extensive stratification of cellular components (Fig. 6). In cells from normal embryos the nuclei are normally centrally located and surrounded by an RNA-rich region. Yolk platelets are distributed throughout the cytoplasm except in the RNA-rich region and the spindle region of dividing cells. In these arrested embryos the nuclei and RNA-containing cytoplasm are displaced to the upper region of the cells and yolk platelets and pigment granules are concentrated in the lower regions (Briggs & Humphrey, 1962). The displaced nuclei enlarge and contain a large amount of DNA, which exists in the form of fine threads. There are excess numbers of nucleoli, ranging from 3 to 14 per nucleus, and although there are few metaphase nuclei, the few that are seen are highly polyploid. These observations indicate that while cytokinesis and karyokinesis are suppressed chromosome replication may be continuing.

The biochemical basis of the gene *v* lesion in embryos from homozygous (*v/v*) females would, at first glance, seem to be quite different from the maternal effect displayed by homozygous embryos derived by mating two heterozygous parents (*V/v*♀ × *V/v*♂). The arrest of cleavage, or the failure to complete gastrulation seem far removed from the vasodilation and pronephric enlargement seen at later stages in *v/v* embryos derived from *V/v* females. In this latter situation a defect in the uptake or utilization of certain ions has been postulated. In the case of the early maternal effect, embryos spawned as eggs from a *v/v* female do not respond as uniformly, or completely, to an increase in

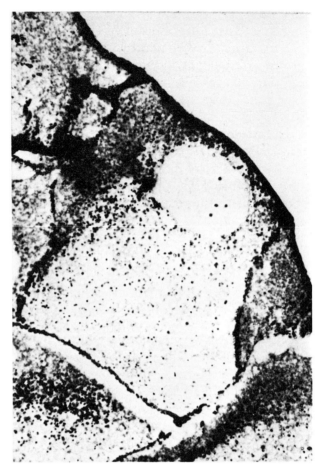

Fig. 6. Section of arrested blastula embryo derived from a v/v female stained with azure b. Note the stratification of the cellular components, with the enlarged nucleus located in the RNA-containing cytoplasm in the upper portion of the cell. The yolk platelets are concentrated in the lower region of the cell. (From Briggs & Humphrey (1962). *Developmental Biology*, **5**, 127–46.)

sodium, potassium and calcium in their water. Embryos from females in which the maternal effect is severe (development arrests in early cleavage stages) do not benefit at all from the higher salt concentrations. There is, however, a clear benefit to the progeny of v/v females in which the maternal effect is, for as yet unexplained reasons, less severe. In these less severe cases the untreated embryos survive beyond cleavage stages. Many of the treated embryos develop to later larval stages and begin to feed. Siblings maintained in tap water all, without exception, die much earlier. Only a very few of the embryos that begin to feed ever reach sexual maturity, whereas a substantial number of *v/v* embryos from heterozygous females develop to adulthood if raised in the higher salt solution (Justus & Briggs, 1964).

Humphrey (1962) emphasized the retardation or cessation of growth as the common factor linking the maternal effect of gene *v* with the usual syndrome appearing in the *v/v* genotype derived from heterozygous parents. The earliest manifestation of the maternal effect is a cessation of development, be it in cleavage, gastrulation or neurulation. If further growth occurs it is at a reduced rate and a complete cessation of growth may occur several days before death and necrosis are observed. The maternal effect may be fundamentally the same as a dwarfing of the larvae seen in the usual *v/v* syndrome, the latter case being only temporary in some instances, which allows for the survival of the individual. Humphrey (1962) suggested further that the vasodilation and pronephric swelling in all *v/v* larvae from heterozygous mothers may be a reflection of physiological problems in certain organs due to their delayed growth and differentiation.

Whether or not the relationship between the maternal effect of gene *v* and the syndrome expressed in *v/v* embryos from hetero- zygous parents proves to be simple, it is now at least known that the maternal effect is not the result of a general metabolic deficiency in the mother. This is evident from the results of ovary grafts where parts of ovarys from *v/v* females were grafted into wild-type (*V/V*) hosts. In such grafts the *V/V* host produced some eggs derived from the graft which continued to express the mutant phenotype. Likewise, it is known that the deficiency leading to the arrest of development associated with the maternal effect is a deficiency which can be corrected by joining *v/v* embryos (from *v/v* mothers) parabiotically to normal embryos, suggesting a transmissible product of the gene *V* (Humphrey, 1962). These experiments have been done only with *v/v* embryos which survived to post-gastrula stages and which could be joined easily with normal embryos. It is assumed that earlier manifestations of the maternal effect during blastula and gastrula stages might also be corrected if similar parabiotic unions could be made.

The one feature of the *v* phenotype which sets it apart from the other known maternal effect mutants in the axolotl is the variability seen in eggs from different v/v females. This variability is evidenced in v/v embryos from v/v mothers as the broad range of developmental stages of which the embryos arrest. Most v/v embryos arrest between cleavage and gastrulation but a few may survive longer. In addition, there is a difference in the viability of heterozygous (V/v) embryos from v/v females. In each case the variability would seem to be due to something intrinsic in the individual v/v females in that the range of variation between different spawnings of an individual female is small. The constancy of the maternal effect in successive spawnings of individual females, regardless of their age or date of spawnings, argues against variability arising as the result of different metabolic states of individual v/v females (Humphrey, 1962). The difference seems, therefore, to have a genetic basis, possibly induced by modifying genes which, if present, would undoubtedly complicate the biochemical and molecular analysis of this lesion.

f: fluid imbalance mutant

A. mexicanum embryos which are homozygous for gene f (fluid imbalance) display a marked suprabranchial swelling (Fig. 7). That swelling is first observed as a substantial enlargement of the head at stages 24–27 (tail-bud). This is later followed by fluid accumulation dorsal to the branchial arches, and rostral to the pronephros. That entire region becomes distended, as a bilateral swelling (Humphrey, 1948). The fluid appears to accumulate between various cell layers (e.g. mesoderm and endoderm). Puncturing the abdomen results in a substantial, albeit temporary, dimunition in the extent of swelling.

Originally, characterization of the phenotype of embryos which display the fluid imbalance syndrome was complicated. When F/f females were mated to F/f males two distinctly different phenotypes were observed to segregate out among the progeny. One phenotype displayed the characteristic swelling shown in Fig. 7 and Fig. 8(a). The other phenotype exhibited various gill abnormalities. Some progeny showed both the fluid accumulation and the gill defects. Virtually all embryos which contained the defects in gill structure died shortly after hatching. The gene g (designating gill lethal: Tompkins, 1970) was eventually recognized

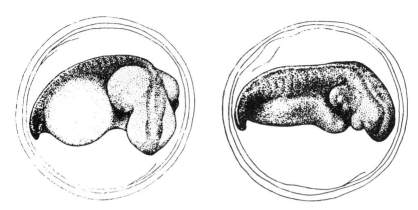

Fig. 7. Swelling of an *f/f* embryo (upper) is readily apparent in the head, suprabranchial and abdominal regions. Normal sibling is shown below. (From Dunson *et al.* (1971). *Comparative Biochemistry and Physiology*, **40**A, 319–37. © Pergamon Press Ltd.)

as usually accompanying the gene *f*, and was ultimately discovered to occupy a locus close to that of gene *f*. The two genes may become separated by crossing-over during meiosis at a rate of approximately 2 % (Humphrey, 1959).

With cross-over stocks of the gene *f* it was possible to perform a precise characterization of the fluid imbalance phenotype. Further studies revealed that the extent of the fluid imbalance, as well as the developmental stage at which it is first recognized, depends upon the genotype of the mother. An elegant description of these effects has been published by Humphrey (1960). What follows represents a summary of that paper. When both the female and the male parents are heterozygous for the gene *f* ($F/f \, ♀ \times F/f ♂$), all the progeny develop normally through early cleavage and gastrula stages of embryogenesis. But by tail-bud stages the swelling in the suprabranchial regions (Fig. 7) is observed in approximately 25 % of the embryos. These embryos usually display a diminution in the swelling as circulation develops. They begin feeding at the usual time, and survive to sexual maturity.

A much more complicated picture of the gene *f* phenotype emerges from an examination of the progeny of a cross in which the female is homozygous for the gene *f* and the male is heterozygous for *f* ($f/f ♀ \times F/f ♂$). All the offspring of this cross display a fluid imbalance as early as the cleavage stage of embryogenesis! By the late blastula/early gastrula stage the blastocoel cavity is enlarged, and the roof of the blastocoel (ectodermal layer) is abnormally thin. As gastrulation proceeds the accumulation of fluid apparently alters the position of the centre of gravity in such a way that the yolk plug is turned downward. With continuing development the embryo comes to lie with its ventral surface up.

As the progeny of these cross develop further and approach the hatching stage, an interesting segregation of the phenotypic effects is observed. The progeny which are genotypically homozygous for gene f (f/f) continue to manifest the severe fluid imbalance. The size of the archenteron is diminished. Fluid does not escape through the blastopore, and the neural folds do not close properly. Subsequent organogenesis is retarded, and the embryos fail to hatch. All the f/f embryos among the progeny of this cross die. Fig. 8(a) and (f) shows the gross morphological features of this 'severe' fluid imbalance syndrome. Surprisingly, f/f embryos joined in parabiosis with normal embryos show a decrease in the swelling, and survive. This apparent rescue is probably accomplished by the action of the circulatory system of the normal parabiont.

Embryos heterozygous for f (F/f) which originated from a cross involving a homozygous female ($f/f \female \times F/f \male$) also show, as mentioned previously, the blastocoel swelling at early embryonic stages. In fact, the heterozygotes are indistinguishable from the homozygotes (f/f) at the gastrulation stage. During neurulation, however, much of the blastocoel fluid escapes since a functional connection develops between the blastocoel, archenteron and blastopore. At organogenesis and subsequent stages of larval development these heterozygous (F/f) progeny can be divided into three broadly characterized groups. In the first group are embryos which appear to be normal in all regards, or at most display some microcephaly (Fig. 8c). These embryos survive to adulthood. In a second group are embryos which, after the initial swelling has regressed, become distended once again (Fig. 8d). Perhaps the development of a normal circulation fails. Ascites follows, and death eventually results. In a third group are embryos which display intermediate effects. Blisters containing fluid frequently appear on the body wall, but in some embryos they regress, leaving only a wrinkled appearance on the larva's surface. A variable proportion of embryos in this group develop to the hatching stage (Fig. 8c), and survive to adulthood.

From the foregoing description it is apparent that the gene f

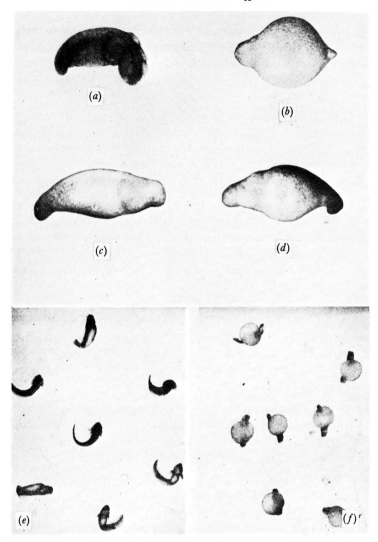

Fig. 8. Gross morphological features of the fluid imbalance syndrome in embryos of various genetic backgrounds. (*a*) 'Mild' fluid imbalance syndrome displayed by *f/f* embryo which was spawned by a heterozygous (*F/f*) female. (*b*) 'Severe' fluid imbalance in *f/f* embryo spawned by a homozygous (*f/f*) female (this embryo shows more fluid accumulation, and an earlier manifestation of the *f* syndrome, than the embryo in (*a*). (*c*) *F/f* embryo from an *f/f* female which – at Harrison stage 31 – is apparently normal, except for a minor degree of microcephaly. (*d*) *F/f* embryo showing second accumulation of fluid, which leads to death. (*e*) *F/f* embryos ready to hatch which show no evidence of prior fluid imbalance. (*f*) Later development of embryos shown in (*b*). These embryos lie ventral side up, and do not continue organogenesis, as do *F/f* embryos in (*e*). (From Humphrey (1960). *Developmental Biology*, **2**, 105–28.)

displays a true maternal effect. Further evidence for the maternal character of the *f* defect was derived from the results of ovary graft experiments performed by Humphrey (1960). Gonad primordia of the *f/f* genotype were implanted into normal hosts at the tail-bud stage. At the age of 3 months the host's ovaries were removed, while the donor's ovaries were permitted to develop to maturity. When spawned, all of the progeny dispayed the characteristic fluid imbalance despite the fact that the eggs had been matured in a normal host.

Experimentation on the nature of the fluid defects generated by the gene *f* has been limited mainly to physiological studies. Dunson, Packer & Dunson (1971) employed progeny from a cross of heterozygous animals (*f/f* ♀ × *F/f* ♂) for their work. They examined swollen embryos histologically and concluded that the extracellular fluid accumulation was largely confined to the endodermal regions of the embryos. This swelling was reported to be diminished somewhat in a high osmotic strength mannitol solution. The swelling could be exaggerated by rearing embryos at low (e.g. 10 °C) or high (e.g. 28–30 °C) temperatures. Sodium influx of *f/f* embryos was measured, and found to be diminished compared with normal embryos. The potassium content of mutant embryos is also altered, as is the Na:K.

The uniform penetrance of the gene *f* in early (e.g. cleavage to gastrulation) stages of post-fertilization development displayed by eggs from an *f/f* female should provide favourable circumstances for further experimentation. Such further work should probably include more embryological as well as biochemical analyses. The authors of this review are of the opinion that one important question should be answered immediately: Is the fluid imbalance syndrome a property of each of the individual cells of the embryo, or a syndrome displayed by the embryo as a whole? Various types of grafting experiments (e.g. Van Deusen, 1973) and cell culture techniques could be employed to answer this question. A second question – Are the surface proteins of *f/f* embryos altered? – might be approached. The expectation here is, of course, that the proteins involved in fluid and ion transport into the embryo would be located at the surface (i.e. the embryonic membranes). Membrane technology (Cuatrecasos & Hollenberg, 1976) has advanced sufficiently far in recent years to permit this type of investigation to stand a good chance of being successful.

We are currently building up our breeding stocks of animals which carry the gene *f* (Bacher & Malacinski, unpublished observation), and have plans for pursuing several of those experimental approaches.

o: ova deficient mutant

Perhaps one of the most interesting maternal effect genes in the axolotl – because of the opportunity for novel experimentation it offers – is the gene *o* (ova deficient) (Humphrey, 1966). Eggs spawned by a female which is homozygous for the gene *o* (*o/o*) uniformly arrest during gastrulation (Fig. 4*d* and *e*), regardless of the genotype of the fertilizing sperm. During cleavage and early blastula stages eggs from *o/o* females are indistinguishable from normal embryos. But both cell division and DNA synthesis become dramatically reduced in mutant embryos at the mid to late blastula stage (Carroll, 1974). The results of a comparative analysis of the mitotic index of normal and mutant embryos from the mid-blastula stage through gastrulation are shown in Fig. 9. At the beginning of Harrison stage 9 the mitotic index in mutant embryos drops precipitously. Tritiated thymidine incorporation studies revealed that the drop in the mitotic index in mutant embryos was accompanied by a substantial decrease in DNA synthesis (Carroll, 1974).

An analysis of the spectrum of new synthesized proteins in mutant embryos revealed alterations in the pattern of protein synthesis. Just prior to the time at which the mitotic index drops a difference in the radioactivity profiles of newly synthesized proteins – separated on acrylamide gels – was detected when extracts of normal embryos were co-electrophoresed with extracts of mutant embryos (Malacinski, 1971). Since more highly resolving gel electrophoresis systems (e.g. O'Farrell, 1975) are now available, those observations should be extended. It is entirely possible that a more precise analysis will yield alterations even earlier in development in protein synthesis. Such alterations apparently represent the first post-fertilization effects manifested by the gene *o*. By the late gastrula stage morphogenesis in mutant eggs comes to a complete halt. There has never been an instance – among the spawnings of dozens of *o/o* females – of an embryo developing beyond gastrulation to the neurula or tail-bud stage.

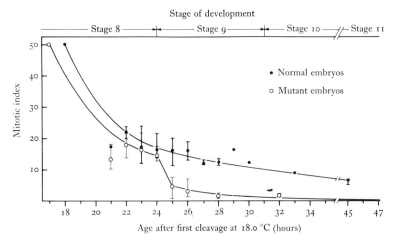

Fig. 9. Mitotic index of normal embryos and embryos spawned by o/o females. Squashes prepared from several embryos were analysed. In each squash preparation at least 200 nuclei were scored. (From Carroll (1974). *Journal of Experimental Zoology*, **187**, 409–22.)

Eggs which are spawned by o/o females, and normally are destined to arrest at gastrulation, can be rescued from that fate by microinjection of normal egg cytoplasm. As long as the egg is injected prior to the first cleavage division, complete correction of the mutant phenotype is achieved (Briggs & Cassens, 1966). If only one of the two blastomeres at first cleavage is injected, a partial (i.e. half-embryo) correction results (Cassens, 1969). Apparently, the active component cannot traverse the cell membrane which partitions the egg at first cleavage. The correction of eggs which are injected prior to cleavage can be complete, for injected eggs can develop to sexual maturity. Indeed, breeding stocks of heterozygous (+/o) animals can be obtained in this manner, if the fertilizing male is homozygous for the wild-type allele of the gene o. These microinjection experiments provide direct proof that the gene o is responsible for a cytoplasmic deficiency. In very few, if any, other experimental systems does such a straightforward case exist for a single mutant gene yielding a developmentally significant deficiency in egg cytoplasm.

The cytoplasmic component which is effective in correcting the gene o defect is actually concentrated in the germinal vesicle (nucleus) of the developing oocyte. Bioassay of the contents of the germinal vesicle, carried out by Briggs (1972), revealed that from early stages of oogenesis onwards, significant amounts of the corrective component are present in the germinal vesicle. During oocyte maturation, when the germinal vesicle breaks down, the corrective component is presumably released to the egg cytoplasm *per se*. This apparently accounts for the ability of injections of normal egg cytoplasm to correct eggs spawned by *o/o* females.

The corrective component has been found in the germinal vesicles of the oocytes of a variety of amphibian species, including *Rana pipiens*, *Rana catesbeiana*, *Xenopus laevis*, *Hyla crucifer* and *Pseudacris triseriata* (Briggs, 1972). This broad distribution of the corrective component is interesting, for these observations suggest that genes resembling the gene *o* might be active in the embryogenesis of a wide variety of species. Perhaps the oocytes of even more diverse species should be assayed, for it is conceivable that the gene *o* represents a major step in a developmental pathway which is fundamental to virtually all organisms!

A limited purification of the active component has been accomplished. The first attempts involved differential centrifugation and ammonium sulphate precipitations (Briggs & Justus, 1968). The active component responded to those treatments as if it were a protein. Further characterization by enzymatic digestion (e.g. trypsin) and heat stability support the expectation that the active component is indeed a protein (Briggs & Justus, 1968). Using molecular sieve column chromatography, our laboratory was able to obtain substantial purification of the corrective 'protein' (see Malacinski & Brothers, 1974). Although the bioassay for the corrective component appears to be relatively straightforward, and the availability of starting material from normal egg cytoplasm should be adequate, the purification of the corrective component is a complicated procedure. On the one hand, the 'Holtfreter' strain of the axolotl which carries the gene *o* is the least vigorous of all the stock maintained in the Indiana University axolotl colony (Malacinski, 1978). Homozygous (*o/o*) females display low survival rates, and are recalcitrant breeders. Hence, suitable numbers of females for the bioassay are very difficult to obtain. On the other hand, the isolation procedures frequently cause the partially purified fractions to be toxic in the bioassay. Microinjection experiments are, therefore, rather cumbersome. Alternative approaches to the assay of the corrective component (e.g. immunological methods) are currently being explored in our laboratory by J. S.

Some insight into the mechanism of action of the corrective

component comes from a set of nuclear transplantation experiments. Normal early-blastulae nuclei (Harrison stages 8–8¼) were transplanted into enucleated eggs spawned by an o/o female (Brothers, 1976). As expected, in the deficient egg cytoplasm these nuclear transplants never developed beyond the stage of development (gastrulation) at which arrest always occurs. Normal mid-blastula or later stage nuclei, however, were capable of supporting complete development to sexual maturity in enucleated recipient eggs spawned by an o/o female. Even after several serial passages through defective eggs, the capacity of normal blastula nuclei to promote development in mutant eggs was retained. Those observations have given rise to the concept that the corrective component might act by effecting a stable (heritable) change in mid-blastula nuclei (Brothers, 1976).

The results of those nuclear transplantation experiments have broad implications for attempts to understand the mode of action of the product of the normal allele of the gene o. We feel, however, that those nuclear transplantation experiments warrant an independent verification. They are technically difficult to perform, and to date represent only a single, albeit elegant, set of experiments carried out on a limited sampling of experimental animals.

In addition to the effects on the egg cytoplasm, two additional effects have been attributed to the gene o. The homozygous (o/o) males are always sterile. They do not display the enlargement of the cloaca which normally accompanies sexual maturity. Their testes are very small (Fig. 10) and their development remains incomplete. Mature sperm seldom, if ever, develop in o/o males (Humphrey, 1966). The relationship between male sterility and a deficiency in the egg cytoplasm remains an enigma. Another effect attributed to the gene o – poor limb regeneration – is much less clear-cut. Most o/o animals display a reduced capacity for limb regeneration (Humphrey, 1966). Several instances have, however, been observed in which o/o animals maintain a completely normal capacity for limb regeneration (Humphrey & Malacinski, unpublished observation). The total genetic background of o/o animals, rather than the gene o itself, might be the most important influence on the capacity for limb regeneration.

The potential usefulness of the gene o is, in our opinion, limited only by the availability of o/o famales (see above), and the imagination of the experimenter. For example, the temporal

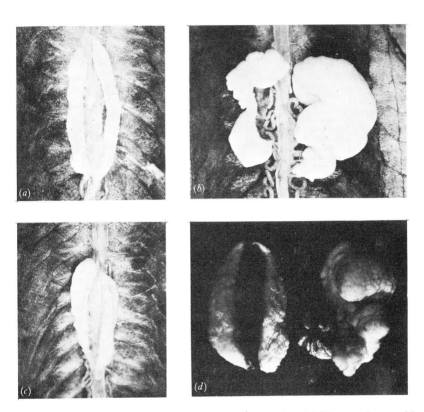

Fig. 10. Development of testes in normal and o/o males. (a) Testes of year-old o/o male. (b) Testes of normal sib of same age as animal shown in (a). (c) Testes of $1\frac{1}{2}$-year-old o/o male do not show further development. (d) Graft of normal testes (right) into peritoneal cavity of o/o male. o/o testes (left) retained juvenile characteristics as in (a), while normal testes continued development and contained growing spermatocytes and differentiating spermatids. (From Humphrey (1966). *Developmental Biology*, **13**, 57–76.)

aspects of the synthesis of the corrective component during oogenesis, the possibility that the corrective component itself interacts directly with the nucleus, and the precise manner in which gastrulation is regulated by the product of the gene o are all issues which warrant a major experimental effort. In our laboratory a long-term outbreeding is being initiated so that in the future a suitable supply of experimental animals will be available.

Other mutant genes

Various other mutant genes in the axolotl can contribute to studies on the maternal control of early development. The pigment markers shown in Fig. 2 have, for example, been extremely valuable as genetic markers in grafting experiments and organ transplantations. As mentioned previously, the true maternal character of several of the axolotl mutants was unequivocally established by grafting presumptive mutant gonads (usually of the wild-type, i.e. dark, pigmentation pattern) into albino hosts.

The persistence of maternally derived amphibian egg pigments has not, to our knowledge, been rigorously analysed. These pigment mutants provide an excellent opportunity for examining the early ontogenesis of pigmentation patterns. The possibility of combining nuclear transplantation techniques and microsurgery with a simple genetic analysis should be exploited.

Other mutant genes in the axolotl may also be useful for studies on gamete development. Animals which are homozygous for the gene e (Fig. 2) not only lack eyes, but are sterile as well. Gonad development may be deficient as a result of a defective hypothalamus (Van Deusen, 1973). The gene s (short toes) gives rise to a complex of abnormalities (limb defects, ascites, etc.), including occasional sterility (Humphrey, 1967). No doubt other axolotl mutants give rise to effects on gamete development. The group of approximately a dozen so-called 'cell lethal' genes might be interesting in this regard. Analyses of their phenotypes is, however, complicated (reviewed by Malacinski, 1978).

Concluding remarks

Since each of the axolotl maternal effect mutants acts through effects on the egg cytoplasm, they are all extremely valuable for studies on the development of the early (post-fertilization) embryo. It is thought that for each of the mutant genes every cell in the embryo is affected. For the genes nc, cl, and probably o, this is certainly the case. For the other two genes (v and f) it remains to be established whether all cells are uniformly affected by the action of the respective mutant genes.

A major question is whether these maternal effect mutations lead to specific deficiencies in mRNA, protein, or other macromolecules (e.g. membrane lipids). In one case – the gene o – a single protein is most probably missing from the cytoplasm of eggs spawned by homozygous (o/o) females. For the other genes direct evidence is lacking which would implicate the absence of a specific macromolecule as the cause of the block to development. Obviously, future experimentation with these genes should be focused on this issue. The clear-cut manner in which some mutants (e.g. cl and o) arrest should facilitate research directed towards this important problem.

For most of these genes the block to development is most probably a direct effect of the action (or lack of function) of the mutant gene. It is especially useful that, rather than representing a secondary effect of the death of a limited number of cells, most of these genes appear to be the primary cause of arrest in embryos spawned by mutant females.

A continuing search for additional maternal effect mutations is being carried out in the Indiana University axolotl colony in laboratory as well as wild strains of the axolotl. The colony has recently acquired several wild *Ambystoma mexicanum* from Lake Xochimilco and other related species from various lakes in Mexico, which are being bred with existing laboratory strains. This extensive breeding programme will hopefully turn up many new maternal effect mutations as well as introducing hybrid vigour into the strains carrying the known mutations. Having these genes in an organism which is so amenable to laboratory research should continue to attract the attention of developmental biologists.

The authors' research is supported by grants from the NSF (PCM 77-04457) and NIH (HD 11295). J. S. is a fellow of the American Cancer Society.

References

ADAMSON, E. D. & WOODLAND, H. R. (1974). Histone synthesis in early amphibian development: histone and DNA synthesis are not coordinated. *Journal of Molecular Biology*, **88**, 263–85.

BRIGGS, R. (1972). Further studies on the maternal effect of the o gene in the Mexican axolotl. *Journal of Experimental Zoology*, **181**, 271–80.

BRIGGS, R. (1977). Genetics of cell type determination. In *Cell Interactions in Differentiation*, ed. L. Saxén & L. Weiss, pp. 22–43. New York & London: Academic Press.

BRIGGS, R. & CASSENS, G. (1966). Accumulation in the oocyte nucleus of a gene product essential for embryonic development beyond gastrulation. *Proceedings of the National Academy of Sciences of the USA*, **55**, 1103–9.

BRIGGS, R. & HUMPHREY, R. R. (1962). Studies on the maternal effect of the semilethal gene, *V*, in the Mexican axolotl. I. Influence of temperature on the expression of the effect. II. Cytological changes in the affected embryos. *Developmental Biology*, **5**, 127–46.

BRIGGS, R. & JUSTUS, J. T. (1968). Partial characterization of the component from normal eggs which corrects the maternal effect of gene *o* in the Mexican axolotl (*Ambystoma mexicanum*). *Journal of Experimental Zoology*, **167**, 105–16.

BROTHERS, A. J. (1976). Stable nuclear activation dependent on a protein synthesized during oogenesis. *Nature, London*, **260**, 112–15.

BROWN, D. D. & GURDON, J. B. (1964). Absence of ribosomal RNA synthesis in the anucleolate mutant of *Xenopus laevis*. *Proceedings of the National Academy of Sciences of the USA*, **51**, 139–46.

CARROLL, C. R. (1974). Comparative study of the early embryonic cytology and nucleic acid synthesis of *Ambystoma mexicanum* normal and *o* mutant embryos. *Journal of Experimental Zoology*, **187**, 409–22.

CARROLL, C. R. & VAN DEUSEN, E. B. (1973). Experimental studies on a mutant gene (*cl*) in the Mexican axolotl which affects cell membrane formation in embryos from *cl/cl* females. *Developmental Biology*, **32**, 155–66.

CASSENS, G. A. (1969). An analysis of the maternal effect of gene *o* in *Ambystoma mexicanum*. PhD thesis, Indiana University.

CHEN, P. S. (1953). The rate of oxygen consumption in the lethal hybrid between *Triton ♀* and *Salamandra ♂*. *Experimental Cell Research*, **5**, 275–87.

CRIPPA, M. & GROSS, P. R. (1969). Maternal and embryonic contributions to the functional messenger RNA of early development. *Proceedings of the National Academy of Sciences of the USA*, **62**, 120–7.

CUATRECASAS, P. & HOLLENBERG, M. D. (1976). Membrane receptors and hormone action. *Advances in Protein Chemistry*, **30**, 252–451.

DUNSON, W. A., PACKER, R. K. & DUNSON, M. K. (1971). Ion and water balance in normal and mutant fluid imbalance (*f/f*) embryos of the axolotl (*Ambystoma mexicanum*). *Comparative Biochemistry and Physiology*, **40A**, 319–37.

HUMPHREY, R. R. (1948). A lethal fluid imbalance in the Mexican axolotl inherited as a simple Mendelian recessive. *Journal of Heredity*, **39**, 255–61.

HUMPHREY, R. R. (1959). A linked gene determining the lethality usually accompanying a hereditary fluid imbalance in the Mexican axolotl. *Journal of Heredity*, **50**, 279–86.

HUMPHREY, R. R. (1960). A maternal effect of a gene (*f*) for a fluid imbalance in the Mexican axolotl. *Developmental Biology*, **2**, 105–28.

HUMPHREY, R. R. (1962). A semilethal factor (*V*) in the Mexican axolotl (*Siredon mexicanum*) and its maternal effect. *Developmental Biology*, **4**, 423–51.

HUMPHREY, R. R. (1966). A recessive factor (*o*, for ova deficient) determining a complex of abnormalities in the Mexican axolotl (*Ambystoma mexicanum*). *Developmental Biology*, **13**, 57–76.

HUMPHREY, R. R. (1967). Genetics and experimental studies on a lethal trait ('short toes') in the Mexican axolotl (*Ambystoma mexicanum*). *Journal of Experimental Zoology*, **164**, 281–96.

HUMPHREY, R. R. (1975). The axolotl, *Ambystoma mexicanum*. In *Handbook of Genetics*, ed. R. C. King, vol. 4, pp. 3–17. Englewood Cliffs, New Jersey: Prentice-Hall.

JUSTUS, J. T. & HUMPHREY, R. R. (1964). The effects of sodium, potassium and calcium ions on certain expressions of the semilethal gene *V* in the Mexican axolotl *Ambystoma* (= *Siredon*) *mexicanum. Developmental Biology*, **9**, 255–68.

MALACINSKI, G. M. (1971). Genetic control of qualitative changes in protein synthesis during early amphibian (Mexican axolotl) embryogenesis. *Developmental Biology*, **26**, 442–51.

MALACINSKI, G. M. (1978). The Mexican axolotl, *Ambystoma mexicanum*: its biology and developmental genetics, and its autonomous cell-lethal genes. *American Zoologist*, **18**, in press.

MALACINSKI, G. M. & BROTHERS, A. J. (1974). Mutant genes in the Mexican axolotl. *Science*, **184**, 1142–7.

MOORE, J. A. (1941). Developmental rate of hybrid frogs. *Journal of Experimental Zoology*, **86**, 405–22.

MOORE, J. A. (1955). Abnormal combinations of nuclear and cytoplasmic systems in frogs and toads. *Advances in Genetics*, **7**, 139–82.

O'FARRELL, P. H. (1975). High resolution two-dimensional electrophoresis of proteins. *Journal of Biological Chemistry*, **250**, 4007–21.

RAFF, E. C. (1977). Microtubule proteins in axolotl eggs and developing embryos. *Developmental Biology*, **58**, 56–75.

RAFF, E. C., BROTHERS, A. J. & RAFF, R. (1976). Microtubule assembly mutant. *Nature, London*, **260**, 615–17.

RAFF, E. C. & RAFF, R. (1978). Tubulin and microtubules in the early development of the axolotl and other amphibia. *American Zoologist*, in press.

SUBTELNY, S. (1974). Nucleocytoplasmic interactions in development of amphibian hybrids. *International Review of Cytology*, **39**, 35–88.

TOMPKINS, R. (1968). Biochemical analysis of the effects of the genes 'g' and 'v' on the development of the Mexican axolotl, *Ambystoma mexicanum*. PhD thesis, Indiana University.

TOMPKINS, R. (1970). Biochemical effects of the gene *g* on the development of the axolotl *Ambystoma mexicanum. Developmental Biology*, **22**, 59–83.

VAN DEUSEN, E. B. (1973). Experimental studies on a mutant gene (*e*) preventing the differentiation of eye and normal hypothalamus primordia in the axolotl. *Developmental Biology*, **34**, 135–58.

WOODLAND, H. R. & ADAMSON, E. D. (1977). The synthesis and storage of histones during oogenesis of *Xenopus laevis. Developmental Biology*, **57**, 118–35.

WRIGHT, D. A. (1975). Expression of enzyme phenotypes in hybrid embryos. In *Isozymes*, ed. C. L. Markert, vol. 4, pp. 649–64. New York & London: Academic Press.

WRIGHT, D. A. & MOYER, F. H. (1968). Inheritance of frog lactate dehydrogenase patterns and the persistence of maternal isozymes during development. *Journal of Experimental Zoology*, **107**, 197–206.

Studies on the maternal effect of the semi-lethal factor *ac* in the salamander *Pleurodeles waltli*

J. C. BEETSCHEN AND M. FERNANDEZ

Laboratoire de Biologie générale, Université Paul Sabatier,
31077 Toulouse, France

Several maternal effects have been discovered in the offspring of various mutant female axolotls (for reviews see Briggs, 1973; Malacinski & Brothers, 1974; Malacinski & Spieth, this volume). In the salamander *Pleurodeles waltli* the principal characteristics of another maternal effect have been briefly described, in the progeny of *ac/ac* females (Beetschen, 1970, 1976). Evidence from crosses between phenotypically normal heterozygous + /*ac* animals indicates that the *ac* factor (= 'caudal ascites') behaves as a semi-lethal recessive gene, the phenotypic expression of which is temperature-sensitive (Beetschen & Jaylet, 1965; Beetschen, 1971; Fernandez & Beetschen, 1975). Adult *ac/ac Pleurodeles* can nevertheless be obtained, though growth is greatly retarded and sexual maturity in females is usually attained only after 3 years instead of the normal 18 months. A small number of females are actually fertile and may even be capable of laying eggs at intervals, several times within a few weeks, after one mating. Adult males are frequently fertile and their sexual behaviour is normal. In both sexes the adult lifespan is usually shorter than in normal animals and the reproductive capacities are irregular, even for fertile animals.

A maternal effect affects the whole progeny of *ac/ac* females crossed with + / +, + /*ac* or *ac/ac* males (Beetschen, 1970). From such crosses, + /*ac* embryos may survive due to their heterozygous constitution, provided the maternal effect does not induce morphogenetic abnormalities that are too severe (Fig. 2*a*), but *ac/ac* embryos always die (Fig. 2*b*). The + /*ac* and *ac/ac* embryos arising from mutant females all normally show the same syndrome from the very beginning of gastrulation. The ectoderm of the

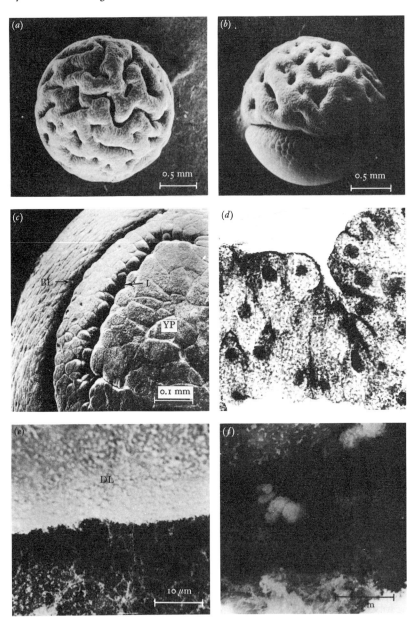

Fig. 1. (*a*), (*b*), (*c*), (*e*) and (*f*) are scanning electron micrographs of gastrulae with maternal effect. (*a*) Animal hemisphere of a stage 11 gastrula from *ac/ac* mother, with a strong maternal effect. (*b*) Another stage 11 gastrula with prominent yolk-plug (exogastrulation) added to ectodermal syndrome. Lateral view, animal pole upward. (*c*) Part of a large yolk-plug (YP) of a third stage 11 gastrula

animal hemisphere becomes furrowed with irregular grooves, the depth and frequency of which depend on the individual. In many embryos, as gastrulation proceeds, the furrows become continuous, giving the animal hemisphere a 'brain-like' appearance (Fig. 1*a*). During this phase epiboly movements (spreading) of the ectoderm become more disturbed as the syndrome becomes more intense. The blastocoele roof remains thicker than in controls of the same age but, on sectioning, the cells lining the furrows show an elongated shape and their migration inwards result in localized thinnings at the bottom of the furrows (Fig. 1*d*). Chordamesoderm and endoderm invagination are hindered and slow down, as can be seen in sagittal sections of young gastrulae 5 hours and 8 hours after the appearance of blastopore (Fig. 3) and, later on, at stages 11 and 12 of the normal developmental table (Gallien & Durocher, 1957), due to an abnormally large sized yolk-plug. The endoderm itself can be affected by irregular cell immigration, as shown in one case by a secondary superficial channel, running parallel to the blastoporal groove (Fig. 1*c*). Under these conditions many embryos exogastrulate and die either at the end of gastrulation or during neurulation without having withdrawn their yolk-plug (Fig. 1*b*). In the least-abnormal neurulae the lateral and ventral epiblast is generally wrinkled and irregular, but neural tube formation can proceed normally. The epidermal anomalies become progressively restricted to the most ventral region in embryos which develop beyond the tail-bud stage, and these embryos may survive after hatching if they are $+/ac$ and if head morphogenesis was not affected (Fig. 2*a*). Such $+/ac$ survivors were even able to grow to adult stage and gave rise to normal progeny.

At the gastrula stage the gravity of the ectodermal syndrome is evaluated as follows: *strong*, when the whole of the animal

displays a localized supernumerary invagination (I), roughly parallel to blastoporal groove (BL). (*d*) Cross-section through a furrow in the animal hemisphere of a young gastrula (stage 8a + 9 hours) with maternal effect. Some elongated invaginating cells are conspicuous in the thickened ectoderm. Methyl green–pyronin staining. (*e*) A filamentous and granular network of cell surface material can be seen in the blastoporal groove of a normal gastrula (stage 9), after post-fixation with lanthanum nitrate added to osmium tetroxide. Microvilli are still visible on dorsal lip cells (DL). (*f*) A similar accumulation of irregular cell surface material occurs in deep ectodermal furrows on a gastrula (stage 9) with maternal effect. Microvilli are visible on a linking cell (upward).

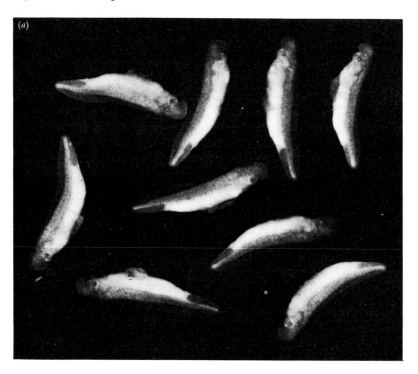

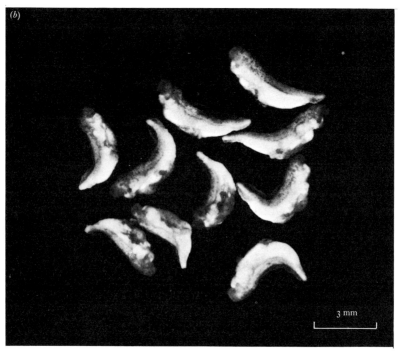

3 mm

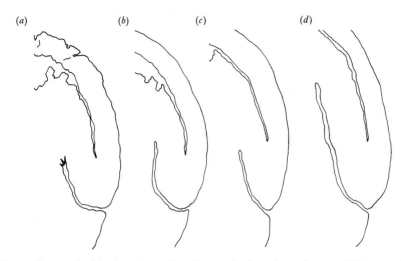

Fig. 3. Camera lucida drawings of median sagittal sections through blastoporal region of various gastrulae. (*a*) Gastrula with maternal effect; stage 8a + 8 hours. (*b*) Gastrula whose ectodermal maternal effect was prevented by pricking the uncleaved egg; stage 8a + 8 hours. (*c*) Control normal gastrula; stage 8a + 5 hours. (*d*) Control normal gastrula; stage 8a + 8 hours.

hemisphere is covered with a series of deep continuous furrows (Fig. 1*a*); *weak*, when small irregular depressions are observed; *average*, when it is between these two conditions, i.e. with clear furrows occasionally interconnected; *nil*, when the surface of the animal hemisphere is regular, a condition which is brought about by correction experiments. The proportions of these categories vary between females and even between spawnings of the same female, but all spawnings display the syndrome. Dejellied embryos, immersed in various saline media (modified Holtfreter, Steinberg, Barth), always show a more severe syndrome than do

Fig. 2. (*a*) A group of advanced + /*ac* embryos (stage 30) displaying a spontaneous recovery from maternal effect syndrome, which is still visible in several cases as an ectodermal pigmented process on the ventral side. General morphogenesis is normal or nearly so, particularly at the head level. These embryos originated from a cross ♀ *ac*/*ac* × ♂ + /*ac*. (*b*) A group of advanced *ac*/*ac* embryos from the same spawning as the + /*ac* embryos of the above group, at the same age. The *ac* mutation is expressed by ascites in dorsal and tail fins but its consequences are strongly enhanced by a maternal effect: growth is hindered, head anomalies are widespread, necrotic areas appear on the tail tip and on the ventral side, abnormal pigmented ectoderm is more developed than in + /*ac* embryos from the same spawning. All these embryos will die.

embryos developing inside the jelly. Until now no inorganic medium has been found to reduce the ectodermal syndrome.

Experimental work was carried out to induce a correction of the maternal effect, by analogy with the findings in the *o* mutant axolotl (Briggs & Cassens, 1966). The previously published data will be summarized and we shall report with more details on the main results which were obtained in new correction experiments and by analysis of some particular aspects of the ectodermal syndrome. Most of these data are unpublished.

Correction of maternal effect

From preliminary work, carried out on spawnings from crosses $♀ac/ac \times ♂ +/+$, it was concluded that a partial correction of the maternal effect could be obtained by injection of a small quantity of normal fertilized egg cytoplasm into the mutant female egg (Romanovsky & Beetschen, 1970). A few injected eggs did not develop any more ectodermal anomalies and many gastrulae showed a less severe syndrome. When eggs were only pricked with the micropipette, their subsequent development was not improved. Such a control experiment could have given different results had it been performed on a larger number of eggs, as it will be now reported. The corrective effect of normal cytoplasm cannot be considered as established at the present time.

Experiments dealing with cytoplasm microinjections were resumed and completed by injection of normal oocyte nuclear sap into eggs of an *ac/ac* female, which had been fertilized by sperms from an *ac/ac* male so that all embryos were *ac/ac*: this genotype, in the case of maternal effect, does not allow of any survival beyond the hatching stage (M. Fernandez, unpublished data).

The first experiments, which were all performed on eggs from the same female, provided clear evidence for a correction of the ectodermal syndrome and developmental anomalies, in particular yolk-plug invagination, after a simple prick of the animal hemisphere with the micropipette. Injection of normal cytoplasm or nuclear sap does not improve the result (Table 1). Nuclear sap injection even seems to be less efficient than a prick alone!

A second series of experiments was performed on three different spawnings and it confirmed the possibility of getting a full correction of ectodermal anomalies in more than 40 % of gastrulae

Table 1. *Experiments on the correction of the maternal effect on eggs of ac/ac females*

	No. of eggs	No. of gastrulae		Ectodermal syndrome[a] at stage 12[b]							
				+++		++		+		o	
		Stage 8a	Stage 12	No.	%	No.	%	No.	%	No.	%
First series											
Dejellied controls	72	62	62	45	72.5	13	21	4	6.5	—	—
Prick in animal hemisphere	79	68	59	13	22	10	17	16	27	20	34
Cytoplasm injection	28	20	20	1	5	2	10	10	50	7	35
Nuclear sap injection	40	36	36	7	19.5	15	41.6	8	22.3	6	16.6
Second series											
Dejellied controls	156	139	139	106	76.2	18	13	12	8.6	3	2.2
Prick in animal hemisphere	249	225	215	23	10.7	30	14	74	34.4	88	40.9

[a] +++, strong; ++, average; +, weak; o, nil (see text).
[b] Percentages calculated on stage 12 gastrulae.

just by pricking the uncleaved egg (Table 1). This prick is also efficient in correcting ectoderm invagination: the yolk-plug persists in more than 55 % of the advanced neurulae of control mutants while it becomes completely withdrawn in more than 70 % of neurulae developing from pricked eggs (Table 2).

Overall, the injury inflicted by pricking 328 uncleaved eggs allowed us to obtain 95 quite normal advanced gastrulae (34.6 %) from 274 which attained stage 12; only one normal gastrula was present in 201 gastrulae from dejellied control mutant eggs. It could be assumed, from the results of the first series of pricking experiments, that the activation of eggs from *ac/ac* females following sperm penetration might be incomplete. As complete activation of virgin amphibian eggs can be induced by an electric shock, especially in *Pleurodeles* (Signoret & Fagnier, 1962), this treatment was given to 179 freshly-laid fertilized mutant eggs, half of them then being subjected to a prick in the animal hemisphere. Correction of the ectodermal syndrome occurred in those eggs which had been either pricked or shocked and pricked, but not in those which had only been shocked. Gastrulation was even better in the batch which was shocked and pricked, the proportion of embryos in which yolk-plug invagination occurred normally (92.5 %) being much higher than that observed for the eggs that were just pricked (79 %), just shocked (68.5 %), or only dejellied (73.5 %). Neurulation was better, too, in embryos from shocked and pricked eggs, 25 normal-looking neurulae developing from 26 normal gastrulae, while only 34 normal neurulae developed from 77 normal-looking 'pricked' gastrulae. On the other hand, withdrawal of the yolk-plug was normal in more than 60 % of the neurulae obtained after the electric shock alone, a percentage which is significantly higher than that obtained in neurulae developing from control dejellied embryos (40 %), though the same electric shock did not improve the ectodermal syndrome.

From these results it appears that the injury inflicted by the prick of a micropipette on the uncleaved egg of an *ac/ac* female corrects the maternal effect in a large number of embryos at the gastrula and neurula stages. Nevertheless, the *ac/ac* embryos do not live beyond the hatching stage. The curative effect of the initial treatment therefore appears to be transitory, perhaps due to the fact that the *ac* mutation is a semi-lethal one and the normal *ac*[+] gene product is not synthesized in *ac/ac* embryos. On the other

Table 2. *Effects of pricking on yolk-plug invagination in gastrulae and neurulae from ac/ac females*

	No. of eggs	No. of gastrulae after stage 12	Normal sized yolk-plug	Abnormal yolk-plug	No. of neurulae after stage 14	Yolk-plug invaginated	Persistent yolk-plug
			Dejellied control eggs				
No.	228	201	154	47	162	72	90
%		100	76.6	23.4	100	44.4	55.6
			Dejellied pricked eggs				
No.	328	274	232	42	246	173	73
%		100	84.6	15.3	100	70.3	29.7

hand, we mentioned that in $+/ac$ embryos from an ac/ac mother spontaneous recovery often occurs after neurulation, the last ectodermal abnormalities being conspicuous on the ventral side of the embryo for a long time (Beetschen, 1970). In that case the abnormal epidermis is progressively rejected and replaced by a normal one (Fig. 2a).

The situation of the ac mutant *Pleurodeles* appears, then, to be quite different from that which is known for various maternal effects in the axolotl. The o maternal effect can be cured by a macromolecular substance which is present in the nuclear sap of the normal oocyte (Briggs & Justus, 1968). A similar situation holds for the nc maternal effect (Raff, Brothers & Raff, 1976). On the other hand, the cl maternal effect is not corrected by injection experiments (Carroll & Van Deusen, 1973). Therefore, as suggested by Beetschen (1976), the ac mutation shows specific characters which could orientate our study towards the properties of the cortical cytoplasm and the cytoplasmic membrane of the oocyte and of the uncleaved egg. The consequences of a wound on those peripheral structures should also be further investigated. We shall now briefly report on a preliminary study of the gastrula cell surface which could support the above interpretation.

Cell surface material in abnormal gastrulae

Deep ectodermal furrows in abnormal gastrulae from mutant females present morphological similarities with a blastoporal groove. In the axolotl, Moran & Mouradian (1975) described a process by which a specific cell surface material (CSM) appears at the blastoporal region and continues to spread throughout gastrulation. This material becomes visible in the scanning electron microscope after treatment with Alcian blue, added to glutaraldehyde fixative, and lanthanum nitrate, added during osmium tetroxide post-fixation. The CSM should therefore contain mucopolysaccharides. According to Moran & Mouradian's brief review of several papers dealing with various embryonic systems, the presence of CSM is linked to the occurrence of morphogenetic movements and it was suggested that CSM corresponds to Holtfreter's 'coat', which is not visible on electron micrographs after normal fixation.

Pleurodeles gastrulae were fixed according to a modified procedure, without Alcian blue. Fixation was carried out with 3 % glutaraldehyde in cacodylate buffer 0.05 M, pH 7.3, adjusted with sucrose to 150 mosmol. Gastrulae were rinsed in the same buffer and vitelline membrane was removed. Post-fixation was carried out in aqueous solution of 1 % osmium tetroxide containing 1 % lanthanum nitrate. The presence of CSM in the blastoporal region of a normal gastrula was demonstrated in the scanning electron microscope, where it appears as an irregular network of fibrillar and granular material (Fig. 1*e*). A similar superficial network seems to be present in the deep ectodermal furrows of gastrulae with the maternal effect (Fig. 1*f*). These investigations have yet to be extended to a complete series of developmental stages throughout gastrulation. Nevertheless, it already seems that a surface material, similar to CSM of normal blastoporal cells, is a concomitant of abnormal behaviour of ectodermal cells. It can also be pointed out that such abnormal immigrating behaviour is not associated with changes in embryonic determination: furrowed ectoderm keeps its neural potencies and does not exhibit mesodermal properties. Abnormal immigrating behaviour is sometimes also visible in endodermal cells (Fig. 1*c*).

Maternal effect and cell division

The occurrence of deep ectodermal furrows could be a consequence of an increase in mitotic activity leading to a superficial wrinkling, a situation which might be similar to that described by Ede (1956) in the blastoderm of X2 mutant *Drosophila* embryos. A comparison between the mitotic indices in animal hemispheres of normal and abnormal gastrulae first at stage 8a (appearance of blastoporal groove), then 5 hours and 9 hours later (ectodermal furrows visible), did not provide any evidence for differences of this sort between the two kinds of gastrulae. But it did seem to indicate a smaller increase in the number of nuclei in abnormal gastrulae at stage 8a + 9 hours, a condition which then could be interpreted as consecutive to an increase in the duration of mitosis in these gastrulae (Beetschen, 1976). After pricking experiments, further comparisons were made on such gastrulae, all of which had an *ac*/*ac* genotype (M. Fernandez, unpublished results). The

Table 3. *Evolution of the mitotic index in ectomesoderm of
early gastrulae*

Eggs	Gastrula stage[a] (at 18 °C)	Mitotic index (MI)[b] (per 1000 nuclei)	Relative decrease of MI between stage 8a and 8a + 8 h (as % of stage 8a value)
Control (+/+)	8a	120.54 ± 12.55	43.5 ± 11.1
	8a + 8 h	68.05 ± 8.07	
Mutant (*ac/ac*)	8a	124.61 ± 34	
1st series	8a + 5 h	76.93 ± 11.29	34.7 ± 18.4
	8a + 8 h	81.32 ± 7.66	
Mutant (*ac/ac*)	8a	81.55 ± 5.83	
2nd series	8a + 5 h	82.78 ± 3.99	21.6 ± 8.3
	8a + 8 h	63.89 ± 6.13	
Mutant (*ac/ac*)	8a	93.52 ± 4.53	
2nd series,	8a + 5 h	80.79 ± 10.32	27.7 ± 5.9
pricked	8a + 8 h	67.66 ± 4.71	

[a] Stage 8a = appearance of blastopore; Stage 8a + 5 h (or + 8 h) = 5 hours
(or 8 hours) later.
[b] ± standard error.

results are summarized in Tables 3 and 4 and Fig. 4. Mitoses and
nuclei were numbered on 10 sagittal sections for each gastrula –
one section in five – and each batch contained 10 gastrulae.

The mitotic index at stage 8a appears to be fairly variable in
different batches of mutant eggs (from 8 to 12 %), though it can
still be compared to that of control eggs (10 to 12 % in two control
batches). A wide range of individual values may also occur in
different gastrulae of the same batch, not only in mutant spawnings
but in control embryos also. Moreover, it appears that the overall
number of nuclei in animal hemisphere plus chordamesoderm
('ectomesoderm') at stage 8a is fairly variable, not only in control
(+/+) and abnormal gastrulae, but also in different batches of
mutant gastrulae (from about 1000 to 1850). This may be due to
egg size, which varies from one female to another, but also to
differences in cell division rhythms, which should now be investigated in cleavage and blastula stages. At later gastrula stages –
5 hours and 8 hours after stage 8a – a differential cell distribution
could also be a difference between normal and abnormal gastrulae,

Table 4. *Evolution of the number of nuclei in ectomesoderm of early gastrulae*

Eggs	Gastrula stage[a] (at 18 °C)	No. of nuclei in ectomesoderm (NNEM)	Relative increase of NNEM between gastrula of stage 8a and 8a + 8 h (in % of stage 8a value)
Control (+/+)	8a	1083 ± 60	87.5 ± 5.8
	8a + 8 h	2031 ± 134	
Mutant (*ac/ac*) 1st series	8a	1322 ± 164	51.0 ± 8.6
	8a + 8 h	2000 ± 97	
Mutant (*ac/ac*) 2nd series	8a	1884 ± 175	51.6 ± 7.8
	8a + 5 h	2296 ± 83	
	8a + 8 h	2857 ± 186	
Mutant (*ac/ac*) 2nd series, pricked	8a	1499 ± 107	72.8 ± 5.4
	8a + 5 h	1858 ± 134	
	8a + 8 h	2591 ± 93	

[a] Stage 8a = appearance of blastopore; stage 8a + 5 h (or + 8 h = 5 hours (or 8 hours) later.

inasmuch as chordamesoderm invagination and archenteron form-ation in abnormal gastrulae are delayed and do not proceed at the same speed as in normal gastrulae (Fig. 3). But it could be that the changes in cell distribution and the increase in cell number are not very closely associated, because in gastrulae from pricked mutant eggs endomesoderm invagination does not proceed faster than in control mutant gastrulae – but ectodermal furrowing is corrected – though a relatively greater increase in cell number does occur.

Trying to eliminate such variables, we compared the relative decrease in mitotic index and the relative increase in the number of ectomesodermal nuclei during the first hours of gastrulation. In spite of the possible limitations mentioned above, such com-parative data can reveal some meaningful facts. From the first results it appears that the endodermal nuclei are not affected by the maternal effect and we shall not report on their situation.

1. The decrease in mitotic index (MI) during the first 8 hours of gastrulation is not significantly greater in mutant than in control eggs when two batches of eggs having the same high MI (about

12 %) at stage 8a are compared. But the relative increase in the number of nuclei is much greater in control gastrulae (87.5 % versus 51 %), a result which confirms earlier data (Beetschen, 1976).

2. In two batches of mutant gastrulae with different numbers of nuclei and different MI at stage 8a, the relative increase in the number of ectomesodermal nuclei is nevertheless identical (51 %).

3. Pricking the uncleaved mutant egg significantly enhances the relative increase in nuclear number in young gastrulae, though, curiously, the mean number of nuclei is significantly lower in young gastrulae from pricked eggs than in control abnormal gastrulae from the same spawning. On the other hand, the MI is significantly higher in 'pricked' gastrulae at stage 8a, but this difference does not persist later on. Finally, after 8 hours, the relative increase in ectomesodermal nuclear number in 'pricked' gastrulae is intermediate between that of control normal gastrulae and that of 'unpricked' abnormal gastrulae. Let us recall that endomesoderm invagination does not proceed faster in 'pricked' gastrulae: at stage 8a + 8 hours it can be compared to that of stage 8a + 5 hours in control normal gastrulae and it is similar to that in abnormal 'unpricked' gastrulae of the same age. No disturbances in the spatial distribution of nuclei should then occur between the two kinds of mutant gastrulae.

4. Does the larger relative increase in ectomesodermal nuclei develop regularly in gastrulae from pricked eggs? In Fig. 4 the histograms indicate the overall increase between stage 8a and stage 8a + 8 hours, between stage 8a and stage 8a + 5 hours, and finally between stage 8a + 5 hours and stage 8a + 8 hours. It can be seen that there is no difference between pricked and unpricked eggs at 5 hours, but there is one at 8 hours. Therefore, cell divisions must be less frequent in abnormal control embryos and mitotic activity should be higher in 'pricked' gastrulae. However, mitotic indices are similar and fairly low in both kinds of mutant gastrulae, and they decrease between 5 and 8 hours.

5. This discrepancy between the greater increase in nuclear number and the similarity in mitotic indices in 'pricked' mutant gastrulae is analogous to the situation that we found when we compared control, normal (+ / +) gastrulae and gastrulae displaying the maternal effect. It seems probable that a lengthening

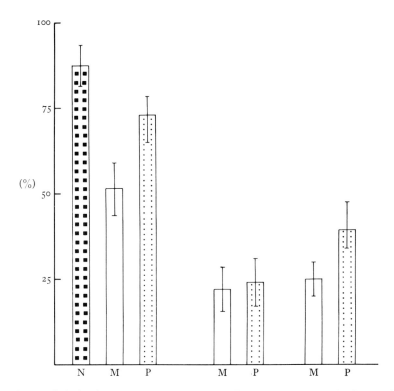

Fig. 4. Relative increase – as a percentage of stage 8a number – in the number of nuclei in ectoderm plus chordamesoderm during the first 8 hours of gastrulation. (*a*) Increase from stage 8a to stage 8a + 8 hours in control normal gastrulae (N), *ac/ac* gastrulae with maternal effect (M) and *ac/ac* gastrulae developing from pricked eggs (P). (*b*) Increase from stage 8a to stage 8a + 5 hours in M and P gastrulae. (*c*) Increase from stage 8a + 5 hours to stage 8a + 8 hours in M and P gastrulae.

of mitotic processes in abnormal gastrulae is responsible. This would mean that more mitoses would be present at any given time independent of the rate of proliferation. Therefore, the actual proliferative activity might be reduced without a decrease in MI. Such a hypothesis needs to be confirmed using more accurate methods and the cell cycle duration should be investigated with classical techniques.

6. In any case, we may conclude that the *ac* maternal effect, apart from its consequences on ectodermal cell movements, induces disturbances of cell proliferation, but that the irregular ectodermal furrowing does not appear to be correlated with an increase in cell

multiplication, which on the contrary is lowered. Both kinds of anomalies can be partly prevented by pricking the uncleaved fertilized egg.

Concluding remarks

Many questions bearing on the consequences of the *ac* maternal effect on gastrulation and neurulation processes remain open. The paradoxical corrective effect brought about by injury of egg cytoplasm should be investigated in connection with changes in either osmotic properties of the egg plasmalemma and cortical structures, or, more probably, selective ionic permeability. The role played by glycoproteins in cell membrane structures and the importance of mucopolysaccharides in cell movements may be a key to the further elucidation of the abnormalities of *ac/ac* oocytes. From work on the *o* mutant of axolotl (Briggs & Cassens, 1966; Briggs & Justus, 1968; Brothers, 1976) it is known that a macromolecular nuclear component, which is synthesized during oogenesis, interacts with cleavage nuclei at an advanced blastula stage and is essential to the initiation of gastrulation. The *Pleurodeles ac* mutation shows that other material prerequisites for gastrulation are established during oogenesis. The normality of cell behaviour at the beginning of gastrulation depends on cytoplasmic properties which are present in the egg before cleavage and were controlled by the maternal genome. One possibility is that these properties are expressed through later control of extracellular cell material formation. Another possibility is that the cytoplasmic anomalies cause a secondary effect on the duration of cell cycle in embryos from *ac/ac* mothers.

More experimental studies are obviously needed before all possible hypotheses could be tested. Investigations on the *ac* mutant oocyte should therefore lead to a better understanding of normal gastrulation processes.

This work was part of the programme of the Equipe de Recherche Associée No. 327 of the Centre National de la Recherche Scientifique.

References

BEETSCHEN, J. C. (1970). Existence d'un effet maternel dans la descendance de femelles de l'Amphibien Urodèle *Pleurodeles waltlii* homozygotes pour le facteur *ac* (ascite caudale). *Comptes Rendus des Séances de l'Académie des Sciences, Paris*, **270**, 855–8.

BEETSCHEN, J. C. (1971). Thermosensibilité de la mutation 'ascite caudale' chez *Pleurodeles waltlii* (Amphibien Urodèle). *Comptes Rendus des Séances de l'Académie des Sciences, Paris*, **273**, 97–100.

BEETSCHEN, J. C. (1976). Observations préliminaires sur les perturbations de la gastrulation consécutives à l'effet maternel lié à mutation *ac* chez l'Amphibien *Pleurodeles waltlii*. *Bulletin de la Société Zoologique de France*, **101**, 57–61.

BEETSCHEN, J. C. & JAYLET, A. (1965). Sur un facteur récessif semi-létal déterminant l'apparition d'ascite caudale (*ac*) chez le Triton *Pleurodeles waltlii*. *Comptes Rendus des Séances de l'Académie des Sciences, Paris*, **261**, 5675–8.

BRIGGS, R. (1973). Developmental genetics of the axolotl. In *Genetic Mechanisms of Development*, ed. F. H. Ruddle, pp. 169–99. New York & London: Academic Press.

BRIGGS, R. & CASSENS, G. (1966). Accumulation in the oocyte nucleus of a gene product essential for embryonic development beyond gastrulation. *Proceedings of the National Academy of Sciences of the USA*, **55**, 1103–9.

BRIGGS, R. & JUSTUS, J. T. (1968). Partial characterization of the component from normal eggs which corrects the maternal effect of gene *o* in the Mexican axolotl (*Ambystoma mexicanum*). *Journal of Experimental Zoology*, **167**, 105–16.

BROTHERS, A. J. (1976). Stable nuclear activation dependent on a protein synthesized during oogenesis. *Nature, London*, **260**, 112–15.

CARROLL, C. R. & VAN DEUSEN, E. B. (1973). Experimental studies on a mutant gene (*cl*) in the Mexican axolotl which affects cell membrane formation in embryos from *cl/cl* females. *Developmental Biology*, **32**, 155–66.

EDE, D. A. (1956). Studies on the effects of some genetic lethal factors in the embryonic development of *Drosophila melanogaster*. II. An analysis of the mutant X2. *Wilhelm Roux' Archiv für Entwicklungsmechanik der Organismen*, **148**, 437–51.

FERNANDEZ, M. & BEETSCHEN, J. C. (1975). Recherches sur le rôle de la température dans la réalisation du phénotype chez des embryons de l'Amphibien *Pleurodeles waltlii* homozygotes pour la mutation thermosensible *ac* (ascite caudale). *Journal of Embryology and Experimental Morphology*, **34**, 221–52.

GALLIEN, L. & DUROCHER, M. (1957). Table chronologique du développement chez *Pleurodeles waltlii* Michah. *Bulletin Biologique de la France et de la Belgique*, **91**, 97–114.

MALACINSKI, G. M. & BROTHERS, A. J. (1974). Mutant genes in the Mexican axolotl. *Science*, **184**, 1142–7.

MORAN, D. & MOURADIAN, W. E. (1975). A scanning electron microscopic study of the appearance and localization of cell surface material during amphibian gastrulation. *Developmental Biology*, **46**, 422–9.

RAFF, E. C., BROTHERS, A. J. & RAFF, R. A. (1976). Microtubule assembly mutant. *Nature, London*, **260**, 615–17.

ROMANOVSKY, A. & BEETSCHEN, J. C. (1970). Correction partielle de l'effet maternel lié à la mutation *ac* chez l'Amphibien *Pleurodeles waltlii* par injection de cytoplasme d'œufs normaux dans les œufs de femelles mutantes homozygotes *ac/ac*. *Comptes Rendus des Séances de la Société de Biologie*, **164**, 440–4.

SIGNORET, J. & FAGNIER, J. (1962). Activation expérimentale de l'œuf de Pleurodèle. *Comptes Rendus des Séances de l'Académie des Sciences, Paris*, **254**, 4079–80.

The impact of pre-fertilization events on post-fertilization development in mammals

ANNE McLAREN

MRC Mammalian Development Unit (University College London),
Wolfson House, 4 Stephenson Way, London NW1 2HE, UK

Fertilization is a developmental node. It may be regarded as the point in time and space from which a new individual develops; but the egg and spermatozoon that meet at that point are themselves each the culmination of a developmental process, the product of its own cell lineage, with an initial genetic endowment developing in interaction with the environment of the parent organism.

This paper will be concerned with those genetic and environmental forces that not only go to fashion the gamete, but also exert an influence beyond fertilization. Such influences may be termed *maternal* or *paternal effects*, according to the sex of the parent in which the causative forces are operating. First we must enquire into some relevant aspects of the development of the gametes. We shall then consider the effect on subsequent development of *environmental forces* acting during gametogenesis, and finally the effect of *genetic* factors.

Germ cell development

Cell lineage studies in the mouse (Falconer & Avery, 1978) suggest that the cellular ancestors of the germ cell population are contained in the embryonic ectoderm or epiblast that is separated off from the primary endoderm on the 5th day of development. Primordial germ cells can first be distinguished on the 9th day, at the base of the allantois (Ozdzenski, 1967). They migrate via the roof of the hindgut to their definitive position in the genital ridges, arriving during the course of the 11th day, and overt sexual differentiation of the gonad becomes evident a couple of days later. Claims that primordial germ cells migrate in the blood-stream of mammals as they do in birds have not yet been confirmed (Ohno & Gropp, 1965). Nothing is known of the influence upon the germ cells of their cellular environment prior to their arrival in the genital ridges.

The subsequent course of germ cell development differs markedly between the two sexes, in ways that affect the contribution that they make to the next generation and the types of perturbation that may be produced in them by environmental agencies.

Spermatogenesis

In the male, the germ cells to a large extent cease mitotic division at about the same stage of fetal life that female germ cells embark on meiosis. Many degenerate, and the survivors remain in a state of dormancy until the first type A spermatogonia appear, a few days after birth in mice and rats, a couple of months after birth in man. After a further period of active mitosis, the first germ cells enter meiosis, about 2 weeks after birth in mice and rats, at puberty in man. The cyclic pattern of spermatogenesis then becomes established, with a proliferating stem cell population of type A spermatogonia, giving off successive spermatogenic waves that over the course of a few weeks undergo mitotic multiplication and then maturation into spermatocytes, meiosis, and finally the transformation from spermatids into mature spermatozoa. The spermatids in particular develop in close association with the surrounding somatic tissue, their heads entirely embedded in the giant Sertoli cells (Nagano, 1968). The condensation of the nuclear material so characteristic of spermatozoa, and the elimination of virtually all the cytoplasmic mass, occurs during the transformation of spermatids into spermatozoa. The various stages of spermatogenesis are distributed progressively from the periphery towards the centre of each seminiferous tubule. The stem cell population in the seminiferous epithelium continues to throw off new cohorts of spermatogenic cells in a regular cycle, throughout the reproductive life of the male.

Although synthesis of both ribosomal and poly(A)$^+$ (presumptive messenger) RNA has been shown to occur during mouse spermatogenesis during the period immediately following meiosis, as late as the haploid 'round spermatid' stage (Geremia, D'Agostino & Monesi, 1978), this does not necessarily mean that the haploid genotype is expressed in the sperm phenotype. The four daughter spermatids resulting from the two meiotic divisions remain connected by cytoplasmic bridges (Dym & Fawcett, 1971); any RNA or protein synthesized during the post-meiotic period could therefore be distributed to all four cells so as to produce a common phenotype. Indeed, examples of haploid effects on sperm phenotype are hard to come by (Beatty, 1970). Apart from a suggestive toxic effect of anti-H(Y) antiserum on sperm carrying a Y chromosome rather than an X (Bennett & Boyse, 1973), they are confined to the *T/t* region of the mouse genome (for review, see Sherman & Wudl, 1977), where differential transmission of haplotypes has so far eluded all attempts at explanation in terms other than of haploid gene expression.

Protein synthesis occurring after the cessation of RNA synthesis, i.e. in the later spermatid, is presumably supported by stable RNA molecules. The remaining RNA is eliminated along with the excess cytoplasm during the transition from spermatid to spermatozoon, so that the mature spermatozoon lacks the capacity to synthesize protein, except perhaps in its mitochondria.

Oogenesis

In the female, the germ cells enter meiosis during fetal life in most mammals, including mouse and man, and shortly after birth in others. They proceed through zygotene and pachytene, the stages during which chromosome pairing takes place and chiasmata are formed, and arrest midway through the diplotene stage of first meiotic prophase. In long-lived mammals such as man, meiotic arrest may last for 40 years or more. The great majority of the primordial germ cells degenerate before birth; the remainder form a finite population of non-replicating oocytes, destined either to be ovulated or to be lost by atresia during the reproductive life of the female.

Some increase in size of the germ cells occurs as they pass through the successive stages of meiotic prophase, but most of the cytoplasmic growth occurs later. After birth, each 'resting' oocyte becomes surrounded by a single layer of flattened cells, forming the primordial follicle. Such follicles, which form the starting point for subsequent follicular growth, constitute the great majority of the germ cell population of the ovary throughout life.

Whether a hormonal stimulus is needed for the initiation of growth in a proportion of the primordial follicles is not known, but the presence of gonadotrophins is certainly not required for the maintenance of follicular growth. Henderson & Edwards (1968) have suggested that the order in which follicles start to grow reflects the order in which the oocytes that they contain entered meiosis in fetal life, but this model is not as yet supported by experimental evidence. The formation of the zona pellucida is initiated in the early stages of follicle growth. Along with the accumulation of cytoplasm, RNA and protein synthesis are extremely active in the oocyte during this period, although from the meiotic point of view it is still 'resting'.

Follicular growth results, after about 16 days in the mouse, in a population of large, multilayered follicles. These may be induced by follicle-stimulating gonadotrophin to develop into mature Graafian follicles, which then undergo maturation and ovulation under the influence of ovulation-inducing gonadotrophin. No further growth of the oocyte occurs during this period (4 to 5 days in the mouse). Meiosis is not resumed until shortly before ovulation, when under the influence of ovulation-inducing gonadotrophic hormone (e.g. HCG) prophase of the first meiotic division is completed, the nuclear membrane breaks down, the first polar body is extruded and the oocyte moves into the second meiotic division. This process takes about 12 hours in the mouse. RNA synthesis continues up to 2 hours before germinal vesicle breakdown (Rodman & Bachvarova, 1976). Some of the newly synthesized transcripts are released into the cytoplasm before the nuclear membrane disappears (Wassarman & Letourneau, 1976). Both nucleolus and nucleoplasm can be labelled with [³H]uridine, suggesting that both ribosomal and heterogeneous RNA are being synthesized. Most of the maternal RNA in the fertilized egg is presumably synthesized at this time; no RNA synthesis would be expected in the unfertilized egg, as the chromosomes are in metaphase and condensed. The pattern of polypeptide synthesis shows both qualitative and quantitative changes during oocyte maturation; each stage of maturation is associated with the appearance of specific polypeptides, some synthesized for very brief periods only (Van Blerkom & McGaughey, 1978a). No information is available as yet on how translation in oocytes is regulated.

In the ovary, processes from the follicle cells around the oocyte penetrate through the zona pellucida and terminate in gap junctions on the surface of the vitellus (Anderson & Albertini, 1976). These junctions persist up to the time that gonadotrophin-induced oocyte maturation begins (Szollosi, Gerard, Menezo & Thibault, 1978). The passage of RNA and other substances from somatic nurse cells into the developing oocytes is well documented in insects; whether an analogous transfer of either informational macromolecules or small signalling molecules such as cyclic AMP takes place from follicle cells into oocytes in the mammalian ovary is not known, but there are clear indications that some maternal proteins are taken into the oocyte by pinocytosis (Glass, 1970). The mature egg shows striking heterogeneity of the vitelline membrane, in that 20% of the cell surface, in the vicinity of the meiotic spindle, is devoid of both microvilli and cortical granules, and is very rarely penetrated by spermatozoa (Nicosia, Wolf & Inoue, 1977).

The female mammal does not have a haploid gamete comparable to the haploid spermatozoon, since the second meiotic division is only completed after sperm entry into the egg. The only opportunity for haploid gene action in the egg would be after fertilization, if RNA synthesis from the maternal genome occurred at a time when the paternal chromosomes were still condensed and transcriptionally inactive. No such RNA synthesis has been detected, and the new polypeptides that appear after fertilization are thought to be coded for by preformed, 'masked' maternal messenger RNA.

Non-genetic effects on gamete development

For convenience of description, I shall distinguish four aspects of the environment of a germ cell: (a) cellular, (b) metabolic, (c) chemical, (d) external.

Cellular and metabolic environment

The cellular environment of the gamete consists of the surrounding somatic cells, with which, as we have seen, the developing germ cells enter into extremely intimate associations in both testis and ovary. The extent to which the genetic constitution of the somatic cells affects gamete development is hard to determine in a normal individual, since germ cells and somatic cells share the same genotype. Some relevant studies have been carried out in chimaeric mice, in which two genetically distinct cell populations co-exist, since in such animals a germ cell of one genotype may be associated with somatic cells of contrasting genotype (McLaren, 1976*a*).

Although some indications to the contrary have been claimed (Kanazawa & Imai, 1974), no clear evidence exists that germ line genes are subject to any form of transformation or transgenosis from their somatic neighbours (McLaren, 1975). Similarly, genetic effects on sperm shape do not appear to be exerted through the somatic cells, but are intrinsic to the germ cells themselves (Burgoyne, 1975). On the other hand the failure of spermatozoa to develop normally in chromosomally male (i.e. XY) mice feminized by the gene for *testicular feminization* (*Tfm*) has been shown to be due to gene action in the cellular environment rather than in the male germ cells, since these can form normal fertile spermatozoa if associated with non-*Tfm* somatic tissue in a chimaeric testis (Lyon, Glenister & Lamoreux, 1975). The gene abolishes androgen sensitivity; evidently the response to androgen required for the completion of spermatogenesis is a response not of the germ cell but of some somatic cell, probably either the Sertoli cell or the tubular epithelium.

Some similar effect of the hormonal environment during oogenesis may be responsible for the small increase in chromosome abnormalities detected in some but not all studies of spontaneous abortions of women who have recently ceased taking oral contraceptives (for references, see Alberman, Creasy, Elliott & Spicer, 1976). In cultures of pig oocytes, addition of progesterone and oestrogen to the medium significantly reduces the incidence of chromosome abnormality (McGaughey, 1977). Indeed, in the sheep it has been shown that inadequate steroid support during oocyte maturation leads not only to anomalies of fertilization, but also to delay in cleavage and failure of blastocyst formation (Moor, 1978).

Not only the hormonal environment but also other aspects of metabolism, such as body temperature or the local pH in the gonad, may affect gamete development. Above a certain temperature spermatogenesis is known to be inhibited. At lower temperatures there may be an increase in the incidence of gene mutation, chromosome breakage or non-disjunction, which would manifest itself as an environmentally induced maternal or paternal effect. Burfening, Elliott, Eisen & Ulberg (1970) subjected male mice to heat stress for 24 hours at intervals of up to 4 weeks before mating; they found the greatest effect on fertilization failure 18 days before mating, during maturation of the spermatids, and the greatest effect on implantation rate 15 days before mating, when late spermatids and mature spermatozoa would have been present in the reproductive tract. Bellvé (1972) reported that the exposure of male and/or female mice to heat stress a few days before mating results in arrest of embryos at the 4- to 8-cell stage, as well as increased post-implantation mortality. The heat stress was applied 2 days before mating, at a time when the spermatozoa would have been in the epididymis or vas deferens. High temperature is thought to inhibit RNA and/or protein synthesis; but as we have seen, RNA synthesis ceases soon after the end of meiosis, and the protein-synthetic machinery is eliminated along with the excess cytoplasm during the transition from spermatid to spermatozoon. An exception may be made for the sperm mitochondria; although they are thought to degenerate soon after fertilization and not to play any role in subsequent embryonic development (Szollosi, 1976), it may be that in the immediate pre-fertilization period they support synthesis of some protein that plays a key role during cleavage.

Other factors, including infectious diseases, local inflammation, alcoholism and dietary deficiencies, have been shown to produce extensive germ cell degeneration, but the rate of germ cell development, i.e. the duration of the spermatogenic cycle, is unaffected, and no effects on post-fertilization development have yet been established.

Chemical environment

Similar effects may be produced by drugs and toxic chemicals, as well as by radiation. Administration of caffeine to male Chinese hamsters for 2 months prior to mating significantly increased the proportion of females in subsequent litters (Weathersbee, Ax & Lodge, 1975). Caffeine and its derivatives are known to be mutagenic, causing translocations, deletions, and fragmentation of chromosomes, at least in bacteria (for a detailed review of chemical mutagenesis, see Auerbach, 1977). The effect on spermatogenesis may have been due to an increased rate of non-disjunction during meiosis, resulting in loss of the small Y chromosome from a proportion of the spermatozoa; the excess female progeny would in this event have consisted of XO females. Alternatively, damage induced in the X chromosome may have proved lethal in hemizygous condition, reducing the number of XY embryos.

Attempts have been made to determine the mutagenic or teratogenic effect of exposure of human populations to various toxic chemicals, such as anaesthetic gases and vinyl chloride, but problems of interpretation abound. An excess of fetal wastage has been reported among the wives of men occupationally exposed to vinyl chloride; this could result from dominant lethal mutations induced during spermatogenesis, but here and with other similar industrial hazards it is hard to rule out the possibility that small amounts of the toxic substance are carried home on clothing, and have a direct teratogenic effect during pregnancy. In the case of anaesthetic gases (Knill-Jones, Newman & Spence, 1975), the apparent increase in incidence of minor fetal abnormalities among anaesthetists' wives may have been affected by reporting bias; the reduced reproductive efficiency of women anaesthetists was more convincing, but the role of other factors such as stress was not adequately controlled, and exposure was not limited to the prefertilization period. Experiments in mice suggest that anaesthesia shortly before ovulation may interfere with meiotic chromosome segregation, leading to aneuploid eggs and embryonic mortality (Kaufman, 1977).

There is also evidence from experiments on cow and sheep oocytes that sulphur dioxide, a common constituent of air pollution, can cause fragmentation and non-disjunction of meiotic chromosomes (Jagiello, Lin & Ducayen, 1975).

Radiation

In general, mammalian germ cells are well protected from the assaults of the external environment. The major exception is ionizing radiation, the mutagenic effects of which have been studied extensively in mammals (for review, see, for example,

Russell, 1965), as well as in lower organisms. Large doses of radiation lead to extensive death of germ cells, resulting in temporary or permanent sterility; lower doses cause chromosome breakage and non-disjunction, which are scored as dominant lethal mutations, resulting in embryonic mortality; still lower doses induce point mutations, which may be either lethal or viable.

In the male, the spermatogenic stages that are most sensitive to X-irradiation, as judged by incidence of dominant lethals, are some of the spermatogonial stages, followed by spermatocytes and spermatids. Mature spermatozoa are relatively resistant to irradiation. Some type A spermatogonia are also highly resistant, and hence are able to regenerate the seminiferous epithelium after other germ cell populations have been eliminated. Chromosome breakage is more readily induced in post-spermatogonial stages, so chromosome aberrations are seen in progeny sired fairly soon after irradiation; point mutations, on the other hand, are characteristic of spermatogonial irradiation, so can be found throughout the subsequent reproductive life of the individual. Different genetic loci may have very different sensitivities to radiation-induced mutation.

In the female, oocytes seem particularly sensitive to irradiation during first meiotic metaphase. In mice, oocytes in primordial follicles are more sensitive to killing than those in growing ones, though the opposite is true for the Rhesus monkey. Chromosome aberrations are common, and the X chromosome is particularly susceptible to loss. The most marked effects of irradiation on the mutation rate in mice, as well as on the rate of non-disjunction (see p. 303), are observed in older females.

Differences in radiation resistance also exist between species: for example, the oocytes of rats and mice are much more sensitive to X-irradiation than are those of guinea-pigs, monkeys and man (Baker, 1971). This may reflect ultrastructural differences between the dictyate resting oocyte of rodents, and the lampbrush diplotene resting oocyte seen, for example, in women. In all species tested, a dose-rate effect exists for oocytes and spermatogonia, though not for post-spermatogonial stages, in that a given amount of X- or gamma-radiation produces fewer mutations if spread out over several small doses than if given in one large dose. Neutron irradiation appears not to show a dose-rate effect.

The chromosome anomalies induced by X-irradiation of human

germ cells will be considered further below, in relation to the
effects of maternal ageing.

Aetiology of chromosome abnormalities

When a known environmental agent acting during gametogenesis,
whether it be radiation, or caffeine ingestion, or heat shock, is
associated with subsequent embryonic mortality or malformation,
the cause of the developmental aberration is not in doubt, though
its mode of action may remain unknown. We know most about
radiation, since its destructive action on the genetic material has
been extensively analysed in other biological systems, and the
resulting chromosome breaks or errors of segregation can often be
directly visualized in the affected embryo. For other agents, if no
chromosome aberrations can be detected the mode of action is still
obscure. An agent acting before fertilization that exerts an adverse
effect in the female but not the male presumably acts through the
cytoplasm, but even this conclusion depends on proof that the
agent does not affect the maternal environment in such a way as
to damage the developing embryo subsequent to fertilization: such
proof is seldom if ever available.

However, a great deal of embryonic mortality and malformation
is observed that is not associated with any known environmental
agent. Much of this must have its genesis during gestation, and
is thus of no concern here. But some involves chromosome
abnormalities that must have arisen during meiosis, though we do
not know their cause. In man, chromosome abnormalities have
been detected in 0.5 % of all live births, 5 % of still-births, and
40 to 60 % of spontaneous abortions, with a particularly high
frequency during the first trimester. Since some 15 % of all
recognized pregnancies end in spontaneous abortion, this implies
that nearly 10 % of all conceptuses are chromosomally abnormal,
even if no allowance is made for those that are lost before the
pregnancy is recognized. Estimates of the frequency of chromosome abnormality among human zygotes range up to 50 %; the
true frequency is unlikely to be less than 15 %.

What are these chromosomal errors, how many of them arise
during gametogenesis, and what causes them?

The chromosomal abnormalities that have been identified may
be classified into structural rearrangements (e.g. translocations),

Table 1. *Relative loss of chromosomally abnormal human embryos*
before birth

	Incidence per 1000 recognized conceptions in:		
	Spontaneous abortions	Newborn babies	% surviving to birth
Triploidy	15	< 0.1	< 0.01
Tetraploidy	5	< 0.1	< 0.01
Sex chromosome trisomies	< 1	1.5	> 99.9
Sex chromosome monosomies	11	0.1	0.5
Autosomal trisomies	40	1.1	3.0
Autosomal monosomies	< 1	< 0.1	—
Structural rearrangements	3	1.6	35.0
No abnormalities detected	75	845.7	—
Total	150	850	—

Spontaneous abortions have been taken as 15 % of all recognized conceptions. The level of abnormalities detected has been taken as 50 % for spontaneous abortions and 0.5 % for newborns. The relative proportions of the various abnormalities are based on those given by de Grouchy (1976).

errors of ploidy (triploidy, tetraploidy), and errors of chromosome number (trisomy, monosomy), affecting either the sex chromosomes or the autosomes (Carr, 1965, 1971). Their relative frequency is shown in Table 1. Structural rearrangements are the most frequent abnormalities to be identified at birth, and are also seen in abortions, though at a lower relative frequency. Triploids constitute about 20 % of all spontaneous abortions, and tetraploid abortuses are also not uncommon, but it is extremely rare for either to survive to birth. Among sex chromosome anomalies, trisomies (XXY, XYY, XXX) make up more than a third of chromosome anomalies seen at birth, but rarely cause death during embryonic life; in contrast, XO monosomies are common among spontaneous abortions but only about 1 in 200 survives to birth (Turner's syndrome). The most common group of abnormalities among abortions is the autosomal trisomies, mainly involving chromosomes of groups C, D, E and G. By the time of birth only the group G trisomies (i.e. Down's syndrome) are at all frequent, and although they constitute about 25 % of the chromosome anomalies seen at birth, this represents less than 5 % of the total group G trisomies. Non-disjunction should give rise to equal numbers of trisomic and monosomic nuclei, and chromosome loss due to anaphase lagging should give monosomies only. One would therefore expect to see at least as many monosomic as trisomic embryos, yet autosomal monosomies have rarely been identified even early in gestation: presumably they are either spermato-gonial cell-lethals (Searle, 1975) or they cause embryonic death before the pregnancy is recognized. Those trisomies that have rarely been reported may also be lethal in the very early embryo. In mice, embryos with autosomal monosomies die at or before the time of implantation, while trisomies survive until later in gestation (Ford, 1972; Gropp, 1976). Monosomy for chromosomes 17 and 5 seems particularly deleterious, since development ceases at the 8-cell and morula stage respectively; embryos trisomic for either of these chromosomes also die earlier than do most other trisomies (Dyban, 1978; Dyban & Baranov, 1978).

Structural rearrangements, especially balanced translocations, may be passed on from an earlier generation, rather than arising *de novo* during gametogenesis. Mosaics presumably arise by non-disjunction or loss of a chromosome during cleavage or subsequent embryonic development. Triploids may originate at fertilization, if two spermatozoa succeed in penetrating the same egg (dispermy). With these exceptions, all the chromosome anomalies listed in Table 1 are thought to arise during gameto-genesis.

Modern HLA typing and chromosome banding techniques, that allow the identification of individual variant chromosomes, can often establish whether the error responsible for an observed chromosome anomaly occurred during maternal or paternal game-togenesis, and even whether it occurred at first or second meiotic division. For triploids, Jacobs *et al.* (1978*a*) conclude that up to 90 % have two paternal chromosome sets; most of these arise as a result of dispermy, but some are likely to be due to diploid spermatozoa, formed by failure of the first meiotic division. XYY triploids are of course of paternal origin, but are rare, presumably because they possess very low viability. Only 10 to 20 % of triploids have two maternal chromosome sets, and these also seem mainly to be caused by failure of the first meiotic division. Using somewhat different methods, Beatty (1978) has come to similar conclusions, though his results put greater weight on errors at fertilization and less on failure of the first meiotic division. The tetraploids analysed so far have proved to be all XXXX and XXYY, and are thought to arise from failure of an early cleavage division in the zygote rather than from any error of gametogenesis. About 75 % of Turner's syndrome patients have a maternal X, but the origin of the X chromosome in lethal XO chromosome constitutions has not yet been determined. The extra chromosome 21 in Down's syndrome has been claimed to originate mostly from non-disjunction at the first maternal meiotic division (e.g. Robinson, 1973), but data summarized in a recent paper suggest that the additional chromosome is often paternal in origin (Erickson, 1978).

When individuals with chromosome abnormalities survive to reproductive age, one may enquire whether pre-fertilization events correct the abnormality in such a way as to allow the production of chromosomally normal gametes, or whether the abnormality is passed on to subsequent generations. Many such individuals are sterile, owing to a failure of gametogenesis (e.g. XXY males, Down's syndrome women, many unbalanced translocation carriers). In other cases gametogenesis is impaired, but some gametes struggle through, and these are usually, though not always, chromosomally normal. For example, meiotic studies on XYY men have shown that most primary spermatocytes contain only one Y chromosome, but there are some reports of the presence of a YY bivalent, or an X and two univalents (for review, see Chandley, 1975). Similarly, most of the progeny of XYY men are normal XX females and XY males, but some XYY spermatocytes have been detected, and there have been reports of XYY sons born to XYY men. In XO female mice, Kaufman (1972) has shown a significant tendency for non-random segregation at meiosis, in that the single X chromosome was transmitted to the oocyte rather than to the first polar body in about 70 % of cases. Non-random segregation of translocations and abnormal X chromosomes has also been reported, with the abnormal chromosome being preferentially segregated to the polar body (see Eicher, 1970).

A bizarre chromosome anomaly with drastic consequences has recently come to light. Hydatidiform moles of the 'complete' category that tend to undergo malignant transformation have been

shown to be 46 XX in chromosome constitution, with two
identical haploid chromosome sets of paternal origin (Kajii &
Ohama, 1977; Jacobs, Hassold, Matsuyama & Newlands, 1978b).
Such trophoblastic tissue presumably originates from an egg that
has lost its maternal chromosome complement and been fertilized
either by a diploid XX sperm formed by failure of the second
meiotic division, or by a haploid X-bearing sperm with subsequent
diploidization. Yet mice with no maternal chromosome complement and two identical paternal sets appear to be phenotypically
normal (Hoppe & Illmensee, 1977).

Maternal age

From this brief survey of the chromosomal causes of developmental abnormality it is evident that the largest contribution of
pre-fertilization events to subsequent mortality and morbidity is
through the production of trisomies and monosomies. It is also
here that the effects of maternal age are most clearly seen (Table
2). Trisomies with an additional X chromosome, or an additional
chromosome from the acrocentric autosomal D and G groups, are
associated with increased maternal age; this is not so for triploids,
nor for monosomies, and much less so for other trisomies (Alberman et al., 1976; Creasy, Crolla & Alberman, 1976). Thus XXX
and XXY babies are more common in older women (Court Brown,
Law & Smith, 1960); the XO constitution, whether viable or
lethal, shows no such association, nor do XYY trisomies. The
importance of maternal age for Down's syndrome has long been
known (Penrose, 1934), and is equally strong for trisomy 21
embryos dying before birth; there is no suggestion of a similar
effect for the rare 'anti-mongolism' syndrome, monosomy 21.
Babies with Paten's syndrome (trisomy 13) and Edwards' syndrome (trisomy 18) also tend to be born to older women; and aged
female mice have an increased number of trisomies among their
embryos, resulting from non-disjunction at the first meiotic
division (Henderson & Edwards, 1968; Yamamoto, Endo &
Watanabe, 1973).

Why should the incidence of certain trisomies increase with
maternal age? An understanding of the maternal age effect might
throw light on the aetiology of the anomaly. Despite claims to the
contrary, a recent exhaustive analysis by Erickson (1978) reveals

Table 2. *The effect of maternal age on the incidence of various chromosome abnormalities*

Karyotype	Maternal age (yr) (mean ± S.E.)
Normal	27.5 ± 0.4
Triploid	27.4 ± 0.8
Tetraploid	26.8 ± 1.4
Translocation	27.0 ± 2.3
XO	27.6 ± 0.9
Trisomy (all)	31.3 ± 0.6
Group A	29.6 ± 2.2
Group B	33.4 ± 7.1
Group C	30.9 ± 1.7
Group D	32.5 ± 1.3
Group E	29.6 ± 0.9
Group F	30.1 ± 5.3
Group G	33.2 ± 1.4

Data from Boué, Boué & Lazar (1975).

no sign of any independent increase of trisomy 21 with paternal age. This suggests that the crucial non-disjunctional event did not take place during a mitotic division, since the number of mitoses in the ancestry of the germ cells increases with age in men but not in women (see p. 289). On the other hand, for the female germ cell the completion of the first meiotic division is delayed for 20 years in a 40-year-old woman compared to a 20-year-old. Perhaps degenerative changes have taken place in the cytoplasm of the non-dividing oocyte that interfere with the formation of a normal spindle and hence increase the incidence of non-disjunction. The differential effect on certain chromosomes could reflect some facet of their attachment to the spindle, since the X chromosome and groups D, E and G, which are most often involved in trisomy, have associated nucleoli; it is hard, however, to see how such a mechanism would not at the same time increase the incidence of the corresponding monosomy.

The suggestion by German (1968), that the association of certain trisomies with increased maternal age might be due to delayed fertilization in older women resulting from reduced frequency of intercourse, was not borne out by subsequent statistical analysis

(Cannings & Cannings, 1968), nor is it consistent with the absence of any effect of paternal age.

A different type of explanation has been put forward by Henderson & Edwards (1968). They report that, with increasing age of the mother, chiasma frequency in mouse oocytes declines, chiasmata are more often located at the ends of the chromosomes, and univalents are more often seen. The decline in chiasma frequency could arise either by chiasmata becoming terminalized and lost during the protracted first meiotic prophase in the older mothers, or because fewer chiasmata were initially formed in the oocytes of older females. This in turn could only arise if oocytes with fewer chiasmata were selectively conserved until later life, perhaps because they initiated meiosis at a relatively late stage of fetal life (the so-called 'production line' theory). Evidence as to whether genetic recombination frequencies in general decline with maternal age is conflicting (Fowler & Edwards, 1973): in man, some decline and some increase with maternal age, while in the mouse some decline with maternal age but at least one increases with paternal age. However, whatever the cause, if fewer chiasmata and more univalents existed in the oocytes of older women, the relationship between trisomy and maternal age would be explained. Here again, the evidence is not encouraging, since the claimed increase in univalents in older mouse oocytes could not be confirmed on human material.

Reports from several laboratories (summarised by Erickson, 1978) suggest that, in a substantial proportion of trisomy 21s, the additional chromosome comes from the father. The crucial observation will be whether the ratio of paternally to maternally derived trisomies stays constant with increasing maternal age, or declines. If it stays constant, the explanation of the maternal age effect can no longer be sought in oogenesis, but must involve the improved survival of trisomic embryos in the older uterus. More trisomy-21 embryos are found among spontaneous abortions of older women, as well as at birth; the hypothesis would be that in the more efficient younger uterus, these would have been recognized as abnormal and rejected sooner, before the pregnancy was recognized, while some that would previously have been aborted now survive to term. Some circumstantial evidence for this view comes from the small but significant *decrease* in frequency of Down's syndrome with increasing birth order in older women (i.e.

women over 40 are more likely to have an affected child if it is their first than if it is their sixth) and from the finding of a decrease in incidence in mothers up to the age of 20 (Erickson, 1978).

The view that the long period that the oocytes of older women have spent in meiotic prophase may allow the accumulation of environmentally induced defects, leading to subsequent non-disjunction, is supported by the findings of Alberman *et al.* (1972*b*). In a retrospective study, they compared the total estimated ovarian dose of diagnostic plus therapeutic radiation received by women having spontaneous abortions, chromosomally normal or abnormal, with that received by control women with liveborn children. The group with chromosomally abnormal abortions had received significantly more X-irradiation than either of the other two groups and the difference was particularly striking if the exposure to radiation had been 10 or more years before conception (105 mrem, as against 43 and 29 mrem respectively). The excess radiation dose was highest for triploid abortuses, but was also evident among mothers of trisomics. A similar relationship was obtained from studies (Uchida, Ray, McRae & Besant, 1968; Alberman *et al.*, 1972*a*) on the radiation exposure of mothers of Down's syndrome babies. No effect of paternal irradiation was detected, although Boué, Boué & Lazar (1975) had found an increased frequency of chromosome anomalies after occupational exposure of the father to radiation. These results suggest that maternal age and exposure to radiation have some combined effect on chromosome behaviour during meiosis, such that non-disjunction is more frequent in oocytes that have been irradiated in the distant past. The 'doubling dose' of radiation for the risk of Down's syndrome is thought to be of the order of 2 rads; for comparison, the risk more than doubles for every 5 years of maternal age over the age of 30.

In partial confirmation of these findings, Yamamoto, Shimada, Endo & Watanabe (1973) have reported a significant increase in the incidence of non-disjunction after the exposure of aged female mice to as little as 5 rads of X-irradiation.

Of the various possible mechanisms that have been put forward to explain the association of trisomy and maternal age, one of the most plausible involves the nucleolus. The association is most striking for acrocentric chromosomes; these remain attached to the nucleolus throughout interphase. The normal breakdown of the nucleolus towards the end of first meiotic prophase in oogenesis might be delayed by some ageing factor (Ford, 1960); the same process might be impeded by irradiation; the nucleolus might then interfere with chromosome pairing and increase the incidence of non-disjunction.

Numerous reports exist of effects of maternal age on the incidence of other congenital anomalies in man and laboratory mammals (for references, see Parsons, 1963, 1964). In man, the incidence typically increases with age; in the mouse, the incidence of polydactyly and many other minor skeletal abnormalities decreases with age, while cleft lip and palate shows a minimum incidence in mothers 3 to 4 months old. Such effects are more likely to be due to physiological influences exerted during pregnancy than to errors of meiosis; the decreasing incidence in the mouse may, as Parsons pointed out, reflect more stringent competition between fetuses *in utero* as the female ages.

Genetic effects on gamete development
Maternal effects: cytoplasmic or uterine?

When different strains of mice or other mammals are crossed, the characteristics of the progeny may differ according to which strain provided the mother and which the father. Differences that are manifest in progeny of both sexes, so that sex linkage is ruled out, are termed matroclinous or patroclinous, according to which parental strain they resemble, and are deemed to be due to maternal or paternal effects. Maternal effects are more common than paternal; they are usually matroclinous, but can also be patroclinous.

Maternal effects may be exerted before fertilization, through the cytoplasm of the egg; during gestation, through the environment of the reproductive tract; or after birth, for example via lactation. These three categories are often referred to as cytoplasmic, uterine and milk effects, and can in principle be distinguished by a combination of egg or embryo transfer and cross-fostering (see McLaren, 1962, for fuller discussion). For example, a maternal effect on number of lumbar vertebrae in mice has shown by embryo transfer to be exerted through the uterine environment (McLaren & Michie, 1958); maternal effects on birth weight in rabbits and mice are uterine, though some cytoplasmic influence has also been claimed in mice (Brumby, 1960). In practice, however, very few maternal effects have been analysed in this way, and we remain in ignorance of the basis of many effects such as the matroclinous influence on number of whiskers in *Tabby* mice (Kindred, 1961), or the patroclinous effect on manifestation of the *Fused* gene (Reed, 1937), whereby more of the genetically *Fused* mice are phenotypically normal when the mother is *Fused* than when the mother is herself normal.

Such evidence as we have suggests that uterine effects, which will be considered in detail in the paper by Morriss in this volume, are very much more common in mammals than are cytoplasmic effects. Although the mammalian egg has vastly more cytoplasm than the mammalian spermatozoon, so that any cytoplasmic effects would be expected to be maternal rather than paternal, it has much less cytoplasm than the larger amphibian egg. In Amphibia, morphogenetic substances in the egg cytoplasm have been shown to control early development (see, for example, Brothers, 1976; and Malacinski & Spieth, this volume); such cytoplasmic effects are common also among invertebrates (see Fischer, 1977, for a well-analysed example in annelids). In Amphibia, however, the embryonic genome is not expressed until relatively late in development, at about the time of gastrulation, while in mammals, it begins to be expressed as early as the 6- to 8-cell stage.

Genetic control of early development

What is the relative contribution of maternal and embryonic gene products to early development in mammals?

We may first enquire when embryonic genes begin transcription. Studies on the amount and rate of synthesis of polyadenylic acid during mouse pre-implantation development (Levy, Stull & Brinster, 1978) suggest that, while significant quantities of maternal messenger RNA are present in early cleavage-stage embryos, polyadenylation of RNA transcribed from the embryonic genome occurs as early as the 2-cell stage. A small amount of uridine incorporation into newly synthesized RNA has been detected 1 to 3 hours after fertilization (Young, Sweeney & Bedford, 1978), but is thought not to be due to transcription from the embryonic

genome, since incorporation of [³H]uridine does not label the
pronuclei. Perhaps the new RNA results from transcription in a
polar body (Moore, 1975) or in mitochondria, though Pikó (1975)
has shown that no substantial RNA synthesis in mitochondria
occurs until the 8-cell stage. In the mouse, some of the inactive
mRNA synthesized during oogenesis (see p. 290) is methylated
immediately after fertilization (Young, 1977), and may control
development to the 2-cell stage. In the rabbit, poly(A)-containing
mRNA in the oocyte is used as a template for protein synthesis
from fertilization up to the 8-cell stage (Schultz, 1975; Schultz &
Tucker, 1977); fertilized eggs cultured in the presence of α-
amanitin cleave to the 8-cell stage, but no further (Manes, 1973,
1975). In the mouse, experiments with α-amanitin and other
inhibitors of RNA synthesis have yielded ambiguous results
(Golbus, Calarco & Epstein, 1973; Epstein, 1975), but there seems
no doubt that transcription from the embryonic genome occurs
at the 2- to 4-cell stage, with synthesis of ribosomal, 4S, 5S and
heterogeneous RNA (Graham, 1973). Studies using virus-induced
cell fusion (Bernstein & Mukherjee, 1973) suggest that this
nuclear RNA synthesis is under the control of cytoplasmic factors,
presumably maternal.

The stage at which embryonic genes are not only transcribed
but also expressed is likely to vary from one locus to another. For
proteins in general, the most extensive changes in synthetic
pattern take place during cleavage, between the 2- and 8-cell stages
in the mouse (Epstein & Smith, 1974; Van Blerkom & Brockway,
1975) and between the 8-cell and early blastocyst in the rabbit
(Van Blerkom & McGaughey, 1978b). It seems likely that these
changes represent the shift from maternal to embryonic gene
products, but this has not so far been unambiguously established.
In both mouse and rabbit some of the proteins synthesized in the
oocyte continued to be synthesized after fertilization, at least to the
2-cell stage (Levinson, Goodfellow, Vadeboncoeur & McDevitt,
1978; Van Blerkom & McGaughey, 1978b). For enzymes with
known genetic variants, paternal gene expression can be taken as
direct evidence that the embryonic genome is active. Paternal
variants have been detected in 8-cell rat embryos for 6-phos-
phogluconate dehydrogenase (Khlebodarova, Serov & Korochkin,
1975), 8-cell mouse embryos for glucose phosphate isomerase
(Brinster, 1973) and as early as the 4-cell stage for β-gluc-

uronidase (Wudl & Chapman, 1976). Cell surface antigens of paternal origin have been demonstrated by immunofluorescence at the 6- to 8-cell stage (Muggleton-Harris & Johnson, 1976). Some recessive lethal genes are expressed during cleavage (for review, see McLaren, 1976*b*).

Thus evidence from the mouse suggests that maternal gene products are entirely responsible for development up to the 2-cell stage, and continue to be involved at least up to the 8-cell stage. Some maternal gene products may play a part in development for a considerably longer period. The mechanism of action of the homozygous lethal *Oligosyndactyly*, in which the cells of the embryonic ectoderm appear to be arrested in mitosis in the course of the eighth cell division after fertilization, is consistent with a maternal gene product, derived from the egg cytoplasm, that maintains cell division throughout cleavage and is only used up or diluted out after implantation (McLaren, 1976*b*; Paterson, 1978). The homozygous mutant embryos are presumably unable to synthesize the required product at the time that the embryonic genome would normally take over, hence the mitotic arrest.

X chromosome activity

The X chromosome provides a useful system for assessing the relative role of maternal and embryonic genomes during early development (see Monk, 1978). During oogenesis, both X chromosomes are active (Epstein, 1969; Gartler, Liskay & Gant, 1973); oocytes from XO female mice therefore show only half the level of activity of X-linked enzymes such as HPRT (hypoxanthine phosphoribosyl transferase) that oocytes from normal XX females show (Epstein, 1972). This relationship is maintained up to early 8-cell stage (Monk & Harper, 1978), suggesting that HPRT activity during the first three cleavage divisions is maternal in origin, due either to pre-formed enzyme stored in the egg cyto-plasm or, more likely since enzyme activity increases nearly tenfold during this period, to a store of maternal mRNA. For the next couple of cell divisions, embryos from XX females show a bimodal distribution of HPRT activity, suggesting that both embryonic X chromosomes are active in XX embryos, which thus have twice the level of enzyme activity of their XY litter-mates; but by the blastocyst stage the distribution is once more unimodal, indicating that X inactivation has occurred in the XX embryos, at least in the trophectoderm (Monk & Harper, 1978).

The reduced level of X-coded maternal gene products during the first few cleavage divisions in embryos from XO females may account for their poor viability and retarded development in culture (Burgoyne & Biggers, 1976). Whether a similar explanation could account for the very high mortality among human XO embryos (see p. 000) is not clear: we do not know for how long the human embryo depends on maternal gene products, nor exactly when X-inactivation takes place, nor whether the X chromosome in XO spontaneous abortions is maternal or paternal in origin. Mouse embryos without even one X chromosome (i.e. OY in chromosome constitution) do not survive beyond the 8-cell stage (Burgoyne & Biggers, 1976; Tarkowski & Rossant, 1976).

The X chromosome in mammals appears to be influenced or imprinted in some way during gametogenesis or fertilization, since its behaviour during early development depends on whether it was introduced by the egg or by the sperm (for reviews, see Lyon, 1977; Monk, 1978). Indeed, in marsupials the paternal X chromosome is usually inactive in somatic cells throughout life. In eutherian mammals preferential inactivation of the paternal X chromosome has been found in those tissues that differentiate early (trophoblast, extra-embryonic endoderm), but in the embryonic tissues that differentiate later, inactivation appears to be random. Evidence for preferential inactivation of the paternal X chromosome comes from chromosome labelling studies in mice and rats (Takagi & Sasaki, 1975; Wake, Talagi & Sasaki, 1976), and from studies on the X-linked enzyme phosphoglycerate kinase in mice (West, Frels, Chapman & Papaioannou, 1977).

Cytoplasmic effects

If the cytoplasm of the egg plays some role in controlling early development, genes should exist that exert maternal effects through modifications of the egg cytoplasm.

The early differentiation of mammalian embryos into inner cell mass and trophectoderm is thought to be due to differences in microenvironment between the interior and periphery of the embryo (see Graham, 1971) rather than to any heterogeneity already existing in the cytoplasm or membrane (see p. 291) of the unfertilized egg. The pattern of early cleavage is presumably determined by the cytoplasm, but no genes affecting it are known. An apparent maternal effect on the timing of cleavage (McLaren & Bowman, 1973) turned out to be due to a strain difference in the timing of fertilization, exerted through the environment of the maternal reproductive tract (Nicol & McLaren, 1974). Similarly, the delay in fertilization of eggs in aged female hamsters appears to be due to an effect of senescence on the maternal environment, rather than to any deterioration of the egg itself with age (Parkening & Chang, 1976).

The possible role of genetic variation in maternally inherited mitochondria is raised by the results of Verrusio, Pollard & Frazer (1968). Two inbred strains of mice, if maintained on a particular diet, show a difference in the frequency of cleft palate after treatment with 6-amino-nicotinamide. The difference persists in reciprocal crosses, and in the backcross progeny of F_1 females of the two reciprocal types. Although formally the results could be explained in terms of a uterine effect stretching over more than one generation, the authors prefer the hypothesis that the diet-dependent difference in response to the teratogen is due to a cytoplasmic factor. They point out that, since 6-amino-nicotinamide forms an inactive nicotinamide adenine dinucleotide analogue that interferes with oxidative phosphorylation in mito-chondria, the cytoplasmic factor may well be associated with a genetically determined difference in mitochondria between the two strains.

Only two well-substantiated cases of single genes apparently acting through the cytoplasm have been reported in mammals, both in mice. One concerns a dominant allele at the T/t locus, *hairpin-tail* (T^{hp}), known to consist of a small deficiency (Bennett, 1975). $T^{hp}/+$ heterozygotes die in late gestation if the mutant gene is transmitted from the mother, but are fully viable in both sexes if it comes from the father (Johnson, 1974, 1975). The defect is not exerted through the female reproductive tract, since $T^{hp}/+$ females mated to heterozygous males successfully carry to term those heterozygous embryos in which the T^{hp} gene is paternally derived. The lethal action of the gene derived from the mother is associated with minor malformations such as spina bifida and polydactyly; although the lethality can be suppressed by aggre-gation of $T^{hp}/+$ and normal embryos (Bennett, 1978), the cytoplasmic effect persists, in that the chimaeric females are unable to transmit T^{hp} to viable offspring.

To explain the maternal effect, Johnson (1974) postulated the existence at the T^{hp} locus of an 'activation centre' that switches on some structural locus required for normal development in late gestation. The 'activation centre' would act only in the female, presumably by coding for some cytoplasmic protein or RNA. When the mother transmits T^{hp}, the activating influence is missing, and the corresponding structural locus fails to be expressed. Lyon & Glenister (1977) have suggested that the low frequency in translocation intercrosses of certain classes of progeny deriving both alleles from the father might be explained in a similar way, since the missing progeny would have had a maternal deficiency of the region of chromosome 17 that includes the *t* complex. Other examples from mouse translocation studies of failures of complementation that might involve a deficiency of certain chromosome regions in the egg and their duplication in the spermatozoon (or vice versa) are given in Searle & Beechey (1978). The extreme example of such a phenomenon would be the finding mentioned on p. 299, that in hydatidiform moles the entire maternal chromosome complement is deleted, and the paternal complement duplicated.

An alternative explanation for the *hairpin-tail* maternal effect would be that the region of chromosome involving the deletion is normally expressed in single dose only, in an analogous fashion to X chromosome inactivation, and that it is always the paternally derived chromosome region that is inactivated (as in marsupial X chromosomes, see p. 308). When the T^{hp} deletion is transmitted from the mother, the embryo would then be essentially nullisomic for the chromosome region in question, and nullisomy for even quite small portions of the genome is known to be lethal in the mouse (see, for example, Kaufman & Sachs, 1975; Lewis, 1978). If this explanation were correct, *hairpin-tail* would no longer qualify as a cytoplasmic effect.

The other mammalian instance of a genetic effect apparently exerted through the cytoplasm of the egg involves the Japanese mouse strain DDK. Female DDK mice mated with males of their own strain have litters of reasonable size, with little embryonic mortality, but if DDK females are mated to other strains litter size drops precipitously and most of the embryos die at about the time of implantation (Wakasugi, Tomita & Kondo, 1967). The reciprocal crosses are fully fertile. Ovary grafts from DDK to F_1 females (Wakasugi, 1973) established that the effect was cytoplasmic rather than uterine, i.e. it was not the reproductive tract of DDK mothers that was reacting in a hostile manner to the genetically foreign material, but the cytoplasm of the egg.

Genetic analysis (Wakasugi, 1974) suggested that a single locus or closely linked pair of loci was involved, determining a cytoplasmic factor in the egg as well as a factor in spermatozoa. Wakasugi gave the symbols *om* (*ovum mutant*) and *Om* to the DDK gene and its wild-type equivalent, and *s*, *S* to the corresponding pair of alleles acting in spermatozoa, *om* and *s* being either identical or closely linked. Homozygous *om/om* and *OM/OM* females are postulated to produce substances *o* and *O* respectively during oogenesis; these are stored in the egg cytoplasm, and interact specifically with the *s* or *S* gene or its products to synthesize some substance necessary for the post-implantation development of the embryo. Heterozygous females (*om/OM*) are presumed to produce *o* and *O* in equal amounts, with one or other interacting irreversibly with the sperm product in any given fertilized egg. Of the four combinations *O/s*, *O/S*, *o/s* and *o/S*, the last is usually (but not always) lethal.

Paternal effects

Paternal effects occur in mammals, but are much less common than maternal effects. This may be due to the much greater volume of cytoplasm in the egg, containing maternal messenger RNA and other informational macromolecules; it may also be due to the very different character of gametogenesis in the two sexes, whereby the male has a self-renewing germ cell population, with the actual process of spermatogenesis not taking place until the few weeks prior to sperm release, while in the female a finite population of germ cells enters meiosis before birth, and may remain thus for many years before the eggs are finally matured and ovulated.

In man, a paternal age effect has been reported on the incidence of acrocephalosyndactyly (Penrose, 1960). In the mouse, certain recessive *t* haplotypes show transmission ratio distortion when the father, but not the mother, is heterozygous (*T/t* or *+/t*). The effect is hard to explain other than by post-meiotic (i.e. haploid)

gene action during spermatogenesis (Sherman & Wudl, 1977). An
effect of the *t* haplotype of the fertilizing sperm on the duration
of the pronuclear stage is suggested by the findings of Olds (1971).
An effect of the paternal genome on the cleavage rate of mouse
embryos has been reported (Whitten & Dagg, 1961), but this could
reflect early expression of the embryonic genome (see p. 306)
rather than any non-chromosomal influence exerted through the
male gamete.

It is hard to escape the conclusion that the spermatozoon contributes something other than its genome to the egg, something
that is essential for continued normal development. Parthenogenetically activated embryos, whether haploid or diploid, have
a high failure rate and never develop much beyond the early somite
stage (Graham, 1974; Kaufman, Barton & Surani, 1977), although
the cells of which they are composed can contribute to normal
development in a chimaeric situation (Surani, Barton & Kaufman,
1977; Stevens, Varnum & Eicher, 1977). Yet fertilized mouse eggs
from which the male (or female) pronucleus has been removed
will, provided the diploid state is restored by cytochalasin treatment, develop normally (Hoppe & Illmensee, 1977), giving rise
to homozygous diploid mice with a chromosome constitution
identical to that of the unsuccessful diploid parthenogenone.
Again, the DDK situation (see p. 000) involves a paternal as well
as a maternal effect, in that there seems to be some element in
the 'alien' spermatozoon, lacking in the DDK spermatozoon, to
which the DDK cytoplasm reacts adversely.

The cause of developmental failure in parthenogenones is
unknown. Cortical granule release may be abnormal after some
activating treatments, but not all (Whittingham & Anderson,
personal communication); ribosomal RNA genes are expressed
at the same time as in normal embryos, though there is some
suggestion that their expression is more variable (Hansmann,
Gebauer & Grimm, 1978). Sperm mitochondria do not appear to
contribute to embryonic development (Szollosi, 1976).

Some extrachromosomal infective sperm constituent would also
need to be invoked to explain telegony. The more dramatic
examples that attracted Darwin's attention, such as the crossing
of a mare with a male zebra that supposedly induced stripes in
the pure horse progeny later born to the same mare, have not been
substantiated. On the other hand males from a high-cancer strain

of mice mated to females from a low-cancer strain give rise to progeny with a high incidence of tumours; the incidence increases in later litters, whether or not the subsequent matings are with a high-cancer strain male, and whether or not the mother herself develops a tumour (Mühlbock, 1952). The lag effect is presumably due to the time required for propagation of the infective agent in the female, which she then transmits to her progeny.

References

ALBERMAN, E., CREASY, M., ELLIOTT, M. & SPICER, C. (1976). Maternal factors associated with fetal chromosome abnormalities in spontaneous abortions. *British Journal of Obstetrics and Gynaecology*, **83**, 621–7.

ALBERMAN, E., POLANI, P. E., FRASER ROBERTS, J. A., SPICER, C. C., ELLIOTT, M. & ARMSTRONG, E. (1972*a*). Parental exposure to X-irradiation and Down's syndrome. *Annals of Human Genetics*, **36**, 195–208.

ALBERMAN, E., POLANI, P. E., FRASER ROBERTS, J. A., SPICER, C. C., ELLIOTT, M., ARMSTRONG, E. & DHADIAL, R. K. (1972*b*). Parental X-irradiation and chromosome constitution in their spontaneously aborted foetuses. *Annals of Human Genetics*, **36**, 185–94.

ANDERSON, E. & ALBERTINI, D. F. (1976). Gap junctions between the oocyte and companion follicle cells in the mammalian ovary. *Journal of Cell Biology*, **71**, 680–6.

AUERBACH, C. (1977). *Mutation Research: Problems, Results and Perspectives.* London: Chapman & Hall.

BAKER, T. G. (1971). Comparative aspects of the effects of radiation during oogenesis. *Mutation Research*, **11**, 9–12.

BEATTY, R. A. (1970). The genetics of the mammalian gamete. *Biological Reviews*, **45**, 73–120.

BEATTY, R. A. (1978). The origin of human triploidy: an integration of qualitative and quantitative. *Annals of Human Genetics*, **41**, 299–314.

BELLVÉ, A. R. (1972). Viability and survival of mouse embryos following parental exposure to high temperature. *Journal of Reproduction and Fertility*, **30**, 71–81.

BENNETT, D. (1975). The *T*-locus of the mouse. *Cell*, **6**, 441–54.

BENNETT, D. (1978). Rescue of a lethal *T/t* locus genotype by chimaerism with normal embryos. *Nature, London*, **272**, 539.

BENNETT, D. & BOYSE, T. (1973). Sex ratio in progeny of mice inseminated with sperm treated with H-Y antiserum. *Nature, London*, **246**, 308–9.

BERNSTEIN, R. M. & MUKHERJEE, B. B. (1973). Cytoplasmic control of nuclear activity in preimplantation mouse embryos. *Developmental Biology*, **34**, 47–65.

BOUÉ, J., BOUÉ, A. & LAZAR, P. (1975). Retrospective and prospective epidemiological studies of 1500 karyotyped spontaneous human abortions. *Teratology*, **12**, 11–26.

BRINSTER, R. L. (1973). Paternal glucose phosphate isomerase activity in three-day mouse embryos. *Biochemical Genetics*, **9**, 187–91.

314 *A. McLaren*

BROTHERS, A. J. (1976). Stable nuclear activation dependent on a protein synthesized during oogenesis. *Nature, London*, **260**, 112–15.

BRUMBY, P. J. (1960). The influence of the maternal environment on growth in mice. *Heredity*, **14**, 1–18.

BURFENING, P. J., ELLIOTT, D. S., EISEN, E. J. & ULBERG, L. C. (1970). Survival of embryos resulting from spermatozoa produced by mice exposed to elevated ambient temperature. *Journal of Animal Science, New York*, **30**, 578–82.

BURGOYNE, P. S. (1975). Sperm phenotype and its relationship to somatic and germ line genotype: a study using mouse aggregation chimeras. *Developmental Biology*, **44**, 63–76.

BURGOYNE, P. S. & BIGGERS, J. D. (1976). The consequences of X-dosage deficiency in the germ line: impaired development *in vitro* of preimplantation embryos from XO mice. *Developmental Biology*, **51**, 109–17.

CANNINGS, C. & CANNINGS, M. R. (1968). Mongolism, delayed fertilization and human sexual behaviour. *Nature, London*, **218**, 481.

CARR, D. H. (1965). Chromosome studies in spontaneous abortions. *Obstetrics and Gynecology*, **26**, 308–26.

CARR, D. H. (1971). Chromosomes and abortion. *Advances in Human Genetics*, **2**, 201–57.

CHANDLEY, A. C. (1975). Human meiotic studies. In *Modern Trends in Human Genetics*, ed. A. E. H. Emery, vol. 2, pp. 31–82. London: Butterworth.

COURT BROWN, W. M., LAW, P. & SMITH, P. G. (1960). Sex chromosome aneuploidy and parental age. *Annals of Human Genetics*, **33**, 1–14.

CREASY, M. R., CROLLA, J. A. & ALBERMAN, E. D. (1976). A cytogenetic study of human spontaneous abortions using banding techniques. *Human Genetics*, **31**, 177–96.

DE GROUCHY, J. (1976). Human chromosomes and their anomalies. In *Aspects of Genetics in Paediatrics*, ed. D. Barltrop, pp. 5–13. London: Fellowship of Postgraduate Medicine.

DYBAN, A. P. (1979). Cytogenetics of early stages of mammalian development. *Proceedings of the 14th International Congress of Genetics*, pp. 66–7 (abstr.).

DYBAN, A. P. & BARANOV, V. S. (1978). *The Cytogenetics of Mammalian Development*, pp. 1–254. Moscow: Nauka. (In Russian.)

DYM, M. & FAWCETT, D. W. (1971). Further observations on the numbers of spermatogonia, spermatocytes and spermatids joined by intercellular bridges in mammalian spermatogenesis. *Biology of Reproduction*, **4**, 195–215.

EICHER, E. M. (1970). X autosome translocations in the mouse: total inactivation versus partial inactivation of the X chromosome. *Advances in Genetics*, **15**, 175–259.

EPSTEIN, C. J. (1969). Mammalian oocytes: X chromosome activity. *Science*, **163**, 1078–9.

EPSTEIN, C. J. (1972). Expression of the mammalian X chromosome before and after fertilization. *Science*, **175**, 1467–8.

EPSTEIN, C. J. (1975). Gene expression and macromolecular synthesis during preimplantation embryonic development. *Biology of Reproduction*, **12**, 82–105.

EPSTEIN, C. J. & SMITH, S. A. (1974). Electrophoretic analysis of proteins synthesized by preimplantation mouse embryos. *Developmental Biology*, **40**, 233–44.

ERICKSON, J. D. (1978). Down syndrome, paternal age, maternal age and birth order. *Annals of Human Genetics*, **41**, 289–98.

FALCONER, D. S. & AVERY, P. J. (1978). Variability of chimaeras and mosaics. *Journal of Embryology and Experimental Morphology*, **43**, 195–219.

FISCHER, A. (1977). Autonomy for a specific gene product in oocytes: experimental evidence in the polychaetous annelid, *Platynereis dumerilii*. *Developmental Biology*, **55**, 46–58.

FORD, C. E. (1960). Chromosomal abnormality and congenital malformation. In *Ciba Symposium on Congenital Malformations*, ed. G. E. W. Wolstenholme & C. M. O'Connor, pp. 32–47. London & Edinburgh: Churchill Livingstone.

FORD, C. E. (1972). Gross genome imbalance in mouse spermatozoa: does it influence the capacity to fertilize? In *The Genetics of the Spermatozoon*, ed. R. A. Beatty & S. Gluecksohn-Waelsch, pp. 359–69. Edinburgh & New York: University of Edinburgh & Albert Einstein College of Medicine.

FOWLER, R. E. & EDWARDS, R. G. (1973). The genetics of early human development. *Progress in Medical Genetics*, **9**, 49–112.

GARTLER, S. M., LISKAY, R. M. & GANT, N. (1973). Two functional X-chromosomes in human fetal oocytes. *Experimental Cell Research*, **82**, 464–6.

GEREMIA, R., D'AGOSTINO, A. & MONESI, V. (1978). Biochemical evidence of haploid gene activity in spermatogenesis of the mouse. *Experimental Cell Research*, **111**, 23–30.

GERMAN, J. (1968). Mongolism, delayed fertilization and human sexual behaviour. *Nature, London*, **217**, 516–18.

GLASS, L. (1970). Translocation of macromolecules. In *Cell Differentiation*, ed. O. A. Schjeide & J. de Vellis, pp. 201–20. New York: Van Nostrand Reinhold.

GOLBUS, M. S., CALARCO, P. G. & EPSTEIN, C. J. (1973). The effects of inhibitors of RNA synthesis (α-amanitin and actinomycin D) on preimplantation mouse embryogenesis. *Journal of Experimental Zoology*, **186**, 207–16.

GRAHAM, C. F. (1971). The design of the mouse blastocyst. In *Control Mechanisms of Growth and Differentiation*, ed. D. Davies & M. Balls, *Symposium of the Society for Experimental Biology 25*, pp. 371–8. Cambridge University Press.

GRAHAM, C. F. (1973). Nucleic acid metabolism during early mammalian development. In *The Regulation of Mammalian Reproduction*, ed. S. J. Segal, R. Crozier, P. A. Corfman & P. G. Condliffe, pp. 286–98. Springfield, Illinois: Thomas.

GRAHAM, C. F. (1974). The production of parthenogenetic mammalian embryos and their use in biological research. *Biological Reviews*, **49**, 339–422.

GROPP, A. (1976). Morphological consequences of trisomy in mammals. In *Embryogenesis in Mammals*, ed. M. O'Connor, *Ciba Foundation Symposium 40*, pp. 155–70. London: Churchill Livingstone.

HANSMANN, I., GEBAUER, J. & GRIMM, T. (1978). Impaired gene activity for 18S and 28S rRNA in early embryonic development of mouse parthenogenones. *Nature, London*, **272**, 377–8.

HENDERSON, S. A. & EDWARDS, R. G. (1968). Chiasma frequency and maternal age in mammals. *Nature, London*, **218**, 22–8.

HOPPE, P. C. & ILLMENSEE, K. (1977). Microsurgically produced homozygous-diploid uniparental mice. *Proceedings of the National Academy of Sciences of the USA*, **74**, 5657–61.

JACOBS, P. A., ANGELL, R. R., BUCHANAN, I. M., HASSOLD, T. J., MATSUYAMA, A. M. & MANUEL, B. (1978a). The origin of human triploids. *Annals of Human Genetics*, **42**, 49–57.

JACOBS, P. A., HASSOLD, T. J., MATSUYAMA, A. M. & NEWLANDS, I. M. (1978b). The chromosome constitution of gestational trophoblastic disease. *Lancet*, **ii**, 49.

JAGIELLO, G. M., LIN, J. S. & DUCAYEN, M. B. (1975). SO_2 and its metabolite: effects on mammalian egg chromosomes. *Environmentral Research*, **9**, 84–93.

JOHNSON, D. R. (1974). Hairpin tail: a case of post-reductional gene action in the mouse egg? *Genetics*, **76**, 795–805.

JOHNSON, D. R. (1975). Further observations on the hairpin-tail (T^{hp}) mutation in the mouse. *Genetical Research*, **24**, 207–13.

KAJII, T. & OHAMA, K. (1977). Androgenetic origin of hydatidiform mole. *Nature, London*, **268**, 633–4.

KANAZAWA, K. & IMAI, A. (1974). Parasexual–sexual hybridization – heritable transformation of germ cells in chimeric mice. *Japanese Journal of Experimental Medicine*, **44**, 227–34.

KAUFMAN, M. H. (1972). Non-random segregation during mammalian oogenesis. *Nature, London*, **238**, 465–6.

KAUFMAN, M. H. (1977). Effect of anaesthetic agents on eggs and embryos. In *Development of Mammals*, vol. 1, ed. M. H. Johnson, pp. 137–63. Amsterdam: Elsevier.

KAUFMAN, M. H., BARTON, S. C. & SURANI, M. A. H. (1977). Normal post-implantation development of mouse parthenogenetic embryos to the forelimb bud stage. *Nature, London*, **265**, 53–5.

KAUFMAN, M. H. & SACHS, L. (1975). The early development of haploid and aneuploid parthenogenetic embryos. *Journal of Embryology and Experimental Morphology*, **34**, 645–55.

KHLEBODAROVA, R. M., SEROV, O. L. & KOROCHKIN, L. I. (1976). Expression of alleles at the *Pgd* locus in early rat development. *Dokladÿ Biologicheski Nauk*, **224**, 428–9.

KINDRED, B. M. (1961). A maternal effect on vibrissae score due to the *Tabby* gene. *Australian Journal of Biological Sciences*, **14**, 627–36.

KNILL-JONES, R. P., NEWMAN, B. J. & SPENCE, A. A. (1975). Anaesthetic practice and pregnancy. *Lancet*, **ii**, 807–9.

LEVEY, I. L., STULL, G. B. & BRINSTER, R. L. (1978). Poly(A) and synthesis of polyadenylated RNA in the preimplantation mouse embryo. *Developmental Biology*, **64**, 140–8.

LEVINSON, J., GOODFELLOW, P., VADEBONCOEUR, M. & McDEVITT, H. (1978). Identification of stage specific polypeptides synthesized during murine pre-implantation development. *Proceedings of the National Academy of Sciences of the USA*, **75**, 3332–6.

LEWIS, S. E. (1978). Developmental analysis of lethal effects of homozygosity for the c^{25H} deletion in the mouse. *Developmental Biology*, **65**, 553–7.

LYON, M. F. (1977). Chairman's address. In *Reproduction and Evolution, 4th International Symposium on the Comparative Biology of Reproduction*, pp. 95–8. Canberra: Australian Academy of Sciences.

LYON, M. F. & GLENISTER, P. H. (1977). Factors affecting the observed number of young resulting from adjacent-2 disjunction in mice carrying a translocation. *Genetical Research*, **29**, 83–92.

LYON, M. F., GLENISTER, P. H. & LAMOREUX, M. L. (1975). Normal spermatozoa from androgen-resistant germ cells of chimaeric mice and the role of androgen in spermatogenesis. *Nature, London*, **258**, 620–2.

McGAUGHEY, R. W. (1977). The culture of pig oocytes in minimal medium, and the influence of progesterone and estradiol-17β on meiotic maturation. *Endocrinology*, **100**, 39–45.

McLAREN, A. (1962). Maternal effects in mammals and their experimental analysis. In *Proceedings of the 1st International Conference on Congenital Malformations, London, 1960*, pp. 211–22. Philadelphia: Lippincott.

McLAREN, A. (1975). The independence of germ-cell genotype from somatic influence in chimaeric mice. *Genetical Research*, **25**, 83–7.

McLAREN, A. (1976a). *Mammalian chimaeras*. Cambridge University Press.

McLAREN, A. (1976b). Genetics of the early mouse embryo. *Annual Review of Genetics*, **10**, 361–81.

McLAREN, A. & BOWMAN, P. (1973). Genetic effects on the timing of early development in the mouse. *Journal of Embryology and Experimental Morphology*, **30**, 491–8.

McLAREN, A. & MICHIE, D. (1958). Factors affecting vertebral variation in mice. IV. Experimental proof of the uterine basis of a maternal effect. *Journal of Embryology and Experimental Morphology*, **6**, 645–59.

MANES, C. (1973). The participation of the embryonic genome during early cleavage in the rabbit. *Developmental Biology*, **32**, 453–9.

MANES, C. (1975). Genetic and biochemical activation in preimplantation embryos. In *The Developmental Biology of Reproduction*, ed. C. L. Markert & J. Papaconstantinou, pp. 133–63. New York & London: Academic Press.

MONK, M. (1978). Biochemical studies on mammalian X chromosome activity. In *Development in Mammals*, ed. M. H. Johnson, vol. 3, pp. 189–225. Amsterdam: Elsevier.

MONK, M. & HARPER, M. (1978). X-chromosome activity in preimplantation mouse embryos from XX and XO mothers. *Journal of Embryology and Experimental Morphology*, **46**, 53–64.

MOOR, R. M. (1978). Role of steroids in the maturation of ovine oocytes. *Annales de Biologie Animale, Biochimie, Biophysique*, **18**, 477–82.

MOORE, G. P. M. (1975). The RNA polymerase activity of the preimplantation mouse embryo. *Journal of Embryology and Experimental Morphology*, **34**, 291–8.

MUGGLETON-HARRIS, A. L. & JOHNSON, M. H. (1976). The nature and distribution of serologically detectable alloantigens on the preimplantation mouse embryo. *Journal of Embryology and Experimental Morphology*, **35**, 59–72.

MÜHLBOCK, O. (1952). Studies on transmission of mouse mammary tumor agent by male parent. *Journal of the National Cancer Institute*, **12**, 819–37.

NAGANO, T. (1968). Fine structural relation between the Sertoli cell and the differentiating spermatid in the human testis. *Zeitschrift für Zellforschung und Mikroskopische Anatomie*, **89**, 39–43.

NICOL, A. & McLAREN, A. (1974). An effect of the female genotype on sperm transport in mice. *Journal of Reproduction and Fertility*, **39**, 421–4.

NICOSIA, S. V., WOLF, D. P. & INOUE, M. (1977). Cortical granule distribution and cell surface characteristics in mouse eggs. *Developmental Biology*, **57**, 56–74.

OHNO, S. & GROPP, A. (1965). Embryological basis for germ cell chimerism in mammals. *Cytogenetics*, **4**, 251–61.

OLDS, P. J. (1971). Effect of the *T* locus on fertilization in the house mouse. *Journal of Experimental Zoology*, **177**, 417–34.

OZDZENSKI, W. (1967). Observations on the origin of primordial germ cells in the mouse. *Zoologica Poloniae*, **17**, 367–79.

PARKENING, T. A. & CHANG, M. C. (1976). In-vitro fertilization of ova from senescent mice and hamsters. *Journal of Reproduction and Fertility*, **48**, 381–3.

PARSONS, P. A. (1963). Congenital abnormalities and competition in man and other mammals at different maternal ages. *Nature, London*, **198**, 316–17.

PARSONS, P. A. (1964). Parental age and the offspring. *Quarterly Review of Biology*, **39**, 258–75.

PATERSON, H. (1978). Effects of lethal factors on the early development of mouse embryos. PhD thesis, University of Edinburgh.

PENROSE, L. S. (1934). The relative aetiological importance of birth order and maternal age in Mongolism. *Proceedings of the Royal Society of London, Series B*, **115**, 431–50.

PENROSE, L. S. (1960). Genetical causes of malformation and the search for their origins. In *Ciba Foundation Symposium on Congenital Malformations*, pp. 22–31. Boston: Little & Brown.

PIKÓ, L. (1975). Expression of mitochondrial and nuclear genes during early development. In *The Early Development of Mammals*, ed. M. Balls & A. E. Wild, *British Society for Developmental Biology Symposium 2*, pp. 167–87. Cambridge University Press.

REED, S. C. (1937). The inheritance and expression of *Fused*, a new mutation in the house mouse. *Genetics*, **22**, 1–13.

ROBINSON, J. (1973). Origin of extra chromosome in trisomy 21. *Lancet*, **i**, 131–3.

RODMAN, T. C. & BACHVAROVA, R. (1976). RNA synthesis in preovulatory mouse oocytes. *Journal of Cell Biology*, **70**, 251–7.

RUSSELL, W. L. (1965). Studies in mammalian radiation genetics. *Nucleonics*, **23**, 53–4, 62.

SCHULTZ, G. A. (1975). Polyadenylic acid-containing RNA in unfertilized and fertilized eggs of the rabbit. *Developmental Biology*, **44**, 270–7.

SCHULTZ, G. A. & TUCKER, E. B. (1977). Protein synthesis and gene expression in preimplantation rabbit embryos. In *Development in Mammals*, ed. M. H. Johnson, vol. 1, pp. 69–97. Amsterdam: Elsevier.

SEARLE, A. G. (1975). Radiation-induced chromosome damage and the assessment of genetic risk. In *Modern Trends in Human Genetics*, ed. A. E. H. Emery, vol. 2, pp. 83–110. London: Butterworth.

SEARLE, A. G. & BEECHEY, C. V. (1978). Complementation studies with mouse translocations. *Cytogenetics and Cell Genetics*, **20**, 282–303.

SHERMAN, M. I. & WUDL, L. R. (1977). *T*-complex mutations and their effects. In *Concepts in Mammalian Embryogenesis*, ed. M. I. Sherman, pp. 136–234. Cambridge: MIT Press.

STEVENS, L. C., VARNUM, D. S. & EICHER, E. M. (1977). Viable chimaeras produced from normal and parthenogenetic mouse embryos. *Nature, London*, **269**, 515–17.

SURANI, M. A. H., BARTON, S. C. & KAUFMAN, M. H. (1977). Development to term of chimaeras between diploid parthenogenetic and fertilized embryos. *Nature, London*, **270**, 601–3.

SZOLLOSI, D. (1976). Oocyte maturation and paternal contribution to the embryo in mammals. *Current Topics in Pathology*, **62**, 9–27.

SZOLLOSI, D., GERARD, M., MENEZO, Y. & THIBAULT, C. (1978). Permeability of ovarian follicle; corona cell–oocyte relationship in mammals. *Annales de Biologie Animale, Biochimie, Biophysique*, **18**, in press.

TAKAGI, N. & SASAKI, M. (1975). Preferential inactivation of the paternally derived X chromosome in the extra-embryonic membranes of the mouse. *Nature, London*, **256**, 640–2.

TARKOWSKI, A. K. & ROSSANT, J. (1976). Haploid mouse blastocysts developed from bisected zygotes. *Nature, London*, **259**, 663–5.

UCHIDA, I. A., RAY, M., MCRAE, K. N. & BESANT, D. F. (1968). Familial occurrence of trisomy 22. *American Journal of Human Genetics*, **20**, 107–18.

VAN BLERKOM, J. & BROCKWAY, G. O. (1975). Qualitative patterns of protein synthesis in the preimplantation mouse embryo. I. Normal pregnancy. *Developmental Biology*, **44**, 148–57.

VAN BLERKOM, J. & MCGAUGHEY, R. W. (1978*a*). Molecular differentiation of the rabbit ovum. I. During oocyte maturation *in vivo* and *in vitro*. *Developmental Biology*, **63**, 139–50.

VAN BLERKOM, J. & MCGAUGHEY, R. W. (1978*b*). Molecular differentiation of the rabbit ovum. II. During the preimplantation development of *in vivo* and *in vitro* matured oocytes. *Developmental Biology*, **63**, 151–64.

VERRUSIO, A. C., POLLARD, D. R. & FRASER, F. C. (1968). A cytoplasmically transmitted, diet-dependent difference in response to the teratogenic effects of 6-aminonicotinamide. *Science*, **160**, 206–7.

WAKASUGI, N. (1973). Studies on fertility of DDK mice: reciprocal crosses between DDK and C57BL/6J strains and experimental transplantation of the ovary. *Journal of Reproduction and Fertility*, **33**, 283–91.

WAKASUGI, N. (1974). A genetically determined incompatibility system between spermatozoa and eggs leading to embryonic death in mice. *Journal of Reproduction and Fertility*, **41**, 85–96.

WAKASUGI, N., TOMITA, T. & KONDO, K. (1967). Differences of fertility in reciprocal crosses between inbred strains of mice: DDK, KK and NC. *Journal of Reproduction and Fertility*, **13**, 41–40.

WAKE, N., TAKAGI, N. & SASAKI, M. (1976). Non-random inactivation of X chromosome in the rat yolk sac. *Nature, London*, **262**, 580–1.

WASSARMAN, P. M. & LETOURNEAU, G. E. (1976). RNA synthesis in fully grown mouse oocytes. *Nature, London*, **361**, 73–4.

WEATHERSBEE, P. S., AX, R. L. & LODGE, J. R. (1975). Caffeine-mediated changes of sex ratio in Chinese hamsters, *Cricetulus griseus. Journal of Reproduction and Fertility*, **43**, 141–3.

WEST, J. D., FRELS, W. L., CHAPMAN, V. M. & PAPAIOANNOU, V. E. (1977). Preferential expression of the maternally derived X chromosome in the mouse yolk sac. *Cell*, **12**, 873–82.

WHITTEN, W. K. & DAGG, C. P. (1961). Influence of spermatozoa on the cleavage rate of mouse eggs. *Journal of Experimental Zoology*, **148**, 173–83.

WUDL, L. & CHAPMAN, V. M. (1976). The expression of β-glucuronidase during preimplantation development of mouse embryos. *Developmental Biology*, **48**, 104–9.

YAMAMOTO, M., ENDO, A. & WATANABE, G. (1973). Maternal age dependence of chromosome anomalies. *Nature New Biology*, **241**, 141–2.

YAMAMOTO, M., SHIMADA, T., ENDO, A. & WATANABE, G. (1973). Effects of low dose X-irradiation on the chromosomal non-disjunction in aged mice. *Nature New Biology*, **244**, 206–8.

YOUNG, R. J. (1977). Appearance of 7-methylguanosine-5′-phosphate in the RNA of mouse 1-cell embryos three hours after fertilization. *Biochemical and Biophysical Research Communications*, **76**, 32–9.

YOUNG, R. J., SWEENEY, K. & BEDFORD, J. M. (1978). Uridine and guanosine incorporation by the mouse one-cell embryo. *Journal of Embryology and Experimental Morphology*, **44**, 133–48.

Maternal immunological factors in embryonic and post-natal development

W. D. BILLINGTON AND A. E. WILD

Dept of Pathology, The Medical School, University of Bristol,
Bristol BS8 1TD, UK
and Dept of Biology, Medical and Biological Sciences Building, University
of Southampton, Southampton SO9 3TU, UK

It is now clear that the mammalian embryo does not develop in immunological isolation within the confines of the uterine environment but is involved in a complex and changing immunological interrelationship with its maternal host throughout the period of gestation and is also to some degree susceptible to interference by experimental and pathological conditions. This review is concerned with the different ways in which maternal immune reactivity may have an influence on the development and survival of both embryo and neonate and will focus on four particular topics in this area.

(1) *The survival of the embryo as an intra-uterine allograft.* The implantation of the blastocyst onto the wall of the uterus heralds the beginning of an intimate association of a genetically dissimilar embryo with the maternal uterine tissues. During subsequent development the mother is confronted with a wide spectrum of alien antigens, especially the paternally inherited transplantation alloantigens, which should be sufficient to provoke immunological rejection reactions leading to the destruction of the embryo. The reason for its survival under these conditions is not yet established but a maternal immunoregulatory system is one of the possible mechanisms that have been proposed.

(2) *Maternal influence on development and differentiation.* The possibility exists that maternal responses to the array of emerging antigen systems of the embryo may provide the basis for a significant control mechanism in development. The evidence for this is in fact rather limited but includes some interesting experimental and clinical observations. There is much clearer evidence for maternal modification of fetal and neonatal immune reactivity.

321

(3) *Transmission of passive immunity.* From the moment of birth the young mammal must be equipped with a highly effective immunological defence system to protect itself against a wide range of pathogenic agents in the environment. Although this system begins to develop during early fetal life with the appearance of specialized organs and the differentiation of immunologically active cells, full immunocompetence is not achieved until some stage in the post-natal period. Prior to this the neonate is protected by an endowment of preformed antibodies transferred from the mother either during intra-uterine existence or at the time of suckling. It is also possible that the neonate may benefit from the acquisition of maternal immunologically active cells secreted into the colostral milk together with the immunoglobulins.

(4) *Disorders of the maternal–fetal immunological relationship.* There are a number of both known and suspected instances of fetal pathology resulting directly or indirectly from abnormal or experimentally induced maternal immune responses.

Before considering these four topics in further detail it must be stated that this is not a field rich in established fact and not a little of what follows will therefore necessarily be of an indirect, unconfirmed, controversial or speculative nature. The literature is so diverse and extensive that no attempt has been made to develop an integrated thesis but rather to present a selective view of maternal immunological factors that may have an influence on various events in embryonic development.

Survival of the embryonic allograft

Recent studies have demonstrated that paternally inherited histo-compatibility antigens are expressed on the cell surface of the embryo from a very early stage of its development (Håkansson, Heyner, Sundqvist & Bergström, 1975; Muggleton-Harris & Johnson, 1976; Searle *et al.*, 1976). Although the precise distribution of these antigens on the differentiating tissues is not yet completely defined (Jenkinson & Billington, 1977) they are known to be present in an immunogenic form because the maternal organism is capable of responding to them by the production of both cell-mediated and humoral immunity (Adinolfi & Billington, 1976; Beer & Billingham, 1976). It is not yet known whether all females in their first pregnancy are sensitized in this way. Although anti-HLA antibody responses cannot be detected in all human females this may be due to kinetic fluctuations in the individual, or even lack of sensitivity of the techniques presently used to measure them, rather than to their intrinsic absence. There is rather clearer indication of a consistent maternal cellular immune response to the fetal antigens. Hence the consensus of current opinion is that the pregnant female possesses sensitized lymphocytes that are potentially capable of a destructive cytotoxic attack on the developing embryo. An explanation for the survival of this intra-uterine allograft must therefore be sought in terms of a failure of the rejection reactions at the effector level. This could result from the involvement of a number of different factors operating in either an immunologically specific or non-specific manner.

Non-specific maternal regulators

Studies on lymphocyte reactivity *in vitro* have shown that a wide variety of protein and steroid hormones and macromolecular components of the serum of pregnant women and laboratory mammals are capable of suppressing immunological responses (Gusdon, 1976; Murgita, 1976). Although it is tempting to assume that this is indicative of their potential to inhibit the effector cell attack on the embryonic target *in vivo* the precise role of these serum factors in this context has yet to be established.

Specific maternal regulators

Cytotoxic or cytostatic effects of immune lymphocytes on fetal target cells *in vitro* can be blocked in a specific manner by the addition of serum from pregnant animal donors (Hellström, Hellström & Brawn, 1969; Tamerius, Hellström & Hellström, 1975; Smith, 1978). This reflects the operation of a phenomenon which is believed to take place *in vivo* and is known as efferent immunological enhancement. It is comparable to a situation described for tumour survival in immune hosts (Hellström & Hellström, 1974). The nature of the blocking factor in the serum and its precise mode of action are currently under investigation in both reproductive and tumour immunology laboratories. The earlier suggestions that antibody acts as the effective agent, by binding to the antigenic sites on the target in a non-cytotoxic way and hindering access of host immunocompetent lymphocytes, are now largely superseded by belief in the involvement of antigen shed from the cell surface which acts either alone or as an antigen–antibody complex to block either the lymphocyte receptors or the antigenic determinants of the target (Price & Robins, 1978).

T lymphocyte suppressor cells

Recent analyses of the complex nature of the mechanisms involved in the regulation of immune responses have demonstrated the role of a T lymphocyte subset population that exerts a suppressive effect in various model systems (Gershon, 1974; Taylor & Basten, 1976). The presence of such suppressor T cells in pregnant mice has now been detected in both in-vitro (G. Chaouat, personal communication; G. Smith, unpublished results) and in-vivo (Chaouat, Voisin, Daeron & Kanellopoulos, 1977; Smith & Powell, 1977; Hamilton, 1977) assays and it would seem reasonable to assume that they play some part in the immunoregulatory mechanisms involved in the protection of the embryo, in either an immunologically specific or non-specific manner. Suppressor cells have also been identified in fetal and neonatal lymphoid organs (Skowron-Cendrzak & Ptak, 1976) and in human umbilical cord blood (Oldstone, Tishon & Moretta, 1977), but we are here concerned only with the maternal aspects of the regulatory systems of pregnancy.

Trophoblast barrier

It is important to point out that following implantation the embryo becomes encapsulated within a rapidly proliferating and differentiating outer trophoblastic material. Those tissues that are destined to form the fetus are thus not in direct cellular contact with the uterus, nor is the fetus itself during the later stages of pregnancy since there is no ingrowth of maternal blood vessels and the placental trophoblast and fetal membranes provide the only surfaces of apposition. Since it appears that there is normally little or no ingress of maternal immunocompetent cells into the fetus (Billington *et al.*, 1969; Adinolfi, 1975) it is at the level of the trophoblastic and fetal membrane barriers that the maternal immune attack must be centred. The antigenic status of these twò tissues is thus of paramount importance.

Rather little is known of the immunobiology of fetal membranes, except in the mouse, where it has been demonstrated that histocompatibility antigens are expressed on the yolk sac and act as targets for sensitized lymphocytes *in vitro* (Jenkinson & Billington, 1974). Here, at least, it would seem that maternal immunoregulatory mechanisms must be called upon to play a protective role, always assuming that lymphocytes are able to enter the uterine lumen.

The situation regarding the expression of histocompatibility antigens on the trophoblast cell surface is less clear since there are many difficulties in providing convincing evidence. These have been detailed in a recent review (Billington, 1979). In the human, the current view is that HLA antigens are absent from the trophoblast (Goodfellow *et al.*, 1976; Faulk, Sanderson & Temple, 1977), although it should be noted that the techniques so far employed are not capable of determining a very low level of expression. In addition, the present findings relate only to the main mass of trophoblast covering the surface of the chorionic villi in the placenta and nothing is known of the antigenic status of the trophoblast in direct tissue contact with the maternal uterine decidual cells. Recent information on the trophoblast of the mouse indicates that although this is antigenically deficient during the early post-implantation stages of development there are some trophoblastic populations in the placenta that undoubtedly do express paternally inherited histocompatibility antigens (Sellens, Jenkinson & Billington, 1978). A reason for the survival of these cells *in vivo* must therefore be sought. Even if the human trophoblast does indeed prove to be different from that of the mouse in this respect, it is considered likely that there are other antigen systems present, of a tissue- or organ-specific rather than histocompatibility, nature (see Billington, 1976), that could render the tissue potentially susceptible to cellular rejection reactions and require some form of maternal immunoregulatory protection. Although much less is known of maternal immune responses to non-histocompatibility antigens there is good evidence for their existence (Baldwin & Vose, 1974) and they may have a greater role to play than is currently recognized, in relation not only to embryonic survival but also to the control of development and differentiation (see next section).

Maternal influence on development and differentiation

This is perhaps the least well defined of the four aspects of maternal immunological effects under consideration. Since there are several different lines of evidence, not all of which are clearly related, these will be presented briefly under separate headings.

Fertilization and implantation

Although the evidence is largely indirect, it is possible that these early reproductive events may be influenced to some extent, in either normal or pathological circumstances, by maternal immunological factors. There are numerous reports of inhibition of fertilization following experimental isoimmunization of females of different mammalian species with spermatozoal or seminal plasma antigens (Jones, Ing & Hobbin, 1976), although the precise site of action is unknown. These findings have provided a basis for the proposal that some cases of otherwise unexplained human female infertility may be caused by pathological sensitization to these antigens. There is as yet, however, no clear evidence of an association between the level of anti-sperm antibodies and the infertile state. The involvement of any maternal immune response as part of the normal fertilization process is as yet only a matter for speculation. It has been claimed that antibody in the reproductive tract is responsible for the selection of particular sperm for fertilization (Cohen & Werrett, 1975), but this has been disputed on the grounds that immunoglobulin binding occurs solely on senescent sperm, as a prelude to their destruction by polymorphs (Hancock, 1978).

There is also a suggestion that maternal immune response to insemination is a normal phenomenon and that this is important in providing an 'immunologically conditioned' maternal environment for successful pregnancy (Beer & Billingham, 1976). Some support for this can be deduced from the finding that prior local sensitization of the rat uterus by inoculation of spermatozoal, or other, cells can increase the subsequent reproductive performance of the female, as determined by number of implantation sites and size of the developing conceptuses (Beer & Billingham, 1977).

An interpretation in terms of maternal immunological control has been placed upon some interesting observations on the phenotypes of the backcross progeny of a number of different inbred rat strains (Palm, 1974). It was shown that there was a consistent and significant excess of *AgB* (major histocompatibility) locus heterozygotes above the expected Mendelian ratio which, taken together with other findings, indicated a selective elimination of homozygotes by a maternal response to the non-*AgB* determined antigens on the embryos. The stage of development when these embryos are lost is not known. Michie & Anderson (1966) in a similar, but less extensive, study had earlier suggested that selective fertilization might be responsible, but now appear to favour a differential pre-natal survival hypothesis (see McLaren, 1975).

Post-implantation development: fetal and placental size

There is conflicting evidence concerning the effect of maternal transplantation immunity on the development of the fetus and placenta in laboratory rodents. The reports of increases in placental and fetal weight following maternal pre-immunization against the paternal histocompatibility antigens, and of decreases in immunologically tolerant mothers (James, 1967; Iles, 1973; Beer, Scott & Billingham, 1975) have been challenged by other workers who have failed to confirm these findings (Clarke, 1971; Hetherington, 1978; Hetherington & Fowler, 1978). The reason for the discrepancy between these results is not readily apparent, especially in those instances where the genetic status of the animals and the experimental protocols seem essentially similar.

Whether or not there is a real effect of maternal immunity on fetal and placental sizes there is some as yet uncontested evidence that the extent of the uterine decidual response to the implanting embryo is determined in this way (Clarke & Hetherington, 1971) and that this is an antibody-mediated maternal effect (Clarke, 1973). Decidual development may be a determinant of fetal size (Clarke, 1971). Under certain circumstances experimental immunization of the female can result in a reduction of the numbers of liveborn young or of those surviving to maturity (Breyere &

Sprenger, 1969; Parmiani & Della Porta, 1973; Beer & Billingham, 1973).

Effects of antibodies on fetal development

Reports claiming the induction of fetal malformation by injection of pregnant animals with anti-tissue antisera have been in evidence since the early part of the century (Guyer & Smith, 1918, 1920). Although these were subsequently refuted by a number of workers the studies of Brent and his colleagues (Brent, Averich & Drapiewski, 1961; Brent, 1966) have since demonstrated that heterologous antisera to a number of different tissues are highly teratogenic in rats, especially when given in high doses and at particular stages of development. The various defects induced in the fetal organ systems are not however specifically related to the antisera used. It is now clear that these are in fact secondary effects of interference with fetal membrane function since the antibodies do not concentrate in the embryo but bind to the yolk sac basement membrane (Slotnick & Brent, 1966). It is pertinent to note that the marked teratogenic effects were found only with antisera to tissues having abundant basement membrane material (Barrow & Taylor, 1971).

There is some evidence for the production of more specific defects in target organs by anti-tissue immunity and antisera. Gluecksohn-Waelsch (1957) claimed that active autoimmunization of pregnant female mice with brain antigens induced malformations of the central nervous system in 10 % of the offspring, although there are clearly some doubts about the involvement of antibody, and even the reproducibility of these findings (see Brent, 1971). Levi-Montalcini (1964) originally reported that antibodies against nerve growth factor reduced the development of sympathetic nervous tissue when injected into neonatal and adult animals but had no effect on the fetus when administered to the pregnant female. Klingman & Klingman (1972) have, however, now demonstrated that fetal immunosympathectomy can be obtained by this means in mice.

Maternal immunization can also apparently affect the behavioural responses of the offspring. The injection of anti-brain antibodies (Auroux, Alnot & Jouvensal, 1967) and immune serum against synaptic membrane fractions (Karpiak & Rapport, 1975) into pregnant rats resulted in inferior test performances in the young animals two months after birth. These studies, together with certain clinical observations and the finding that the blood–cerebrospinal fluid barrier is not fully developed until the postnatal period, have provided further support for the controversial suggestion of an association between maternal antibodies to brain antigens and congenital mental retardation and other neurological handicaps (Adinolfi, Beck, Haddad & Seller, 1976; Adinolfi, 1976).

An antiserum with specific affinity for the contractile proteins of the heart has been shown to produce degenerative changes in the myocardium of the rat embryo cultured *in vitro* (Berry, 1971). Although the yolk sac is retained in these explant preparations the antibody is clearly able to traverse this membrane since it was shown to be localized at the myocardial target site. If this can occur *in vivo*, the difference from the results obtained by Brent, and those workers using similar approaches, may lie in the use of purified antigen, with consequent high specificity of antibody, and lack of basement membrane component.

Not all attempts to demonstrate a specific effect of maternal antibody on the appropriate fetal organ have succeeded. Chandler, Kyle, Hung & Blizzard (1962), for example, raised high titres of antibody against thyroid tissue in female rabbits but found no evidence of abnormalities in the fetal thyroid despite the transplacental transfer of the antibody. An association has been proposed between maternal thyroid antibody and infant cretinism, but the evidence is by no means convincing (Hall, Owen & Smart, 1964). Numerous studies have failed to demonstrate any clear effect on eye development following injection of heterologous lens antiserum into pregnant rabbits (see Brent, 1971).

Russian workers have claimed that certain experimental and clinical data provides evidence for a maternal influence on embryonic development by the shedding of substances (antigens?) from the adult organ and their stimulation of growth and differentiation in the corresponding fetal tissue (Vyasov, Volkova, Titova & Murashova, 1962). It is not readily apparent in what way this system could operate but their assertion that maternal antibodies are produced against various, if not all, fetal tissues at different stages of pregnancy (Volkova & Maysky, 1969) might be seen to provide support for the idea of maternal regulation of embryogenesis as a normal phenomenon. This field is clearly in need of more decisive experimentation before any such conclusions can be made. It will require, at least, the identification of the individual phase-specific tissue antigens and the demonstration of production of specific maternal humoral antibody together with the transmission of this antibody into the embryo and its localization at the tissue site. The role of any fetal serum proteins that gain access to the maternal circulation and induce immune responses will also have to be considered.

Modification of fetal and neonatal immunity

Apart from the possibility of a maternal immunological influence on the physical growth and differentiation of other tissues of the embryo there is also evidence for maternal modification of the fetal immunological system, although there is as yet little or no clear indication of the relevance of most of the observed phenomena to normal developmental processes.

Dray (1961) described a situation in which maternally transmitted anti-allotypic antibody affected the immunoglobulin secreting cells of the rabbit fetus. Immunization of does with immunoglobulin of the paternal allotype produced a long-lasting suppression of cells capable of synthesizing this allotype and a compensatory increase in the synthesis of the maternally inherited type. The effect, termed allotype suppression, is a consequence of allelic exclusion, which results in the maternal and paternal allotype being present in separate cells. Allotype suppression also occurs in heterozygous mice but is of shorter duration (Herzenberg, Herzenberg, Goodlin & Rivera, 1967). For reasons that are not clear, it does not occur in homozygous mice, as studied by embryo transfer to allotype-immunized recipients (Tuffrey *et al.*, 1976). The effect could possibly arise in humans as a result of a woman becoming immunized to allotypes of her husband through an appropriate blood transfusion. This would seriously deplete the immunological repertoire of any progeny.

Several studies have demonstrated that immunization of female rodents against viral and other antigens depresses the ability of the offspring to respond to these antigens. The effect is usually ascribed to the transmission of the antibody into the fetus and its persistence into the post-natal period to interfere with the immune response to the subsequent challenge. Antibody-induced suppression appears not to be restricted to a lymphoid cell product. Goldman & Goldman (1978) have now presented evidence for suppression of the fifth component of complement in mice heterozygous for C_5 deficiency following exposure to anti-C_5 antibody during gestation in immunized mothers. The degree of suppression was considerably enhanced if the neonate was given a booster injection of antibody.

In a study in mice assessing the effect of maternal sensitization on the response of the offspring to nitrophenyl haptens, Kindred & Roelants (1974) found a restricted response in terms of circulating anti-DNP antibody titres and the number of DNP-specific clones expressed. These authors, however, suggested that the age of the offspring when tested precluded the likelihood of persisting maternal antibody being responsible and thus favoured the involvement of antigen transferred across the placenta.

Maternal immunization does not always result in depression of responsiveness in the offspring. Gill & Kunz (1971) had previously reported an *enhanced* antibody response to a chemical antigen in the progeny of sensitized rats, but had also implicated transferred antigen. These findings were extended by Stern (1976) who showed in a hapten-carrier system that mice have an enhanced anti-DNP antibody response following maternal sensitization to bovine serum albumin (BSA). He interpreted the results in terms of an increased ability of the offspring to recognize BSA and the expression of this recognition by a secondary response to the hapten DNP when presented on BSA as a carrier protein. The transferred agent was considered to be either antigen or primed cells.

Further work is clearly required to determine the precise mechanisms involved in these maternal effects on immunological responses. The variability reported may lie in the differing protocols employed, especially in relation to the timing of both the maternal sensitization and the assessment of the responses of the offspring. These studies could lead to important clinical application by indicating potential approaches to the prophylactic immunization of young children.

From the results of a series of complex experiments, Uphoff (1973) has concluded that there are maternal influences not only upon immunological responsiveness of the offspring but also on the phenotypic expression of antigenicity. Her approach involved the production of various genetic substrains of mice by mating or egg-transfer procedures and assessment of the capacity of bone marrow grafts from these sources to elicit graft-versus-host responses in X-irradiated recipients. Although the basis for these altered responses, which were permanent and vertically transmitted from mother to offspring for several generations, has not been established it is of interest to note that they can apparently also be induced by some factor (antibody, antigen or cells?) in the milk from appropriate foster mothers. The possibility of maternal modification of the neonatal immune system by passive transfer of immunocompetent cells in the colostrum during suckling has been the subject of other recent enquiries (see next section).

There is some evidence for maternal immunological information transfer in man. Using the controversial macrophage electrophoresis technique, Field & Caspary (1971) claimed that lymphocyte sensitization to PPD (purified protein derivative of *Mycobacterium tuberculosis*) shown in normal or pathologically sensitized mothers was also evident in their newborn children. They concluded that the passage of lymphocyte subcellular elements was responsible, rather than the intact cells or free antigen. It would be valuable to have these findings confirmed and extended to other antigens. Human fetal and neonatal lymphocytes in an in-vitro one-way mixed lymphocyte reaction exhibit a reduced reactivity specifically to the maternal transplantation antigens (Ceppellini *et al.*, 1971). This effect may well provide an explanation for the reported extended survival of maternal but not paternal skin grafts on neonatal rabbits (Demant, 1968).

Transmission of passive immunity
The need for maternal antibody

In mammals (and for analogous reasons the same is probably true of oviparous vertebrates) the placenta and fetal membranes, in conjunction with maternal immune responses, present efficient barriers to the ingress of infectious organisms so that few of the antigenic stimuli experienced by the mother are likely to be experienced by the fetus. (For some notable exceptions, due to spirochaete, toxoplasmosis and viral infections, see Solomon,

1971.) Thus the necessary immunological education of the lymphoid system which is needed to combat infection normally only commences after birth when the newborn mammal is exposed to pathogens for the first time. Were it not for the fact that this education takes place under the cover of protective antibodies acquired from the mother the newly born mammal would almost certainly succumb to serious infection since, at best, all it can produce is a limited primary response which may take several days to become effective and during which time the pathogen may have multiplied to an unsurmountable level. For example, newborn human infants of mothers immune to measles, hepatitis A, poliomyelitis, rubella and diphtheria present little if any symptoms of these diseases when infected, whereas the symptoms can be most severe if the mother is non-immune (Vahlquist, 1960). In domestic livestock, such as pigs and calves, survival barely extends to 24 hours in a conventional environment if the newborn animal is deprived of colostrum, a condition largely attributable to failure to gain maternal antibody protective against enteric colibacillosis (Porter, Parry & Allen, 1977).

Selective immunoglobulin transport

Considerable interest has centred on the physiological and cellular processes involved in the transport of antibody (reviewed in Brambell, 1970; Wild, 1973, 1974). To some extent any understanding of these processes is complicated by the fact that different species vary in the time at which this transport occurs relative to birth and by the fact that different cellular sites are involved (see Table 1). Nevertheless, a number of common features have emerged, and of much importance is that of selection.

Antibodies belong to different classes of immunoglobulin molecules, but only those of the IgG class are transported intact across the human chorioallantoic placenta and suckling rat and mouse gut to any appreciable extent and largely to the exclusion of other serum proteins (Brambell, 1970). The situation is more complex in the rabbit since there is evidence that IgM may also be transported (Hemmings, 1973; Shek, & Dubiski, 1975), but probably not normally by the same route as IgG (Hemmings, 1974a). In the human, at least, selection against IgM has functional significance in that it precludes entry of anti-A and anti-B iso-

Table 1. *Time and route of transmission of IgG from mother to fetal and newborn mammals*

Species	Pre-natal	Post-natal	Route where known	Duration of post-natal transfer
Ox, goat, sheep	o	+ + +	Gut	24 hours
Pig	o	+ + +	Gut	24–36 hours
Horse	o	+ + +	Gut	24–36 hours
Wallaby (*Setonix*)	o	+ + +	Gut	180 days
Dog, cat	+	+ +	Gut	1–2 days
Hedgehog	+	+ +	Gut	40 days
Mouse	+	+ +	Gut and yolk sac	16 days
Rat	+	+ +	Gut and yolk sac	20 days
Guinea-pig	+ + +	o	Yolk sac	—
Rabbit	+ + +	o	Yolk sac	—
Grey squirrel	+ + +	Unknown	Yolk sac	—
Man, rhesus monkey	+ + +	o	Chorio-allantoic placenta	—

Adapted from Brambell (1970) and Wild (1973).

agglutinins of this class, thus preventing possible deleterious combination with appropriate blood group antigens on fetal erythrocytes. However, the consequence of this exclusion is that for certain pathogens which evoke principally IgM responses (*Bordetella pertussis*, streptococcus, staphylococcus and *Escherichia coli*) immune mothers provide little protection for their newborn babies (Vahlquist, 1960), at least by the transplacental route.

Analysis of IgG subclasses in paired human maternal and cord blood sera at different stages of gestation indicates that there may be some preference for IgG1 over the other three subclasses, all of which are transported to the fetus (Virella, Nunes & Tamagnini, 1972; Schur, Alpert & Alper, 1973; Chandra, 1976), but subclass selection, if it occurs, is not nearly so marked in man as it is in the mouse. In this species, IgG3 has been estimated to be transported to the fetus some 20–50 times more readily than

IgG1 and IgG2 (Gitlin, 1971); it represents 40 % of the fetus's total immunoglobulin content at birth but only about 9 % of the mother's (Grey, Hirst & Cohn, 1971). Since mouse IgG3 does not bind complement it has been suggested that besides conferring passive immunity this subclass might also act as a blocking antibody, protecting fetal cells from cytotoxic IgM and cell-mediated immune responses at the maternofetal interface in the placenta (Ralph, Nakoinz & Cohn, 1971). The same sort of inferences cannot easily be made in the human situation since IgG1, the preferred subclass, binds complement better than IgG2 and IgG4 (Morell, Terry & Waldman, 1970). However, since IgG antibody–complement complexes need high epitope density for target cell killing (Linsett, 1970), HLA and H-2 density, especially within the pacenta, may be of over-riding importance in determining which subclasses need to be reduced or preferred during transport. The evidence for histocompatibility antigen expression on the placenta was discussed in the section on the survival of the embryonic allograft (p. 325).

All four subclasses of human IgG are selectively transported across the gut of the suckling rat but there is no significant transport of IgM, IgA, IgD or IgE (Waldman & Jones, 1976). Thus the gut of the suckling rat mimics the situation operating in the human placenta and provides reassurance that the mechanism underlying selection is basically the same even though the cell types engaged in transport are so different. Similarly, all four subclasses of human IgG are transported across the yolk sac splanchnopleur to the blood of the rabbit fetus, but with a preference for IgG1 (Hemmings, 1974*b*). During transport of homologous antibody the suckling rat gut transmits IgG1 very much better than IgG2 (Morris, 1976), but the biological significance of this is not yet known.

Selection during transport of immunoglobulins from mother to young is independent of molecular size, but related to the species of origin of the immunoglobulin molecules and to immunoglobulin structure (Brambell, 1970; Wild, 1973). There is now considerable evidence indicating that what determines the ability of immunoglobulins to be selectively transported across the placenta, yolk sac splanchnopleur and neonatal gut is resident, along with other adjunctive properties, in the Fc region of the molecule (p. 337).

Non-selective immunoglobulin transport and significance of colostrum

In contrast to the situation in man, rabbit, rat and mouse, there is no comparable selective process operating during transport of immunoglobulin across the gut of suckling ruminants and pig (Morris, 1968; Porter, 1969; Brandon, 1976) or, in all probability, of the foal (Jeffcott, 1972). All immunoglobulins, whatever the class (but with the exception of 11S IgA in the pig; see Porter, 1976), are absorbed with equal facility when they are fed as components of colostrum to newborn unsuckled pigs and calves. Selection is obviated in the gut of these species because the mammary gland is able to concentrate immunoglobulins in the colostrum (in particular IgG1 in the cow and IgG2 in the pig) by selectively transporting them from the serum (Dixon, Weigle & Vasquez, 1961; Pierce & Feinstein, 1965; Porter, 1969). Thus passive immunity in these species is transferred immediately the animal suckles, in a large dose and for a short period of time (Table 1), since there is virtually no transport of antibody before birth and concentrations of antibody need to be built up quickly to high levels in the blood to be effective against pathogens.

Colostrum and/or milk, depending upon the species, is richly supplied with IgA compared to the concentrations present intravascularly, and secretory IgA has properties which make it resistant to proteolysis within the gut lumen (Heremans, 1974). It appears to be capable of binding to enterocytes but, unlike IgG, does not become internalized and transported. Retention of this maternal immunoglobulin within the lumen of the gut to form the so-called 'antiseptic paste' over enterocytes can be looked upon as an adaptation to provide local immunity, particularly against *E. coli* (Hill & Porter, 1974) and viruses (Ogra, Weintraub & Ogra, 1977), during this vulnerable period of post-natal development and until such time as the neonatal animal can synthesize and transport its own secretory IgA from plasma cells in the lamina propria. In this context, the importance of immediate breast feeding of newborn human infants, despite there being little evidence of transport of antibody across the gut, cannot be emphasized enough, since the first-formed colostrum is richest in IgA (Beer & Billingham, 1976).

Mechanism of selective immunoglobulin transport

Brambell (1966, 1970) suggested that selective transport could result from the presence of specific receptors for the Fc region of immunoglobulin molecules in or on rabbit yolk sac endoderm and enterocytes of suckling rodent gut (and by implication on syncytiotrophoblast of the human chorioallantoic placenta). The receptors were envisaged by Brambell to provide protection for attached immunoglobulin molecules against proteolysis when transporting endocytic vesicles that also had non-selectively endocytosed proteins subsequently fused with lysosomes. Resulting phagolysosomes were then presumed to exocytose their contents, enabling protected immunoglobulin to then diffuse to the underlying blood capillaries.

As a result of binding studies made with intact cells, tissue sections, or cell membrane preparations and appropriately tagged immunoglobulins, much evidence has now accumulated to support the idea that specific receptors for IgG(Fc) are present on rabbit (Schlamowitz, 1976; Hillman, Schlamowitz & Shaw, 1977; Wild & Dawson, 1977) and mouse (Elson, Jenkinson & Billington, 1975) yolk sac endoderm, on human syncytiotrophoblast (Matre, Tonder & Endresen, 1975; Jenkinson, Billington & Elson, 1976; McNabb, Koh, Dorrington & Painter, 1976; Thomas, MacArthur & Humphrey, 1976; Matre & Johnson, 1977; Balfour & Jones, 1977; Wood, Reynard, Krishnan & Racela, 1978) and on suckling rat gut proximal enterocytes (Jones & Waldman, 1972; Borthistle, Kubo, Brown & Grey, 1977). However, there is no evidence that phagolysosomes (in rabbit yolk sac endoderm at least) ever exocytose their contents (Wild, 1974, 1976) and another mechanism has therefore been proposed (Wild, 1975) in which selective transport functions are ascribed to coated vesicles. Rather than being protective in function, Fc receptors are considered to segregate, at the cell surface, immunoglobulin destined for transport within coated vesicles from that (together with other proteins) destined for proteolysis within phagolysosomes. Goodness of fit with the receptor and availability of receptor sites will determine which, and how much, immunoglobulin is transported intact. Ultrastructural studies of immunoglobulin transport support such a mechanism in rabbit (Moxon, Wild & Slade, 1976) and guinea-pig (King, 1977a) yolk sac endoderm and in proximal enterocytes of suckling rat gut (Rodewald, 1973, 1976; Nagura, Nakane & Brown, 1978), although in the latter case selective and non-selective uptake seems to be operative in different regions of the intestine rather than being operative within the same cell. Coated vesicles may well be the vehicles for selectively transporting IgG across human syncytiotrophoblast (King, 1977b; Ockleford & Whyte, 1977) and their association with specific receptors for other proteins needing to be selectively endocytosed by other cells is becoming increasingly well established (Wild, 1979).

Selective transport of IgG across the acinar epithelium in the bovine mammary gland during formation of colostrum also appears to be based on a specific receptor mechanism (Kemmler et al., 1975) but again, contrary to the Brambell hypothesis, is independent of proteolysis (Brandon, 1976).

A complete understanding of the selective transport mechanism is an important goal. In humans, placental insufficiency can reduce the endowment of maternal IgG to the fetus (Chandra, 1976) and may be related to defects in the mechanism of transport itself. It is perhaps also worth pointing out that the survival advantage of maternally acquired antibody would be seriously reduced if the immunoglobulins were polymorphic and immunogenic by virtue of carrying allotypic variants, since fetal antibody responses to them would lead to their elimination. Studies on mice exposed to maternal immunoglobulins of different allotypes through various cross-matings and foster nursing have in fact demonstrated the induction of tolerance to these molecules, which would be expected to continue so long as the immunoglobulin antigen persisted (Warner & Herzenberg, 1970).

Maternal leucocyte transmission in milk

Beer, Billingham & Head (1975) have reported that foster nursing of neonatal rats onto a female of a different strain can influence their survival and immunological competence. Depending upon the time of fostering the neonates either developed a fatal graft-versus-host disease or survived to show altered reactivity to test skin allografts from the maternal strain. From these and subsequent investigations (Head, Beer & Billingham, 1977) it was concluded that the effects were produced by the passive transfer of maternal immunocompetent leucocytes in the milk and that this could represent an important additional cellular form of immune protection during suckling of the normal neonate. However, these findings have not been confirmed in an independent similar study on both rats and mice (Silvers & Poole, 1975) and the discrepancy has yet to be resolved. There is little doubt that rodent, and human, milk contains substantial numbers of antigen-reactive lymphocytes (and macrophages) but their survival and incorporation into the lymphoid system of the neonate remains to be established. The one attempt so far to demonstrate this directly by chromosomal marker methods has not succeeded (Trentin, Gallagher & Priest, 1977).

Disorders of the maternal–fetal immunological relationship

There is a rather extensive literature in this field and it is the intention here only to indicate the varied nature of the possible maternal immunological factors involved in disturbances to the normal course of pregnancy. Many of the experimental approaches, involving immunization and injection of antibody, have been considered earlier in this review and this section will therefore be concerned more with the clinical aspects.

Pathological sensitization of the human female by a number of different fetal antigens has been documented and effects of varying severity on the fetus or neonate identified. These have been reviewed by Brent (1966) and, more extensively, by Scott (1976).

The best-known example of adverse maternal immunological effect is that of Rhesus disease. Here, Rh(D)-positive red cells of the fetus leak into the maternal circulation, most commonly at or near the time of parturition, and initiate anti-Rh(D) antibody formation in Rh(D)-negative females who are usually ABO compatible. The maternal antibody is predominantly of the IgG1 and IgG3 subclasses and is capable of traversing the placenta and bringing about the elimination of fetal erythrocytes to produce a range of symptoms from anaemia to death *in utero*. The normal delay in producing a full immune response usually results in the effects of the isoimmunization being seen only in the second pregnancy, although significant fetal bleeds into the maternal circulation early in gestation can occasionally occur and the antibodies produced cause damage to the existing fetus. The biological and clinical aspects of this disease have recently been reviewed by Tovey & Maroni (1976). Blood group incompatibility is also known to lead to similar haemolytic disease in mules as a consequence of the newborn foal receiving isoantibodies present in the colostrum of its horse mother. Induction of haemolytic disease by passive antibody has also been achieved experimentally in a number of mammals (Brambell, 1970).

Isoimmunization of the human female by other fetal elements is also known. Antibody production against platelets (sometimes produced also in a maternal autoimmune reaction) causes fetal thrombocytopaenia. Neonates with the isoimmune form of platelet deficiency occur with a frequency of 1 or 2 per 10 000 births (Scott, 1976). Leucocyte antibodies, apparently resulting from fetal white cell leakage (though there is as yet rather little evidence for this), are a common occurrence in pregnant women (see below) and it is claimed that in rare instances these may enter the fetal circulation and be responsible for the known cases of neonatal neutropaenia (Halvorsen, 1965). Lupus erythematosis, with characteristic dis-coid skin lesions, occurs transiently in the newborn offspring of a number of affected mothers and is believed to result from the passive transfer of anti-nuclear antibodies. Mothers with the autoimmune conditions of Graves' disease and myasthaenia gravis similarly may have babies presenting with a transient form of thyrotoxicosis and muscle weakness respectively. Although the aetiology of these diseases has not been finally established it is likely that they are immunologically induced. Other diseases of the mother which may have an immunological basis and produce mild clinical symptoms in the newborn include diabetes mellitus and pernicious anaemia (see Scott, 1976).

The frequent occurrence of antibodies against the paternally inherited HLA antigens of the fetus appears normally to have no adverse effect and has actually been considered to have a role in the *protection* of the fetal allograft (see first section). Although these antibodies are maternal IgG globulins it now seems that they do not cross the placental barrier (Tongio, Mayer & Lebec, 1975; Jeannet *et al.*, 1977). It has been hypothesized that the placenta acts as a filter for these potentially deleterious anti-fetal antibodies by effecting their binding, either directly to specific antigenic determinants on the trophoblast (Billington *et al.*, 1977; Wegmann & Carlson, 1977) or chorionic fetal mesenchyme (Jeannet *et al.*,

1977), or as antigen–antibody complexes to Fc receptors on trophoblast (Jenkinson *et al.*, 1976) or placental stem vessel endothelium (Johnson, Faulk & Wang, 1976). It is possible, however, that under pathological conditions the antibody may overflow the absorptive capacity of the placenta (Jeannet *et al.*, 1977) and be responsible for such fetal disorders as thrombocyto-paenia (discussed above) and malformation (Terasaki, Mickey, Yamazaki & Vredevoe, 1970) and possibly even for some cases of abortion (Burke & Johansen, 1974).

Maternal sensitization to incompatible fetal ABO blood group antigens has for some time been implicated as a cause of abortion (Matsunaga & Itoh, 1957) and there is also a suggestion that immune reaction against the male-specific H–Y antigen may explain the selective elimination of male zygotes implied by analysis of human sex ratio data (Renkonen, Mäkelä & Lehtovaara, 1962). On the basis of a demonstration of fetal mortality in pregnant rabbits injected with antibody against α-foetoprotein, Slade (1973) has also speculated that some congenital abnormalities and abortions may result from maternal isoimmunization to this fetal serum component.

An exceedingly rare form of deleterious maternal immunological effect on human fetal development may be evident in 'runt disease', purported to result from the intrusion into the fetus of large numbers of maternal immunocompetent cells that set up a graft-versus-host reaction (see Brent, 1971). There is some experimental support for this suggestion since there is a report (although not yet confirmed) that maternal immunization can in certain circumstances induce runt disease in a number of laboratory rodent species (Beer & Billingham, 1973).

Clearly the complex immunological interrelationship between the developing embryo and its mother is susceptible to interference by both pathological conditions and experimental insult. That the course of pregnancy is, however, apparently so relatively rarely prejudiced by maternal immune reactions attests to the success of the mammals in solving the immunological problems brought about by the development of the embryo within the intra-uterine environment during the evolution of viviparity.

References

ADINOLFI, M. (1975). The human placenta as a filter for cells and plasma proteins. In *Immunobiology of Trophoblast*, ed. R. G. Edwards, C. W. S. Howe & M. H. Johnson, *Clinical and Experimental Immunoreproduction 1*, p. 193. Cambridge University Press.

ADINOLFI, M. (1976). Neurological handicap and permeability of the blood–cerebrospinal fluid barrier during foetal life to maternal antibodies and hormones. *Developmental Medicine and Child Neurology*, **18**, 243–6.

ADINOLFI, M., BECK, S. E., HADDAD, S. A. & SELLER, M. J. (1976). Permeability of blood–cerebrospinal fluid barrier to plasma proteins during foetal and perinatal life. *Nature, London*, **259**, 140–1.

ADINOLFI, M. & BILLINGTON, W. D. (1976). Ontogeny of acquired immunity and foeto-maternal immunological interactions. In *Fetal Physiology and Medicine*, ed. R. W. Beard & P. W. Nathanielsz, p. 17. Philadelphia: W. B. Saunders.

AUROUX, M., ALNOT, M. O. & JOUVENSAL, C. (1967). Perturbations tardives du système nerveux central compatibles avec la vie. I. Baisse de la capacité d'apprentissage chez le rat per injection d'anticorps hétérospecifiques. *Comptes Rendus de la Société de Biologie (Paris)*, **161**, 1917.

BALDWIN, R. W. & VOSE, B. M. (1974). The expression of a phase-specific foetal antigen on rat embryo cells. *Transplantation*, **18**, 525.

BALFOUR, A. H. & JONES, E. A. (1977). The binding of plasma proteins to human placental cell membranes. *Clinical Science and Molecular Medicine*, **52**, 383–94.

BARROW, M. V. & TAYLOR, W. J. (1971). The production of congenital defects in rats using antisera. *Journal of Experimental Zoology*, **176**, 41–60.

BEER, A. E. & BILLINGHAM, R. E. (1973). Maternally acquired runt disease. *Science*, **179**, 240–3.

BEER, A. E. & BILLINGHAM, R. E. (1976). *The Immunobiology of Mammalian Reproduction*. New Jersey: Prentice-Hall.

BEER, A. E. & BILLINGHAM, R. E. (1977). Histocompatibility gene polymorphisms and maternal–foetal interactions. *Transplantation Proceedings*, **9**, 1393.

BEER, A. E., BILLINGHAM, R. E. & HEAD, J. R. (1975). Natural transplantation of the leucocytes during suckling. *Transplantation Proceedings*, **7**, 399–402.

BEER, A. E., SCOTT, J. R. & BILLINGHAM, R. E. (1975). Histoincompatibility and maternal immunological status as determinants of foetoplacental weight and litter size in rodents. *Journal of Experimental Medicine*, **142**, 180.

BERRY, C. L. (1971). The effects of an antiserum to the contractile proteins of the heart on the developing rat embryo. *Journal of Embryology and Experimental Morphology*, **25**, 203–12.

BILLINGTON, W. D. (1976). The immunobiology of trophoblast. In *Immunology of Human Reproduction*, ed. J. S. Scott & W. R. Jones, p. 81. New York & London: Academic Press.

BILLINGTON, W. D. (1979). The placenta and the tumour: variations on an immunological enigma. In *Placenta – A Neglected Experimental Animal?*, ed. P. Beaconsfield & J. Ginsburg. Oxford: Pergamon Press (in press).

BILLINGTON, W. D., JENKINSON, E. J., SEARLE, R. F. & SELLENS, M. H. (1977)

Alloantigen expression during early embryogenesis and placental ontogeny in the mouse: immunoperoxidase and mixed haemadsorption studies. *Transplantation Proceedings*, **9**, 1371–7.

BILLINGTON, W. D., KIRBY, D. R. S., OWEN, J. J. T., RITTER, M. A., BURTONSHAW, M. D., EVANS, E. P., FORD, C. E., GAULD, I. K. & McLAREN, A. (1969). Placental barrier to maternal cells. *Nature, London*, **224**, 704–6.

BORTHISTLE, B. K., KUBO, R. T., BROWN, W. R. & GREY, H. M. (1977). Studies on receptors for IgG on epithelial cells of the rat intestine. *Journal of Immunology*, **119**, 471–6.

BRAMBELL, F. W. R. (1966). The transmission of passive immunity from mother to young and the catabolism of immunoglobulins. *Lancet*, **ii**, 1087–93.

BRAMBELL, F. W. R. (1970). *The Transmission of Passive Immunity from Mother to Young. Frontiers of Biology 18.* Amsterdam: North Holland.

BRANDON, M. R. (1976). Selective transfer and catabolism of IgG in the ruminant. In *Maternofoetal Transmissioan of Immunoglobulins*, ed. W. A. Hemmings, *Clinical and Experimental Immunoreproduction 2*, pp. 437–48. Cambridge University Press.

BRENT, R. L. (1966). Immunologic aspects of developmental biology. *Advances in Teratology*, **1**, 82–129.

BRENT, R. L. (1971). The effect of immune reactions on fetal development. *Advances in the Biosciences*, **6**, 421–53.

BRENT, R. L., AVERICH, E. & DRAPIEWSKI, V. A. (1961). Production of congenital malformations using tissue antibodies. I. Kidney antisera. *Proceedings of the Society for Experimental Biology and Medicine*, **106**, 523–6.

BREYERE, E. J. & SPRENGER, W. W. (1969). Evidence of allograft rejection of the conceptus. *Transplantation Proceedings*, **1**, 71–5.

BURKE, J. & JOHANSEN, K. (1974). The formation of HL-A antibodies in pregnancy. The antigenicity of aborted and term foetuses. *Journal of Obstetrics and Gynaecology of the British Commonwealth*, **81**, 222–8.

CEPPELLINI, R., BONNARD, G. D., COPPO, F., MIGGIANO, V. C., POSPISIL, M., CURTONI, E. S. & PELLEGRINO, M. (1971). Mixed leucocyte cultures and HL-A antigens. I. Reactivity of young foetuses, newborns and mothers at delivery. *Transplantation Proceedings*, **3**, 58–63.

CHANDLER, R. W., KYLE, M. A., HUNG, W. & BLIZZARD, R. M. (1962). Experimentally induced autoimmunization disease of the thyroid. I. The failure of transplacental transfer of anti-thyroid antibodies to produce cretinism. *Pediatrics*, **29**, 961–7.

CHANDRA, R. K. (1976). Levels of IgG subclasses, IgA, IgM and tetanus antitoxins in paired maternal and foetal sera: findings in healthy pregnancy and placental insufficiency. In *Maternofoetal Transmission of Immunoglobulins*, ed. W. A. Hemmings, *Clinical and Experimental Immunoreproduction 2*, pp. 77–90. Cambridge University Press.

CHAOUAT, G., VOISIN, G. A., DAERON, M. & KANELLOPOULOS, J. (1977). Anticorps facilitants et cellules suppressives dans la réaction immunitaire maternelle antifoetale. *Annales d'Immunologie*, **128**, 21.

CLARKE, A. G. (1971). The effects of maternal preimmunization on pregnancy in the mouse. *Journal of Reproduction and Fertility*, **24**, 369.

CLARKE, A. G. (1973). Studies on the immunological interaction between mother and foetus. PhD thesis, University of Edinburgh.

CLARKE, A. G. & HETHERINGTON, C. M. (1971). Effects of maternal preimmunization on the decidual cell reaction in mice. *Nature, London*, **230**, 114–15.

COHEN, J. & WERRETT, D. J. (1975). Antibodies and sperm survival in the female tract of the mouse and rabbit. *Journal of Reproduction and Fertility*, **42**, 301–10.

DEMANT, P. (1968). Histocompatibility antigens in rabbits and their significance for the survival of maternal skin grafts in newborn rabbits. *Folia Biologica, Prague*, **14**, 9–20.

DIXON, F. J., WEIGLE, W. O. & VASQUEZ, J. J. (1961). Metabolism and mammary secretion of serum proteins in the cow. *Laboratory Investigation*, **10**, 216–37.

DRAY, S. (1962). Effect of maternal isoantibodies on the quantitative expression of two allelic genes controlling γ-globulin allotype specificity. *Nature, London*, **195**, 677–80.

ELSON, J., JENKINSON, E. J. & BILLINGTON, W. D. (1975). Fc receptors on mouse placenta and yolk sac cells. *Nature, London*, **255**, 412–14.

FAULK, W. P., SANDERSON, A. R. & TEMPLE, A. (1977). Distribution of MHC antigens in human placental chorionic villi. *Transplantation Proceedings*, **9**, 1379.

FIELD, E. J. & CASPARY, E. A. (1971). Is maternal lymphocyte sensitization passed to the child? *Lancet*, **ii**, 337–42.

GERSHON, R. K. (1974). T cell control of antibody responses. *Contemporary Topics in Immunology*, **3**, 1.

GILL, T. G. & KUNZ, H. W. (1971). Enhanced antibody response in the offspring of immunized rats. *Journal of Immunology*, **106**, 274–5.

GITLIN, D. (1971). Development and metabolism of the immune globulins. In *Immunological Incompetence*, ed. B. M. Kagan & E. R. Steihm, pp. 3–13. Chicago: Year Book Medical Publishers.

GLUECKSOHN-WAELSCH, S. (1957). The effect of maternal immunization against organ tissues on embryonic differentiation in the mouse. *Journal of Embryology and Experimental Morphology*, **5**, 83–92.

GOLDMAN, J. N. & GOLDMAN, M. B. (1978). Antibody induced suppression of the fifth component of complement in mice. *Journal of Immunology*, **120**, 400–7.

GOODFELLOW, P. N., BARNSTAPLE, C. J., BODMER, W. F., SNARY, D. & CRUMPTON, M. J. (1976). Expression of HLA system antigens on placenta. *Transplantation*, **22**, 595.

GREY, H. M., HIRST, J. W. & COHN, M. (1971). A new mouse immunoglobulin IgG3. *Journal of Experimental Medicine*, **133**, 289–304.

GUSDON, J. P. (1976). Maternal immune responses in pregnancy. In *Immunology of Human Reproduction*, ed. J. S. Scott & W. R. Jones, p. 103. New York & London: Academic Press.

GUYER, M. F. & SMITH, E. A. (1918). Studies on cytolysins. I. Some prenatal effects of lens antibodies. *Journal of Experimental Zoology*, **26**, 65–85.

GUYER, M. F. & SMITH, E. A. (1920). Studies on cytolysins. II. Transmission of induced eye defects. *Journal of Experimental Zoology*, **31**, 171–80.

HÅKANSSON, S., HEYNER, S., SUNDQVIST, K. G. & BERGSTROM, S. (1975). The presence of paternal H-2 antigens on hybrid mouse blastocysts during experimental delay of implantation and the disappearance of these antigens after onset of implantation. *International Journal of Fertility*, **20**, 137–40.

HALL, R., OWEN, S. G. & SMART, G. A. (1964). Paternal transmission of thyroid autoimmunity. *Lancet*, **i**, 115.

HALVORSEN, K. (1965). Neonatal leucopaenia due to fetomaternal leucocyte incompatibility. *Acta Pediatrica Scandinavica*, **54**, 86–90.

HAMILTON, M. S. (1977). Suppressor cells in pregnant mice. *Federation Proceedings*, **36**, 1183.

HANCOCK, R. J. T. (1978). Sperm antigens and sperm immunogenicity. In *Sperms, Antibodies and Infertility*, ed. J. Cohen & W. F. Hendry, pp. 1–9. Oxford: Blackwell.

HEAD, J. R., BEER, A. E. & BILLINGHAM, R. E. (1977). Significance of the cellular component of the maternal immunologic endowment in milk. *Transplantation Proceedings*, **9**, 1465–71.

HELLSTRÖM, K. E. & HELLSTRÖM, I. (1974). Lymphocyte-mediated cytotoxicity and blocking serum activity to tumour antigens. *Advances in Immunology*, **18**, 209–77.

HELLSTRÖM, K. E., HELLSTRÖM, I. & BRAWN, J. (1969). Abrogation of cellular immunity to antigenically foreign mouse embryonic cells by a serum factor. *Nature, London*, **224**, 914–15.

HEMMINGS, W. A. (1973). Transport of IgM antibody to the rabbit foetus. *Immunology*, **25**, 165–6.

HEMMINGS, W. A. (1974a). Transport of maternal antibodies to the rabbit foetus. In *Proceedings of the First International Congress on Immunology in Obstetrics and Gynecology*, ed. A. Centara & N. Carretti, pp. 252–64. Amsterdam: Excerpta Medica.

HEMMINGS, W. A. (1974b). Transport of human IgG subclasses to the rabbit foetus across the yolk sac. *International Research Communications System*, **2**, 1124.

HEREMANS, J. F. (1974). Immunoglobulin A. In *The Antigens*, vol. 2, ed. M. Sela, pp. 365–522. New York & London: Academic Press.

HERZENBERG, L. A., HERZENBERG, L. A., GOODLIN, R. C. & RIVERA, E. C. (1967). Immunoglobulin synthesis in mice. Suppression by anti-allotype antibody. *Journal of Experimental Medicine*, **126**, 701–13.

HETHERINGTON, C. M. (1978). Absence of effect of maternal immunisation to paternal antigens on placental weight, foetal weight and litter size in the mouse. *Journal of Reproduction and Fertility*, **53**, 81.

HETHERINGTON, C. M. & FOWLER, H. (1978). Effect of tolerance to paternal antigens on placental and fetal weight in the mouse. *Journal of Reproduction and Fertility*, **52**, 113–17.

HILL, I. R. & PORTER, P. (1974). Studies of bactericidal activity to *Escherichia coli* of porcine serum and colostral immunoglobulins and the role of lysozyme with secretory IgA. *Immunology*, **25**, 645–7.

HILLMAN, K., SCHLAMOWITZ, M. & SHAW, A. R. (1977). Characterisation of IgG receptors of the fetal rabbit yolk sac membranes: localisation to subcellular fraction and effects of chemical agents and enzymes on binding. *Journal of Immunology*, **118**, 782–88.

ILES, S. A. (1973). Immunological aspects of mammalian pregnancy. D.Phil. thesis, University of Oxford.

JAMES, D. A. (1967). Some effects of immunological factors on gestation in mice. *Journal of Reproduction and Fertility*, **14**, 265–75.

JEANNET, M., WERNER, C., RAMIREZ, E., VASSALLI, P. & FAULK, W. P. (1977). Anti-HLA, anti-human 'Ia-like' and MLC blocking activity of human placental IgG. *Transplantation Proceedings*, **9**, 1417–22.

JEFFCOTT, L. B. (1972). Passive immunity and its transfer with special references to the horse. *Biological Reviews*, **47**, 439–64.

JENKINSON, E. J. & BILLINGTON, W. D. (1974). Studies on the immunobiology of mouse foetal membranes: the effect of cell-mediated immunity on yolk sac cells *in vitro*. *Journal of Reproduction and Fertility*, **41**, 403–12.

JENKINSON, E. J. & BILLINGTON, W. D. (1977). Cell surface properties of early mammalian embryos. In *Concepts in Mammalian Embryogenesis*, ed. M. I. Sherman, p. 235. London: MIT Press.

JENKINSON, E. J., BILLINGTON, W. D. & ELSON, J. (1976). Detection of receptors for immunoglobulins on human placenta by EA rosette formation. *Clinical and Experimental Immunology*, **23**, 456–61.

JOHNSON, P. M., FAULK, W. P. & WANG, A.-C. (1976). Immunological studies of human placentae: subclass and fragment specificity of binding of aggregated IgG by placental endothelial cells. *Immunology*, **31**, 659–64.

JONES, W. R., ING, R. M. Y. & HOBBIN, E. R. (1976). Approaches and perspectives in the development of anti-sperm immunity as a contraceptive principle. In *Development of Vaccines for Fertility Regulation, WHO Symposium Report*, pp. 17–35. Copenhagen: Scriptor.

JONES, E. A. & WALDMAN, T. A. (1972). The mechanism of intestinal uptake and transcellular transport of IgG in the neonatal rat. *Journal of Clinical Investigation*, **51**, 2916–27.

KARPIAK, S. E. & RAPPORT, M. M. (1975). Behavioural changes in 2-month-old rats following prenatal exposure to antibodies against synaptic membranes. *Brain Research*, **92**, 405.

KEMMLER, R., MOSSMAN, H., STROHMAIER, U., KICKHOFEN, B. & HAMMER, D. K. (1975). *In vitro* studies on the selective binding of IgG from different species to tissue sections of the bovine mammary gland. *European Journal of Immunology*, **5**, 603–8.

KINDRED, B. & ROELANTS, G. E. (1974). Restricted clonal response to DNP in adult offspring of immunized mice: a maternal effect. *Journal of Immunology*, **113**, 445–58.

KING, B. F. (1977a). An electronmicroscopic study of absorption of peroxidase-conjugated immunoglobulin G by guinea-pig visceral yolk sac *in vitro*. *American Journal of Anatomy*, **148**, 447–56.

KING, B. F. (1977b). *In vitro* absorption of peroxidase conjugated IgG by human placental villi. *Anatomical Record*, **187**, 624–5.

KLINGMAN, G. I. & KLINGMAN, J. D. (1972). Immunosympathectomy as an ontogenic tool. In *Immuno-Sympathectomy*, ed. G. Steiner & E. Schobaum, p. 91. Amsterdam: Elsevier.

Levi-Montalcini, R. (1964). The nerve growth factor. *Annals of the New York Academy of Sciences*, **118**, 149–56.

Linsett, W. D. (1970). Effect of cell surface antigen density on immunological enhancement. *Nature, London*, **228**, 824–7.

McLaren, A. (1975). Antigenic disparity: does it affect placental size, implantation or population genetics? In *Immunobiology of Trophoblast*, ed. R. G. Edwards, C. S. Howe & M. H. Johnson, *Clinical and Experimental Immunoreproduction 1*, pp. 255–73. Cambridge University Press.

Matre, R. & Johnson, P. M. (1977). Multiple Fc receptors on the human placenta. *Acta Pathologica et Microbiologica Scandinavica, Section C*, **85**, 314–16.

Matre, R., Tonder, O. & Endresen, C. (1975). Fc receptors in human placenta. *Scandinavian Journal of Immunology*, **4**, 741–5.

Matsunaga, E. & Itoh, S. (1957). Blood groups and fertility in a Japanese population, with special reference to intra-uterine selection due to maternal–foetal incompatibility. *Annals of Human Genetics*, **22**, 111–31.

McNabb, T., Koh, T. Y., Dorrington, K. J. & Painter, R. H. (1976). Structure and function of immunoglobulin domains. V. Binding of immunoglobulin G and fragments to placental membrane preparations. *Journal of Immunology*, **117**, 882–8.

Michie, D. & Anderson, N. F. (1966). A strong selective effect associated with a histocompatibility gene in the rat. *Annals of the New York Academy of Sciences*, **129**, 88–93.

Morell, A., Terry, W. D. & Waldman, T. (1970). Metabolic properties of IgG subclasses in man. *Journal of Clinical Investigation*, **49**, 673–80.

Morris, I. G. (1968). Gamma globulin absorption in the newborn. In *Handbook of Physiology : Alimentary Canal*, ed. C. F. Code, sect. 6, vol. 3, pp. 1491–1512. Baltimore: American Physiological Society.

Morris, I. G. (1976). Intestinal transmission of IgG subclasses in suckling rats. In *Maternofoetal Transmission of Immunoglobulins*, ed. W. A. Hemmings, *Clinical and Experimental Immunoproduction 2*, pp. 341–57. Cambridge University Press.

Moxon, L. A., Wild, A. E. & Slade, B. S. (1976). Localisation of proteins in coated micropinocytotic vesicles during transport across rabbit yolk sac endoderm. *Cell and Tissue Research*, **117**, 175–93.

Muggleton-Harris, A. L. & Johnson, M. H. (1976). The nature and distribution of serologically detectable alloantigens on the preimplantation mouse embryo. *Journal of Embryology and Experimental Morphology*, **35**, 59.

Murgita, R. A. (1976). The immunosuppressive role of alpha-foetoprotein during pregnancy. *Scandinavian Journal of Immunology*, **5**, 1003.

Nagura, H., Nakane, P. K. & Brown, W. R. (1978). Breast milk IgA binds to jejunal epithelium in suckling rats. *Journal of Immunology*, **120**, 1333–9.

Ockleford, C. D. & Whyte, A. (1977). Differentiated regions of human placental cell surface associated with exchange of materials between maternal and foetal blood: coated vesicles. *Journal of Cell Science*, **25**, 293–312.

Ogra, S. S., Weintraub, D. & Ogra, P. L. (1977). Immunological aspects of human colostrum and milk. III. Fate and absorption of cellular and soluble components in the gastrointestinal tract of the newborn. *Journal of Immunology*, **119**, 245–8.

OLDSTONE, M. B., TISHON, A. & MORETTA, L. (1977). Active thymus derived suppressor lymphocytes in human cord blood. *Nature, London*, **269**, 333.

PALM, J. (1974). Maternal–fetal histoincompatibility in rats: an escape from adversity. *Cancer Research*, **34**, 2061–5.

PARMIANI, G. & DELLA PORTA, G. (1973). Effects of antitumour immunity on pregnancy in the mouse. *Nature New Biology*, **241**, 26–7.

PIERCE, A. E. & FEINSTEIN, A. (1965). Biophysical and immunological studies on bovine immune globulins with evidence for selective transport within the mammary gland from maternal plasma to colostrum. *Immunology*, **8**, 106–23.

PORTER, P. (1969). Transfer of immunoglobulins IgG, IgA and IgM to lacteal secretions in the parturient sow and their absorption by the neonatal piglet. *Biochimica et Biophysica Acta*, **181**, 381–92.

PORTER, P. (1976). Intestinal absorption of colostral IgA anti-*E. coli* antibodies by the neonatal piglet and calf. In *Maternofoetal Transmission of Immunoglobulins*, ed. W. A. Hemmings, *Clinical and Experimental Immunoreproduction 2*, pp. 397–407. Cambridge University Pres.

PORTER, P., PARRY, S. H. & ALLEN, W. D. (1977). Significance of immune mechanisms in relation to enteric infections of the gastrointestinal tract in animals. in *Immunology of the Gut, Ciba Foundation Symposium 46*, pp. 55–75. Amsterdam: Elsevier/Excerpta Medica/North-Holland.

PRICE, M. R. & ROBINS, R. A. (1978). Circulating factors modifying cell-mediated immunity in experimental neoplasia. In *Immunological Aspects of Cancer*, ed. J. E. Castor, p. 155. Lancaster: MTP Press.

RALPH, P., NAKOINZ, I. & COHN, M. (1972). IgM–IgG1, 2, 3 relationship during pregnancy. *Nature, London*, **238**, 344–5.

RENKONEN, K. O., MÄKELÄ, O. & LEHTOVAARA, R. (1962). Factors affecting the human sex ratio. *Nature, London*, **194**, 308–9.

RODEWALD, R. B. (1973). Intestinal transport of antibodies in the newborn rat. *Journal of Cell Biology*, **58**, 189–211.

RODEWALD, R. B. (1976). Intestinal transport of peroxidase-conjugated IgG fragments in the neonatal rat. In *Maternofoetal Transmission of Immunoglobulins*, ed. W. A. Hemmings, *Clinical and Experimental Immunoproduction 2*, pp. 137–53. Cambridge University Press.

SCHLAMOWITZ, M. (1976). Membrane receptors in the specific transfer of immunoglobulins from mother to young. *Immunological Communications*, **5**, 481–500.

SCHUR, P. H., ALPERT, E. & ALPER, C. (1973). Gamma G subgroups in human fetal, cord, and maternal sera. *Clinical Immunology and Pathology*, **2**, 62–6.

SCOTT, J. S. (1976). Immunological diseases in pregnancy. In *Immunology of Human Reproduction*, ed. J. S. Scott & W. R. Jones, pp. 229–95. New York & London: Academic Press.

SEARLE, R. F., SELLENS, M. H., ELSON, J., JENKINSON, E. J. & BILLINGTON, W. D. (1976). Detection of alloantigens during preimplantation development and early trophoblast differentiation in the mouse by immunoperoxidase labelling. *Journal of Experimental Medicine*, **143**, 348–59.

SELLENS, M. H., JENKINSON, E. J. & BILLINGTON, W. D. (1978). Major histocompatibility complex and non-major histocompatibility complex antigens on mouse ectoplacental cone and placental trophoblastic cells. *Transplantation*, **25**, 173–9.

SHEK, P. N. & DUBISKI, S. (1975). Materno-foetal transfer of normal IgM in the rabbit. *Immunology*, **29**, 365–9.

SILVERS, W. K. & POOLE, T. W. (1975). The influence of foster nursing on the survival and immunologic competence of mice and rats. *Journal of Immunology*, **115**, 1117–21.

SKOWRON-CENDRZAK, A. & PTAK, W. (1976). Suppression of local GVH reactions by mouse foetal and newborn spleen cells. *European Journal of Immunology*, **6**, 451.

SLADE, B. (1973). Antibodies to α-foetoprotein cause foetal mortality in rabbits. *Nature, London*, **246**, 493–4.

SLOTNICK, V. & BRENT, R. L. (1966). The production of congenital malformations using tissue antisera. V. Fluorescent localization of teratogenic antisera in the maternal and fetal tissue of the rat. *Journal of Immunology*, **96**, 606–10.

SMITH, G. (1978). Inhibition of cell-mediated microcytotoxicity and stimulation of mixed lymphocyte reactivity by mouse pregnancy serum. *Transplantation*, **26**, 278–83.

SMITH, R. N. & POWELL, A. E. (1977). The adoptive transfer of pregnancy-induced unresponsiveness to male skin grafts with thymus dependent cells. *Journal of Experimental Medicine*, **146**, 899.

SOLOMON, J. B. (1971). *Foetal and Neonatal Immunology. Frontiers of Biology 20.* Amsterdam: North-Holland.

STERN, C. M. M. (1976). The materno-foetal transfer of carrier protein sensitivity in the mouse. *Immunology*, **30**, 443–8.

TAMERIUS, J., HELLSTRÖM, I. & HELLSTRÖM, K. E. (1975). Evidence that blocking factors in the sera of multiparous mice are associated with immunoglobulins. *International Journal of Cancer*, **16**, 456.

TAYLOR, R. B. & BASTEN, A. (1976). Suppressor cells in humoral immunity and tolerance. *British Medical Bulletin*, **32**, 152–7.

TERASAKI, P. I., MICKEY, M. R., YAMAZAKI, J. N. & VREDEVOE, D. (1970). Maternal–fetal incompatibility. I. Incidence of HL-A antibodies and possible association with congenital anomalies. *Transplantation*, **9**, 538–43.

THOMAS, J. H., MACARTHUR, R. I. & HUMPHREY, L. J. (1976). Fc receptors on the human placenta. *Obstetrics and Gynecology*, **48**, 170–1.

TONGIO, M. H., MAYER, S. & LEBEC, A. (1975). Transfer of HL-A antibodies from the mother to the child. *Transplantation*, **20**, 163–6.

TOVEY, L. A. D. & MARONI, E. S. (1976). Rhesus isoimmunization. In *Immunology of Human Reproduction*, ed. J. S. Scott & W. R. Jones, pp. 187–227. New York & London: Academic Press.

TRENTIN, J. J., GALLAGHER, M. T. & PRIEST, E. L. (1977). Failure of functional transfer of maternal lymphocytes to F_1 hybrid mice. *Transplantation Proceedings*, **9**, 1473–5.

TUFFREY, M., BARNES, R. D., LUND, P., CATTY, D. & KING, T. D. (1976). Manipulation of mouse embryos to study the role of maternal antibody upon the IgG2a levels in the progeny of immunised maternal recipients using allotypic markers. In *Maternofoetal Transmission of Immunoglobulins*, ed. W. A. Hemmings, *Clinical and Experimental Immunoreproduction 2*, pp. 253–60. Cambridge University Press.

UPHOFF, D. E. (1973). Maternal influences on the immune response. *Biomedicine*, **18**, 13–22.

VAHLQUIST, B. (1960). Transfert d'immunité de la mère à l'enfant par la voie transplacentaire et par l'allactement. *Revue Immunologie*, **24**, 3–12.

VIRELLA, G., NUNES, M. & TAMAGNINI, G. (1972). Placental transfer of human IgG subclasses. *Clinical and Experimental Immunology*, **10**, 475–8.

VOLKOVA, L. S. & MAYSKY, I. N. (1969). Immunological interaction between mother and embryo. In *Immunology and Reproduction*, ed. R. G. Edwards, pp. 211–30. London: International Planned Parenthood Federation.

VYASOV, O. E., VOLKOVA, L. S., TITOVA, I. I. & MURASHOVA, A. I. (1962). Humoral relationship between mother and foetus in clinic and experiment. *Vestnik Akademii Med. Nauk SSSR*, **17**, 23.

WALDMAN, T. A. & JONES, E. A. (1976). The role of IgG-specific cell surface receptors in IgG transport and catabolism. In *Maternofoetal Transmission of Immunoglobulins*, ed. W. A. Hemmings, *Clinical and Experimental Immunoreproduction 2*, pp. 123–36. Cambridge University Press.

WARNER, N. L. & HERZENBERG, L. A. (1970). Tolerance and immunity to maternally-derived incompatible IgG$_{2a}$-globulin in mice. *Journal of Experimental Medicine*, **132**, 440–7.

WEGMANN, T. G. & CARLSON, G. A. (1977). Allogeneic pregnancy as immunoabsorbant. *Journal of Immunology*, **119**, 1659.

WILD, A. E. (1973). Transport of immunoglobulins and other proteins from mother to young. In *Lysosomes in Biology and Pathology*, vol. 3, ed. J. T. Dingle, pp. 425–64. Amsterdam: North-Holland.

WILD, A. E. (1974). Protein transport across the placenta. In *Transport at the Cellular Level*, ed. M. A. Sleigh & D. H. Jennings, *Society for Experimental Biology Symposium 28*, pp. 521–46. Cambridge University Press.

WILD, A. E. (1975). Role of the cell surface in selection during transport of proteins from mother to foetus and newly born. *Philosophical Transactions of the Royal Society of London, Series B*, **271**, 395–410.

WILD, A. E. (1976). Protein transport across the rabbit yolk sac endoderm. In *Maternofoetal Transmission of Immunoglobulins*, ed. W. A. Hemmings, *Clinical and Experimental Immunoreproduction 2*, pp. 155–67. Cambridge University Press.

WILD, A. E. (1979). Coated vesicles – a morphologically distinct subclass of endocytic vesicles. In *Coated Vesicles*, ed. C. D. Ockleford & A. Whyte. Cambridge University Press (in press).

WILD, A. E. & DAWSON, P. (1977). Evidence for Fc receptors on rabbit yolk sac endoderm. *Nature, London*, **268**, 443–5.

WOOD, G., REYNARD, J., KRISHNAN, E. & RACELA, L. (1978). Immunobiology of the human placenta. I. IgGFc receptors in trophoblastic villi. *Cellular Immunology*, **35**, 191–204.

Teratogenesis

GILLIAN M. MORRISS

Dept of Human Anatomy, University of Oxford, South Parks Road,
Oxford OX1 3QX, UK

Only a small proportion of the congenital malformations which
occur spontaneously in man and other mammals have a clearly
identifiable genetic or environmental origin. The majority probably have a multifactorial background, involving an interaction
between genetic and environmental cofactors. This review will
look briefly at some of the many maternal factors which are known
to interfere with normal organogenesis in mammalian embryos.
The first section deals mainly with the maternal aspects, and the
second is concerned with studies on the response of the embryo.

Maternal factors in teratogenesis

Introduction

Exposure of the pregnant maternal body to a wide variety of
external environmental factors can alter the internal environment
in ways which affect embryonic and fetal development. Internal
maternal factors without obvious external cause can also affect
development. Some of these factors, each of which represents a
major research area, are listed in Table 1. Only a few of them will
be represented in the examples used to illustrate this review, and
further information may be obtained from the following sources:
Stevenson (1960), Torpin (1968), Berry & Poswillo (1975), Persaud (1977), Wilson & Fraser (1977).

The emergence of abnormal mammalian development (teratology) as a field of scientific investigation has been largely separate
from that of experimental embryology. This has been due partly
to the peculiar inaccessibility of post-implantation mammalian

Table 1. *Maternal factors in teratogenesis*

Diet	Extremes of temperature
Drugs	Radiations
Mutagenic agents	Viral infections
Cytotoxic agents	Metabolic abnormalities
Antimitotic agents	Stress
Agents affecting gene expression	Intra-uterine pressure effects
Gaseous environment	Maternal genotype

embryos, and partly because the teratogenicity of various procedures was usually discovered by workers who were using these procedures in a different scientific context. It is only in recent years that experimental embryologists working with non-mammalian vertebrates, or with pre-implantation mammalian embryos, have begun to investigate post-implantation mammalian development. At the same time, teratologists have moved from a descriptive to a mechanistic approach. However, the research effort which is being applied to elucidating the mechanisms of teratogenesis is still very small compared with the social importance of the problem.

Since the reasons for dichotomy between the two approaches to mammalian development are historical, it will be useful to review briefly some of the major landmarks in teratology.

Historical background

The use of X-rays as a diagnostic procedure began on a significant scale during the First World War. In 1920, Aschenheim reported a case of congenital microcephaly and mental retardation, and attributed it to maternal exposure to X-rays during pregnancy. Murphy (1929) reported the outcome of 625 pregnancies in women exposed to pelvic radiation, and stated that irradiation of the pregnant woman is extremely likely to be followed by the birth of serious defective offspring. von Hippel (1907) had induced ocular anomalies in rabbits by maternal X-irradiation, but more detailed experimental work was not carried out until 1947, when Warkany & Shraffenberger exposed female rats to X-rays on days 13 and 14 of pregnancy; the pups were subsequently found to have skeletal malformations.

Another research area which made an important contribution

to teratology was that of nutrition biochemistry. From the turn of the century onwards, a great deal of progress was being made in the search for the long-recognized 'accessory growth factors', whose deficiency caused night blindness, scurvy, beri-beri, xerophthalmia, and lack of growth. Animal experiments, using mainly rats and pigs, were carried out using carefully controlled diets composed of proteins, carbohydrates, fats, and salts from natural and synthetic sources. The adequacy of a diet was assessed in terms of both growth and reproduction. The accessory growth factors, which we now call vitamins, were divided into fat-soluble and water-soluble components by McCollum & Davis in 1915.

The earliest report of malformed young resulting from a deficient diet was by Zilva, Golding, Drummond & Coward (1921). A sow was reared from birth on a diet restricted in fat-soluble factor. Near maturity it ceased to grow, and the diet was supplemented with cod-liver oil. As soon as normal growth resumed, the supplement was withdrawn, and the sow was mated to a boar which had been reared and maintained on a normal diet. She produced eight piglets, all of which were stillborn or died soon after birth. All were oedematous, and had hydronephrosis, sometimes with hydration of other internal organs; four had abnormal hindlimbs.

This observation was not followed up, and more detailed experiments were not carried out until 12 years later, when Hale (1933) carried out a study which he reported under the title 'Pigs born without eye balls'. This was the first of three classic papers on the teratogenicity of hypovitaminosis A. The first sow was reared on a diet deficient in vitamin A from 4 months of age, and mated at 160 days. She gave birth to 11 piglets, all of which had severe microphthalmia.

At that time, in spite of the earlier reports of malformations resulting from X-rays, congenital anomalies were generally regarded as being due to heredity alone. Hale (1933, 1935) pointed out that for this to be the case in his experiment both parents would have to have been heterozygous for a recessive character, and all of the 11 young homozygous for that character; the chances of this are approximately one in 4 million. He then carried out further mating experiments: a blind boar was mated to his blind sister, who was fed an adequate diet. The young, which on the heredity theory should all have been eyeless, were all normal (Hale, 1935,

1937). Hale's experiments are regarded as the beginning of experimental mammalian teratology.

In 1946, Warkany & Schraffenberger found that the ocular anomalies induced by Hale in pigs could be induced in rats by the same treatment. Warkany, Roth & Wilson (1948) examined the abnormal A-deficient rats more carefully, and discovered that in addition to the eye defects there were soft tissue malformations of the genito-urinary and cardiovascular systems and in the diaphragm, but no skeletal abnormalities.

Following Hale's results, other maternal factors were found to be teratogenic in experimental animals. Warkany & Nelson (1940, 1941) produced skeletal malformations by means of a diet designed to produce endemic goitre (which it failed to do), and found that the malformations could be prevented by supplementation with dried liver. In 1943, Warkany & Schraffenberger were able to show that the preventive factor in the liver was riboflavin, thereby demonstrating the teratogenicity of riboflavin deficiency. The teratogenicity of trypan blue was discovered by Gillman, Gilbert, Gillman & Spence (1948) in their search for a substance which would enable them to study the development of the central nervous system. In 1950 maternal hypoxia (Ingalls, Curley & Prindle, 1950) and the administration of cortisone (Baxter & Fraser, 1950) were reported as teratogenic procedures.

In 1940, crystalline vitamin A was isolated, and extensive studies were subsequently carried out on the effects of vitamin A excess. Its effect on cartilage matrix, particularly of growing long-bones, has been an important and productive research area for over 25 years. Cohlan (1953) was studying this effect when he made the chance discovery of the teratogenicity of excess maternal vitamin A. The abnormalities produced were mainly of the central nervous system, face, palate and eyes.

X-irradiation, hypoxia, trypan blue, cortisone, excess vitamin A and some drugs have all proved to be useful tools in experimental teratology, since they can be administered at specific times during pregnancy. Dietary deficiences are less amenable, since the maternal deficiency must be built up over a considerable period of time in order to deplete body stores.

Teratogenic response and stage of pregnancy

The developing embryo is vulnerable to procedures which cause structural abnormalities only during the period of organogenesis. Pre-implantation embryos are not susceptible, and nor are embryos which have completed organogenesis. The response of the embryo to excess maternal vitamin A illustrates this principle (Fig. 1).

Another of the basic principles of teratology is that the pattern of malformations induced by a teratogenic procedure varies with stage of pregnancy (Fig. 2). This does not mean that two different teratogens acting at the same stage will bring about similar malformations, since they may exert their effects through interference with different developmental mechanisms; hence the skeletal malformations induced by maternal riboflavin deficiency differ from those induced by X-rays, and with excess vitamin A they are almost entirely confined to the neural-crest-derived skeleton (Warkany & Nelson, 1940, 1941; Warkany & Schraffenberger, 1947; Morriss, 1972).

The observation that the teratogenic response varies with stage of pregnancy, and that certain malformations cannot be induced before or after a particular stage, has led to the concept of the 'sensitive period' or 'critical period', i.e. the period during which

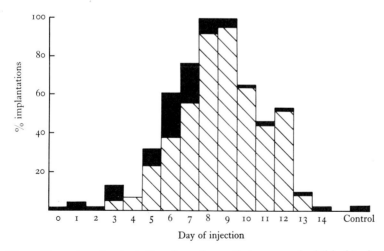

Fig. 1. Percentage implantations in embryos that are resorbed (black), abnormal (cross-hatched) and normal (remainder), according to the day of maternal vitamin A administration. (From Morriss, 1973a.)

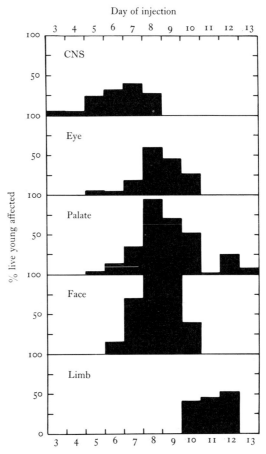

Fig. 2. Types of malformation according to the day of maternal vitamin A administration. (From Morriss, 1973*a*.)

a particular structure or organ is susceptible to a teratogen. The onset of this period may precede overt differentiation; e.g. irradiation of rat embryos on day 9 produced anomalies of the metanephric kidney, even though the primordium of the metanephros does not appear until day 12 (Wilson, Jordan & Brent, 1953). The critical period for a particular organ may last for several days, and there may be more than one period. For example, cleft palate induced by vitamin A excess has two critical periods (Fig. 2): the first of these is associated with facial malformations, and is presumably brought about by a different mechanism from the second, when cleft palate occurs in an otherwise normal head.

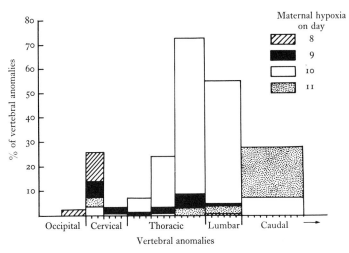

Fig. 3. Position of vertebral anomalies in fetal mice following maternal hypoxia on one of the days 8–11 of gestation. Cervical region divided into upper (3 vertebrae) and lower (4); thoracic region divided into upper (4), middle (4) and lower (5); caudal region: first 10 vertebrae only. Expressed as percentage of affected fetuses. (After Murakami & Kameyama, 1963.)

Experimental or retrospective studies on teratogenic response in relation to stage of pregnancy have been carried out for several teratogens, e.g. X-rays in the mouse (Russell & Russell, 1954) and man (Brill & Forgotson, 1964); thalidomide (Nowack, 1965); and rubella (Mayes, 1957). These studies can provide useful information on the mode of action of the teratogen. For instance, Murakami & Kameyama (1963) found that short-term maternal hypoxia in mice on days 8–11 produced axial anomalies that showed a chronological antero-posterior progression with different times of treatment; susceptibility of a particular segmental level can be correlated with the period of migration and condensation of the sclerotomal cells (Fig. 3).

Problems in interspecies extrapolation

The most important aim of experimental teratology is to gain sufficient understanding of the causes of congenital defects to be able to reduce the large proportion of human malformations that are at present not attributable to any known cause. Advances towards this aim will come through the identification of genetic and environmental factors which cause abnormal development, either alone or through interaction with other factors, and through the ability to predict, and therefore to avoid, new environmental hazards. Since experimental work cannot be carried out on human embryos except by unfortunate accidents, most of the new information is coming from work on the common laboratory species, particularly the rat, mouse, rabbit and guinea pig. Screening programmes for the safety of new drugs and other environmental additives also rely heavily on these species, at least as a first filter.

One of the problems of recognition and prediction is that there is considerable variation of response, not merely between different mammalian species, but between individuals of the same species. The influence of intraspecies differences (maternal genotype) on the teratogenic response will be discussed in a subsequent section. Possible sources of error in interspecies extrapolation are differences in maternal factors, such as enzymes involved in drug metabolism, dietary requirements, binding or storage factors; placental factors, such as the presence or absence of a yolk sac placenta during susceptible stages of development; and embryonic factors which, in addition to the items listed as maternal factors, include interspecies differences in the order of appearance of new structures, and hence differences in the overlapping of critical periods for different organs.

The best-known example of the invalidity of interspecies extrapolation is the case of thalidomide. This drug, which is a derivative of glutamic acid, is not teratogenic in rats and mice, but malformations similar to those occurring in man can be induced by it in New Zealand White rabbits (Heine, Kirchmair, Fiedler & Stuewe, 1964). Monkeys react in the same manner as man, as a result of which strong arguments have been put forward for the use of primates in teratogenicity testing (Wilson, 1971). In contrast to thalidomide, many chemical agents that are highly teratogenic in rats, mice and rabbits have little or no effect in primates, or cause growth retardation and abortion rather than malformation (Wilson, 1972).

Maternal vitamin E (α-tocopherol) deficiency, with supplementation during early post-implantation development, is a teratogenic procedure in rats (Cheng & Thomas, 1954), but does not appear to affect development in other species. In contrast, vitamin A has been found to be teratogenic in both deficiency and excess in all species which have been tested, and in man (Giroud & Martinet, 1959; Wiersig & Swenson, 1967; Pilotti & Scorta, 1965). One could predict that it would not be teratogenic in the polar bear, because of the enormous storage capacity of the adult liver – an essential adaptation to a diet rich in vitamin A (Rodahl & Moore, 1943).

Interspecies differences in teratogenic susceptibility may also be due to differences in placental structure and function. The teratogenicity of trypan blue in the rat, for instance, is due to its interference with yolk-sac-mediated nutrition (Beck, Lloyd, & Griffiths, 1967), a process that does not occur in primates.

Possible sources of extrapolation errors at the embryonic level include differences in the order of occurrence of different morphogenetic events (see species-comparative data in Nishimura & Shiota, 1977). Differences in the amount of growth which is concomitant with differentiation and morphogenesis may also be important; the longer time-scale of human development compared with that of mouse can be correlated with the larger size of organs during their early stages of development.

The problem of distinguishing 'maternal-mediated' and 'direct' effects

Since abnormal morphogenesis is the response of the embryo to a teratogenic procedure, it would seem logical to study the mechanism of action of teratogens *in vitro* wherever possible. However, teratogenic mechanisms may involve different factors *in vivo* and *in vitro*. The maternal organism, placenta and embryo form a complex interacting system, and many procedures have a maternal effect which may itself either bring about or influence the response of the embryo. It is theoretically possible for a procedure to be teratogenic *in vitro* but not *in vivo*, and vice versa.

Some teratogens probably act almost entirely as a direct effect on the embryo, so that in-vitro studies have the potential to provide a relatively complete picture of the mechanism *in vivo*. These include radiations and, possibly, extremes of temperature, although in both cases the maternal effects may have some influence on the embryonic response. Trypan blue is eminently suited to in-vitro study, since the yolk sac is retained with the embryo in whole rat embryo cultures.

Virus infections such as rubella, cytomegalovirus and herpes almost certainly exert their major effect directly on the embryo. But maternal facors may also be important; these include route of infection (e.g. via the placenta or as an ascending genital tract infection), the maternal immune status, the degree of fever, and vascular lesions affecting the placental blood supply.

The problem of distinguishing between maternal-mediated and direct teratogenic effects has been studied in relation to malformations induced by excess vitamin A (Morriss, 1973*a*, *b*; Morriss & Steele, 1974, 1977). The malformations induced in rat embryos following exposure to excess vitamin A on days 8 or 9 of pregnancy were very similar in in-vivo and in-vitro studies, and in both situations they were preceded by characteristic ultrastructural changes in the embryonic cells. However, these observations do not rule out the possibility that the teratogenic activity of vitamin A *in vivo* has a maternal component. Five hours after subcutaneous injection, the decidual tissue immediately surrounding the embryo showed the swollen cell membrane effects characteristic of vitamin-A-induced damage, particularly in cells bordering the capillaries. Within the capillaries, and in the blood space which surrounds Reichert's membrane (the only placental barrier in day 8 embryos), degenerating cells and organelles were observed, effectively narrowing the lumen of the vessels (Fig. 4). The blood cells were frequently distorted and vacuolated, resembling those observed by Glauert, Daniel, Lucy & Dingle (1963) following exposure to vitamin A *in vitro*, where vacuolation was followed by the loss of haemoglobin.

The appearance of these maternal effects coincided in time with the earliest fine structural alterations in the embryonic cells. The observed effect on the blood supply to the embryo could affect development through a reduction in the availability of nutrients and oxygen, and a reduction in the rate of removal of waste products. One indication in the embryonic ultrastructural response that in-vivo and in-vitro conditions were not identical was the presence of condensed, often ring-shaped mitochondria in the embryos *in vivo*; the normal response of mitochondria to excess vitamin A is swelling (Lucy, Luscombe & Dingle, 1963), and condensed mitochondria have been observed elsewhere in embryonic cells in pathological conditions (Bellairs, 1961).

Other maternal factors may also have an effect on the amount

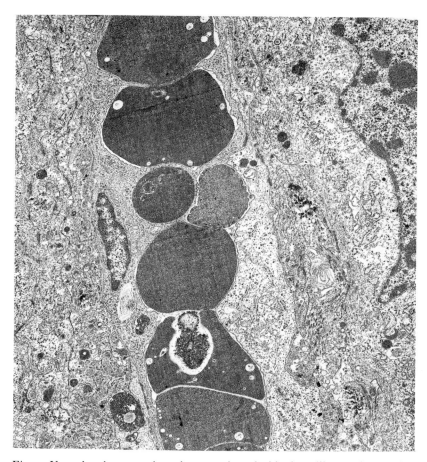

Fig. 4. Vacuolated maternal erythrocytes in a decidual capillary, 5 hours after vitamin A injection on day 8 of gestation. Damaged endothelial cells have collapsed against the erythrocytes, effectively obliterating the lumen at this point. (From Morriss, 1973*b*.)

of vitamin A to which a developing embryo is exposed. Vitamin A is stored in the liver in the ester form, and transported in the plasma attached to specific and non-specific binding proteins (Roels, 1969; Raz, Kanai & Goodman, 1968). Alterations in the levels of available vitamin A to which the embryo is exposed are therefore difficult to assess. Woollam & Millen (1960) reported that the incidence of vitamin-A-induced cleft palate in rats was increased when cortisone was administered, although cortisone alone does not cause cleft palate in this species. This effect is probably due to cortisone-induced release of vitamin A from the liver stores (McGillivray, 1961).

Release of abnormally large quantities of vitamin A from the liver stores may be brought about in the absence of direct intake. In man, high serum vitamin A levels due to decreased liver storage capacity have been associated with liver injuries such as cirrhosis and carbon tetrachloride poisoning (Baker & Frank, 1968). Maternal hepatic damage caused by carbon tetrachloride poisoning has been used to induce malformations in rabbits (Heine *et al.*, 1964), but unfortunately serum vitamin A levels were not measured.

Maternal genotype

The relative importance of genetic and environmental factors in human teratogenesis is impossible to assess. Examples of clearly genetic or clearly environmental causes are well known, but it is probable that most congenital defects are the result of complex interactions between the two. The variable incidence and severity of hereditary malformations in a given population may be due to the influence of environmental factors. Two groups of human malformations which appear to have this type of multifactorial background are anencephaly and spina bifida, and cleft lip and palate (Carter, 1969). Epidemiological studies of the incidence of these malformations takes many variable factors into account: time, place, sex, ethnic group, family, social class, parental age and parity, maternal chronic diseases, and reproductive wastage (M. Klingberg, personal communication). In this way it has been demonstrated that anencephaly and spina bifida are alternative expressions of the same genetic background (Leck, 1974). On the other hand, cleft lip with cleft palate, and cleft lip alone, are more closely related to each other than cleft palate alone (Fogh-Anderson, 1942).

The environmental influences that may be responsible for variations in the expressivity and penetrance of these malformations in human populations are not known; the suspicion that blighted potatoes might be associated with anencephaly and spina bifida (Renwick, 1972) was an interesting hypothesis although it did not receive subsequent support (Clarke, McKendrick & Sheppard, 1973; Lorber, Stewart & Milford-Ward, 1973). The interaction between genetic predisposition and environmental factors is more easily studied in experimental animals, where an influence of maternal genotype on teratogenic response can be demonstrated by reciprocal crosses between strains which differ in their response (Biddle & Fraser, 1977). A clear example of this is provided by the different incidence of glucocorticoid-induced cleft palate in two strains of mice. Fraser & Fainstat (1951) found that the A/J strain (A) is more susceptible to maternally administered cortisone than C57BL/6J (B6). In reciprocal crosses, F_1 embryos developing in A females showed a higher incidence of cortisone-induced cleft palate than did those developing in B6 females, although the embryonic genotype was identical. Back-crosses of the two different types of F_1 females with A males produced offspring which did not differ in teratogenic suscepti-bility to cortisone, thus ruling out cytoplasmic inheritance as a possible mechanism (Kalter, 1954; Fraser, Kalter, Walker & Fainstat, 1954). The strain differences were therefore considered to be due to genetic differences in the maternal or uterine environment, i.e. to differences in the maternal metabolism, storage, or transport of cortisone.

More recently, Dostal & Jelinek (1973) have demonstrated that strain differences in glucocorticoid sensitivity also exist in the embryos at the time of palatal shelf formation and closure. They used four strains, whose susceptibility to maternally administered cortisone was in the order A > B10 > B6 > CBA. When hydro-cortisone was injected directly into the amniotic fluid, the strain difference in response was different: A,B6 > B10 > CBA. Unfor-tunately, since two different glucocorticoids were administered by two different routes, the two results cannot be regarded as comparable. Other studies have shown strain differences in the number of glucocorticoid binding proteins in the maternal and/or embryonic palatal shelves (Salomon & Pratt, 1976; Goldman *et al.*, 1976, 1977).

Maternal genotype has also been shown to influence the terato-genicity of 6-aminonicotinamide (Goldstein, Pinsky & Fraser, 1963), ethyl nitrosourea (Diwan, 1974), hypoxia (Ingalls *et al.*, 1953), and excess vitamin A (Nolen, 1969).

Studies on mechanisms of abnormal morphogenesis

Introduction

The fact that the stage-related distribution of malformations is similar but not identical for different teratogenic procedures, suggests that although there are 'critical periods' for vulnerability to a particular developmental abnormality, each procedure exerts its effect through a rather specific interference with normal developmental events.

The elucidation of mechanisms of abnormal organogenesis is being derived from two major approaches, which may be broadly considered as analytic and synthetic. The 'analytic' approach entails a careful study of the pathogenesis of induced malformations, *in vivo* or *in vitro*. The 'synthetic' approach involves controlled interference with specific developmental events over a short period of time, using whole embryo culture or organ culture. Neither of these approaches is ideal, but taken together they can provide useful information about some of the abnormal developmental mechanisms. Information derived from a pathogenesis study can indicate which morphogenetic events are primarily affected, so that more specific questions can be asked through the experimental approach.

Pathogenesis

Before it is worthwhile carrying out a study of the pathogenesis of the malformations brought about by a teratogen, it must be established that the procedure under investigation is a reliable tool, i.e. the maternal treatment must result in 100 % abnormalities in the young at term, without increasing the incidence of embryonic death. If these criteria are fulfilled, embryos can be examined at all stages after exposure to the teratogen, with the assumption that they would have developed predictable malformations if left undisturbed.

The pathogenesis of craniofacial abnormalities induced by excess maternal vitamin A will be used as an example of this approach. Abnormalities of the brain tube were found to occur subsequent to the presence of an abnormally small number of primary mesenchyme cells under the cranial neural folds at an early stage in neural fold elevation (Marin-Padilla, 1966; Morriss,

1972). These cells migrate into the cranial region from the primitive streak. Abnormalities of the face were correlated with an earlier defect of neural crest cell migration (Poswillo, 1975; Morriss & Thorogood, 1978).

These findings were taken as the starting point for further studies on the effect of vitamin A on migrating cells. Inhibition of migration was demonstrated when primary mesenchyme (Morriss, 1975) and limb-bud mesenchyme (Kwasigroch & Kochhar, 1975) was explanted and cultured *in vitro* in the presence of vitamin A. Although this teratogen has ultrastructural effects on other parts of the embryo (Morriss, 1973a), the combination of pathogenesis and in-vitro studies suggests that the most important component of its teratogenic action is an effect on cell migration. The affected migrating cells showed blebbing, and occasionally swelling followed by death. These phenomena may be attributable to the expansion of biological membranes which is brought about by vitamin A (Dingle *et al.*, 1962; Bangham, Dingle & Lucy, 1964). However, it is possible that an effect on the synthesis of cell surface glycosaminoglycans is also involved in the mechanism of abnormal migration. Vitamin A is known to affect glycosaminoglycan synthesis (Wolf & Varandani, 1960; Kochhar & Johnson, 1965; Solursh & Meier, 1973), and sulphated glycosaminoglycans are essential for the migration of primary mesenchyme cells in the sea urchin (Karp & Solursh, 1974). Cell surface glycoprotein has also been implicated in the cell adhesion component of cell migration (Rees, Lloyd & Thom, 1977), and the synthesis of this component of the cell membrane is altered by vitamin A excess (Lewis, Pratt, Pennypacker & Hassell, 1978).

Interference with specific developmental events

The above speculations on the mechanism by which vitamin A may bring about its effects on cell migration, and thereby at least part of its effects on morphogenesis, reflect the kind of difficulties that are inherent in an approach which takes as its starting point a particular teratogenic procedure. In contrast, the 'synthetic' approach begins with a morphogenetic event, failure of which brings about an easily recognizable abnormality. The role of a substance which has been implicated |in| the morphogenetic |event| in question, is investigated through the use of chemicals which are competitors or antagonists of the substance, or which interfere with its synthesis or post-synthetic organization.

In practice such precise interference is rarely possible, since most of the biochemical substances which have been used as tools in this way have secondary effects in addition to the effect for which they were chosen. Nevertheless, this approach has proved to be very useful in providing insight into normal and abnormal morphogenetic mechanisms. Studies on the mechanism of palatal shelf adhesion will serve as an example.

Following rotation of the palatal shelves from a vertical to a horizontal position, their medial edges adhere; adhesion is followed by fusion, which occurs through pre-programmed cell death of the apposed epithelial cells (see Greene & Pratt, 1976, for review). Cell surface carbohydrates have been implicated in the process of adhesion (Pourtois, 1972), through studies using concanavalin A binding (Pratt, Gibson & Hassell, 1973), ruthenium red staining (Greene & Kochhar, 1974), and [^{3}H]glucosamine incorporation (Pratt & Hassell, 1974).

Cell surface carbohydrates exist in the form of glycosamino-glycans and glycoproteins, both of which contain significant levels of the amino sugar, glucosamine. The formation of glucosamine-containing macromolecules can be blocked by the use of diazo-oxo-norleucine (DON), which inhibits glutamine synthesis and amino transfer reactions (Telser, Robinson & Dorfman, 1965). Culture of apposed palatal shelves in medium containing DON did not allow fusion, and this effect could be reversed by the addition of D-glucosamine to the medium (Pratt, Greene, Hassell & Greenberg, 1975). DON-cultured shelves did not bind as much concanavalin A at the medial epithelial surface as shelves in control cultures (Greene & Pratt, 1976). These results suggest that amino-sugar-containing macromolecules are responsible for adhesion of the palatal shelves prior to fusion.

At earlier stages of development rat embryos can be studied *in vitro* by the technique of whole embryo culture (New, 1978). We are currently using this technique to investigate some aspects of the mechanism of brain tube formation. We have used β-D-xyloside to investigate the role of proteoglycans in neural fold formation (Crutch & Morriss, unpublished data) and cytochalasin B to investigate the role of microfilaments in curvature of the neural epithelium prior to closure. (β-D-xyloside interferes with the linkage of sulphated glycosaminoglycans onto the protein core in proteoglycan synthesis (Schwartz, 1977); cytochalasin B appears to inhibit a number of biological processes thought to be dependent on the contractile properties of actin-containing microfilaments (Wessells *et al.*, 1971).) Both of these procedures resulted in exencephaly in the cultured embryos, and further investigation of the mechanisms is continuing.

Conclusions

The most important aim of teratology is to find ways of decreasing the incidence of human congenital malformations. In order to do this, it is necessary to view the developing embryo or fetus as part of a materno-placento-embryonic unit within its own fluctuating environment. It is impossible to carry out experimental work on such a broad basis, but the specific aspect under scrutiny must be set within this total context before interpretations and predictions are made. Two useful approaches to human teratology are being undertaken at present: epidemiological studies and comparative studies on normal and abnormal development. An example of the former is the work of Smithells & Sheppard (1978) which has eliminated bendictin, a common drug taken for pregnancy sickness, as a possible teratogen. An example of the latter is the work which led to the discovery that open neural tube defects can be identified by high alpha-fetoprotein levels in the amniotic fluid and maternal plasma (Brock & Sutcliffe, 1972; Brock, Bolton & Monaghan, 1973). It is to be hoped that further progress in pre-conception or pre-natal prediction of human malformations will be made through these methods, and through application of the results of experimental work in other species, in the near future.

The original work described in the last section was carried out in collaboration with Beth Crutch, and was supported by an MRC project grant. I wish to thank Marcus Klingberg for helpful comments on the manuscript, and Edgar and Violet Morriss for providing a congenial environment during its composition.

References

ASCHENHEIM, E. (1920). Schädigung einer menschlichen Frucht durch Röntgenstrahlen. *Archiv für Kinderheilkunde*, **68**, 131-40.

BAKER, H. & FRANK, O. (1968). *Clinical Vitaminology*. New York: Interscience Publishers.

BANGHAM, A. D., DINGLE, J. T. & LUCY, J. A. (1964). Studies on the mode of action of excess of vitamin A. IX. Penetration of lipid monolayers by compounds in the vitamin A series. *Biochemical Journal*, **90**, 133-40.

BAXTER, H. & FRASER, F.C. (1950). Production of congenital defects in offspring of female mice treated with cortisone. *McGill Medical Journal*, **19**, 245.

BELLAIRS, R. (1961). Cell death in chick embryos as studied by electron microscopy. *Journal of Anatomy*, **95**, 54-60.

BECK, F., LLOYD, J. B. & GRIFFITHS, A. (1967). Lysosomal enzyme inhibition by trypan blue: a theory of teratogenesis. *Science*, **157**, 1180-2.

BERRY, C. L. & POSWILLO, D. E. (1975). *Teratology: Trends and Applications*. Berlin: Springer-Verlag.

BIDDLE, F. G. & FRASER, F. C. (1977). Maternal and cytoplasmic effects in experimental teratology. In *Handbook of Teratology*, vol. 3, ed. J. G. Wilson & F. C. Fraser, pp. 3-33. New York: Plenum Press.

BRILL, A. B. & FORGOTSON, E. H. (1964). Radiation and congenital malformations. *American Journal of Obstetrics and Gynecology*, **90**, 1149-68.

BROCK, D. J. H., BOLTON, A. E. & MONAGHAN, J. M. (1973). Prenatal diagnosis of anencephaly through maternal serum alpha-fetoprotein measurements. *Lancet*, **ii**, 923-4.

BROCK, D. J. H. & SUTCLIFFE, R. G. (1972). Alpha-fetoprotein in the antenatal diagnosis of anencephaly and spina bifida. *Lancet*, **ii**, 197-8.

CARTER, C. O. (1969). Genetics of common disorders. *British Medical Bulletin*, **25**, 52-7.

CHENG, D. W. & THOMAS, B. H. (1953). Relationship of time of therapy to teratogeny in maternal avitaminosis E. *Proceedings of the Iowa Academy of Sciences* **60**, 290-9.

CLARKE, C. A., MCKENDRICK, O. M. & SHEPPARD, P. M. (1973). Spina bifida and potatoes. *British Medical Journal*, **3**, 251-4.

COHLAN, S. Q. (1953). Excessive intake of vitamin A as a cause of congenital anomalies in the rat. *Science*, **177**, 535-6.

DINGLE, J. T., GLAUERT, A. M., DANIEL, M. & LUCY, J. A. (1962). Vitamin A and membrane systems. I. The action of the vitamin on the membranes of cells and intracellular particles. *Biochemical Journal*, **84**, 76P.

DIWAN, B. A. (1974). Strain-dependent teratogenic effects of 1-ethyl-1-nitrosourea in inbred strains of mice. *Cancer Research*, **34**, 151-7.

DOSTAL, M. & JELINEK, R. (1973). Sensitivity of embryos and intraspecies differences in mice in response to prenatal administration of corticoids. *Teratology*, **8**, 245-52.

FOGH-ANDERSON, P. (1942). *Inheritance of Harelip and Cleft Palate*. Copenhagen: Busck.

FRASER, F. C. & FAINSTAT, T. D. (1951). Production of congenital defects in the offspring of pregnant mice treated with cortisone. *Pediatrics*, **8**, 527-33.

FRASER, F. C., KALTER, H., WALKER, B. G. & FAINSTAT, T. D. (1954). The experimental production of cleft palate with cortisone and other hormones. *Journal of Cellular and Comparative Physiology, Supplement 1*, **43**, 237–59.

GILLMAN, J., GILBERT, C., GILLMAN, T. & SPENCE, I. (1948). A preliminary report on hydrocephalus, spina bifida, and other congenital anomalies in the rat produced by trypan blue. *South African Journal of Medical Sciences*, **13**, 47–90.

GIROUD, A. & MARTINET, M. (1959). Extension à plusieurs éspeces de mammifères des malformations embryonnaires par hyperitaminose A. *Compte rendu des Séances de la Société de Biologie*, **153**, 201–2.

GLAUERT, A. M., DANIEL, M. R., LUCY, J. A. & DINGLE, J. T. (1963). Studies on the mode of action of excess of vitamin A. VII. Changes in the fine structure of erythrocytes during haemolysis. *Journal of Cell Biology*, **17**, 111–21.

GOLDMAN, A. S., KATSUMATA, M., YAFFE, S. J. & GASSER, D. L. (1977). Palatal cytosol cortisol-binding protein associated with cleft palate susceptibility and H-2 genotype. *Nature, London*, **265**, 643–5.

GOLDMAN, A. S., KATSUMATA, M., YAFFE, S. & SHAPIRO, B. H. (1976). Correlation of palatal cortisol receptor levels with susceptibility to cleft palate teratogenesis. *Teratology*, **13**, 22A.

GOLDSTEIN, M., PINSKY, M. F. & FRASER, F. C. (1963). Genetically determined organ specific responses to the teratogenic action of 6-aminonicotinamide in the mouse. *Genetical Research*, **4**, 258–65.

GREENE, R. M. & KOCHHAR, D. M. (1974). Surface coat on the epithelium of developing palatine shelves in the mouse as revealed by electron microscopy. *Journal of Embryology and Experimental Morphology*, **31**, 683–92.

GREENE, R. M. & PRATT, R. M. (1976). Developmental aspects of secondary palate formation. *Journal of Embryology and Experimental Morphology*, **36**, 225–45.

HALE, F. (1933). Pigs born without eye balls. *Journal of Heredity*, **24**, 105–6.

HALE, F. (1935). Relation of vitamin A to anophthalmos in pigs. *American Journal of Ophthalmology*, **18**, 1087–93.

HALE, F. (1937). Relation of maternal vitamin A deficiency to microphthalmia in pigs. *Texas State Journal of Medicine* **33**, 228–32.

HASSELL, J. R. (1975). The development of rat palatal shelves *in vitro*. *Developmental Biology*, **45**, 90–102.

HEINE, W., KIRCHMAIR, H., FIEDLER, M. & STUEWE, W. (1964). Thalidomid-Embryopathie im Tierversuch (I). *Zeitschrift für Kinderheilkunde*, **91**, 213–21.

INGALLS, T. H. L., AVIS, F. R., CURLEY, F. J. & TEMIN, H. M. (1953). Genetic determination of hypoxia-induced congenital abnormalities. *Journal of Heredity*, **44**, 185–94.

INGALLS, T. H. L., CURLEY, F. J. & PRINDLE, R. A. (1950). Anoxia as a cause of fetal death and congenital defect in the mouse. *American Journal of Diseases of Children*, **80**, 34–45.

KALTER, H. (1954). The inheritance of susceptibility to the teratogenic action of cortisone in mice. *Genetics*, **39**, 185.

KARP, G. C. & SOLURSH, M. (1974). Acid mucopolysaccharide metabolism, the cell surface, and primary mesenchyme cell activity in the sea urchin embryo. *Developmental Biology*, **41**, 110–23.

KOCHHAR, D. M. & JOHNSON, E. M. (1965). Morphologic and autoradiographic studies of cleft palate induced in rat embryos by maternal hypervitaminosis A. *Journal of Embryology and Experimental Morphology*, **14**, 223–38.

KWASIGROCH, T. E. & KOCHHAR, D. M. (1975). Locomotory behavior of limb bud cells: effect of excess vitamin A *in vivo* and *in vitro*. *Experimental Cell Research*, **95**, 269–78.

LECK, I. (1974). Causation of neural tube defects: clues from epidemiology. *British Medical Bulletin*, **30**, 158–63.

LECK, I. (1977). Correlations of malformation frequency with environmental and genetic attributes in man. In *Handbook of Teratology*, vol. 3, ed. J. G. Wilson & F. C. Fraser, pp. 243–324. New York: Plenum Press.

LEWIS, C. A., PRATT, R. M., PENNYPACKER, J. P. & HASSELL, J. R. (1978). Inhibition of limb chondrogenesis *in vitro* by vitamin A: alterations in cell surface characteristics. *Developmental Biology*, **64**, 31–47.

LORBER, J., STEWART, C. R. & MILFORD-WARD, A. (1973). Alpha-fetoprotein in antenatal diagnosis of anencephaly and spina bifida. *Lancet*, **i**, 1187.

LUCY, J. A., LUSCOMBE, M. & DINGLE, J. T. (1963). Studies on the mode of action of excess of vitamin A. VIII. Mitochondrial swelling. *Biochemical Journal*, **89**, 419–25.

McCOLLUM, E. V. & DAVIS, M. (1915). The nature of the dietary deficiencies of rice. *Journal of Biological Chemistry*, **23**, 181–230.

McGILLIVRAY, W. A. (1961). Some factors influencing the release of vitamin A from the liver. *British Journal of Nutrition*, **15**, 305–12.

MARIN-PADILLA, M. (1966). Mesodermal alterations induced by hypervitaminosis A. *Journal of Embryology and Experimental Morphology*, **15**, 261–9.

MAYES, B. (1957). Rubella and the unborn. *Triangle*, **3**, 10–16.

MORRISS, G. M. (1972). Morphogenesis of the malformations induced in rat embryos by maternal hypervitaminosis A. *Journal of Anatomy* **113**, 241–50.

MORRISS, G. M. (1973a). The ultrastructural effects of excess maternal vitamin A on the primitive streak stage rat embryo. *Journal of Embryology and Experimental Morphology*, **30**, 219–42.

MORRISS, G. M. (1973b). An electron microscopic study of the effects of hypervitaminosis A on the maternal–embryonic relationship of the rat at 8 days of pregnancy. *Journal of Reproduction and Fertility*, **33**, 451–6.

MORRISS, G. M. (1975). Abnormal cell migration as a possible factor in the genesis of vitamin A-induced craniofacial anomalies. In *New Approaches to the Evaluation of Abnormal Mammalian Embryonic Development*, ed. D. Neubert & K.-H. Merker, pp. 678–87. Berlin: Geo. Thieme Verlag.

MORRISS, G. M. & STEELE, C. E. (1974). The effects of excess vitamin A on the development of rat embryos in culture. *Journal of Embryology and Experimental Morphology*, **32**, 505–14.

MORRISS, G. M. & STEELE, C. E. (1977). Comparison of the effects of retinol and retinoic acid on postimplantation rat embryo *in vitro*. *Teratology*, **15**, 109–20.

MORRISS, G. M. & THOROGOOD, P. V. (1978). An approach to cranial neural crest cell migration and differentiation in mammalian embryos. In *Development in Mammals*, vol. 3, ed. M. H. Johnson, pp. 363–412. Amsterdam: North-Holland.

MURAKAMI, N. & KAMEYAMA, Y. (1963). Vertebral malformation in the mouse foetus caused by maternal hypoxia during early stages of pregnancy. *Journal of Embryology and Experimental Morphology*, **11**, 107–18.

MURPHY, D. P. (1929). The outcome of 625 pregnancies in women subjected to pelvic radium or roentgen irradiation. *American Journal of Obstetrics and Gynecology*, **18**, 179–87.

NEW, D. A. T. (1978). Whole-embryo culture and the study of mammalian embryos during organogenesis. *Biological Reviews*, **53**, 81–122.

NISHIMURA, H. & SHIOTA, K. (1977). Summary of comparative embryology and teratology. In *Handbook of Teratology*, vol. 3, ed. J. G. Wilson & F. C. Fraser, pp. 119–54. New York: Plenum Press.

NOLEN, G. A. (1969). Variations of the teratogenic response to hypervitaminosis A in 5 strains of the albino rat. *Food and Cosmetics Toxicology*, **7**(3), 209–14.

NOWACK, E. (1965). Die sensible Phase bei der Thalidomid-Embryopathie. *Humangenetik* **1**, 516–36.

PERSAUD, T. V. N. (1977). *Problems of Birth Defects*. Lancaster: MTP Press.

PILOTTI, G. & SCORTA, A. (1965). Ipervitaminosi A gravidica e malformazioni neonatali dell'apparto urinario. *Minerva Ginecologia* **17**, 1103–8.

POSWILLO, D. (1975). The pathogenesis of the Treacher–Collins syndrome (mandibulofacial dysostosis). *British Journal of Oral Surgery*, **13**, 1–26.

POURTOIS, M. (1972). Morphogenesis of the primary and secondary palate. In *Developmental Aspects of Oral Biology*, ed. H. C. Slavkin & L. A. Bavetta, pp. 81–108. New York & London: Academic Press.

PRATT, R. M., GIBSON, W. A. & HASSELL, J. R. (1973). Concanavalin A binding to the secondary palate of the embryonic rat. *Journal of Dental Research*, **52**, 111.

PRATT, R. M., GREENE, R. M., HASSELL, J. R. & GREENBERG, J. (1975). Epithelial cell differentiation during secondary palate development. In *Extracellular Matrix Influences on Gene Expression*, ed. H. C. Slavkin & R. C. Greulich, pp. 561–5. New York & London: Academic Press.

PRATT, R. M. & GREENE, R. M. (1976). Inhibition of palatal epithelial cell death by altered protein synthesis. *Developmental Biology*, **54**, 135–45.

PRATT, R. M. & HASSELL, J. R. (1974). Prefusion synthesis of a carbohydrate-rich surface coat in the rat palatal epithelium. *Journal of Dental Research*, **53**, 64.

RAZ, A., KANAI, A. & GOODMAN, DE W. S. (1968). Retinol binding protein (RBP): a newly recognized plasma lipoprotein. *Federation Proceedings*, **27**, 594.

REES, D. A., LLOYD, C. W. & THOM, D. (1977). Control of grip and stick in cell adhesion through lateral relationships of membrane glycoproteins. *Nature, London*, **267**, 124–8.

RENWICK, J. H. (1972). Hypothesis: anencephaly and spina bifida are usually preventable by avoidance of a specific but unidentified substance present in certain potato tubers. *British Journal of Preventive and Social Medicine*, **26**, 67–88.

RODAHL, K. & MOORE, T. (1943). The vitamin A content and toxicity of bear and seal liver. *Biochemical Journal*, **37**, 166–8.

ROELS, O. A. (1969). The fifth decade of vitamin A research. *American Journal of Clinical Nutrition*, **22**, 903–7.

RUSSELL, L. B. & RUSSELL, W. L. (1954). An analysis of the changing radiation response of the developing mouse embryo. *Journal of Cellular and Comparative Physiology, Supplement 1*, **43**, 103–50.

SALOMON, D. S. & PRATT, R. M. (1976). Glucocorticoid receptors in embryonic facial mesenchyme cells. *Nature, London*, **264**, 174–7.

SCHWARTZ, N. B. (1977). Regulation of chondroitin sulfate synthesis. Effect of beta-xylosides on synthesis of chondroitin sulfate proteoglycan, chondroitin sulfate chains, and core protein. *Journal of Biological Chemistry*, **252**, 6316–21.

SMITHELLS, R. W. & SHEPPARD, S. (1978). Teratogenicity testing in humans: a method demonstrating safety of bendictin. *Teratology*, **17**, 31–5.

SOLURSH, M. & MEIER, S. (1973). The selective inhibition of mucopolysaccharide synthesis by vitamin A treatment of cultured chick embryo chondrocytes. *Calcified Tissue Research*, **13**, 131–42.

STEVENSON, A. C. (1960). The association of hydramnios with congenital malformations. In *Congenital Malformations, Ciba Foundation Symposium*, ed. G. E. W. Wolstenholme & C. M. O'Connor, pp. 241–63. London: Churchill.

TELSER, A., ROBINSON, H. C. & DORFMAN, A. (1965). The biosynthesis of chondroitin-sulphate protein complex. *Proceedings of the National Academy of Sciences of the USA*, **54**, 912–19.

TORPIN, R. (1968). *Fetal Malformations caused by Amnion Rupture during Gestation*. Springfield, Illinois: Thomas.

VON HIPPEL, E. (1907). Über experimentelle Erzeugung von angeborenen Star bei Kaninchen nebst bemerkungen über gleichzeitig Beobachteten Mikrophthalmus und Lidcolobom. *Archives of Ophthalmology*, **65**, 325–60.

WARKANY, J. & NELSON, R. C. (1940). Appearance of skeletal abnormalities in the offspring of rats reared on a deficient diet. *Science*, **92**, 383–4.

WARKANY, J. & NELSON, R. C. (1941). Skeletal abnormalities in offspring of rats reared on deficient diets. *Anatomical Record*, **79**, 83–100.

WARKANY, J., ROTH, C. B. & WILSON, J. G. (1948). Multiple congenital malformations: a consideration of etiological factors. *Pediatrics*, **1**, 462–71.

WARKANY, J. & SCHRAFFENBERGER, E. (1943). Congenital malformations induced in rats by maternal nutritional deficiency. V. Effects of a purified diet lacking riboflavin. *Proceedings of the Society for Experimental Biology and Medicine*, **54**, 92–4.

WARKANY, J. & SCHRAFFENBERGER, E. (1946). Congenital malformations induced in rats by maternal vitamin A deficiency. I. Defects of the eye. *Archives of Ophthalmology*, **35**, 150–69.

WARKANY, J. & SCHRAFFENBERGER, E. (1947). Congenital malformations induced in rats by roentgen rays. *American Journal of Roentgenology*, **57**, 455–63.

WESSELLS, N. K., SPOONER, B. S., ASH, J. F., BRADLEY, M. O., LUDUENA, M. A., TAYLOR, E. L., WRENN, J. T. & YAMADA, K. M. (1971). Microfilaments in cellular and developmental processes. *Science*, **171**, 135–43.

WIERSIG, D. O. & SWENSON, M. J. (1967). Teratogenicity of vitamin A in the canine. *Federation Proceedings*, **26**, 486.

WILSON, J. G. (1971). Use of rhesus monkeys in teratological studies. *Federation Proceedings*, **30**, 104–9.

WILSON, J. G. (1972). Abnormalities of intrauterine development in non-human primates. In *The Use of Non-human Primates in Research on Human Reproduction*, ed. E. Diczfalusy & C. C. Standley, pp. 261–92. Stockholm: WHO Research and Training Center on Human Reproduction.

WILSON, J. G. & FRASER, F. C. (1977). *Handbook of Teratology*, vols. 1–4. New York: Plenum Press.

WILSON, J. G., JORDAN, H. C. & BRENT, R. L. (1953). Effects of irradiation on embryonic development. II. X-rays on the 9th day of gestation in the rat. *American Journal of Anatomy*, **92**, 153–87.

WILSON, J. G., ROTH, C. B. & WARKANY, J. (1953). An analysis of the syndrome of malformations induced by maternal vitamin A deficiency. Effects of restoration of vitamin A at various times during gestation. *American Journal of Anatomy*, **92**, 189–217.

WOLF, G. & VARANDANI, P. T. (1960). Studies on the function of vitamin A in mucopolysaccharide biosynthesis. *Biochimica et Biophysica Acta*, **43**, 501–12.

WOOLLAM, D. H. M. & MILLEN, J. W. (1960). The modification of the activity of certain agents exerting a deleterious effect on the development of the mammalian embryo. In *Congenital Malformations, Ciba Foundation Symposium*, ed. G. E. W. Wolstenholme & C. M. O'Connor, pp. 158–72. London: Churchill.

ZILVA, S. S., GOLDING, J., DRUMMOND, J. C. & COWARD, K. J. (1921). The relation of the fat-soluble factor to rickets and growth in pigs. *Biochemical Journal*, **15**, 427–37.

Effects of maternal diabetes on embryonic development in mammals

ELIZABETH M. DEUCHAR

Dept of Biological Sciences, Hatherly Laboratories, University of Exeter,
Exeter EX4 4PS, UK

Introduction: the evidence in man

Diabetes mellitus was first recognized as a disease in humans. It
was also in humans that the possibility that a mother's diabetes
might have ill-effects on the development of her offspring *in utero*
was first suspected. It is only in the last 10 to 14 years, however,
that sufficient data have been collected to show that there is
definitely an association between maternal diabetes and increased
incidences of congenital abnormalities in human infants. A more
frequent outcome of diabetic pregnancies used to be death of
infants in the perinatal period, owing sometimes to ketosis or
pre-eclampsia in the mother and sometimes to a low lecithin/
sphingomyelin ratio in the fetal lungs, causing abnormally high
surface tension in the alveolar fluid so that it was not easily cleared
from the lungs and replaced by air when the first breaths were
taken. Obstetricians are now able to foresee and to prevent such
complications as these, however, so that the incidence of perinatal
deaths among infants of diabetic mothers is very much lower than
it used to be. As a result, the higher incidence of developmental
abnormalities among these infants than in the general population
has become apparent (Malins, 1978). Pedersen and his collaborators (1964, 1967, 1971, 1974) published some of the first extensive
data on the higher incidence of abnormalities among children of
diabetic mothers, collected in Copenhagen over a period of 45
years. Since then further data have been collected in Holland,
Belgium and the United States (Bergstein, 1979; Naeve, 1979;
Bennett, Webner & Miller, 1979) which all show this preponder-
ance of abnormalities among children of diabetic mothers, and

at the same time rule out any connection with diabetes in the
father. So this is definitely a maternal effect. The present estimate
is that infants of diabetic mothers (IDMs) show three times as
many serious abnormalities as infants of non-diabetic mothers
(INDMs) (Beard & Hoet, 1979).

When the kinds of abnormalities occurring in IDMs are sur-
veyed, there appears to be no one type that is specific to them, but
simply an increase in the incidence of the commoner serious
abnormalities such as cardiac defects, anencephaly and spina
bifida, urinogenital defects and anorectal atresia (Gabbe, 1977).
There is, however, one rare condition, the so-called 'caudal
regression syndrome' which has been seen more often in IDMs
than in INDMs (Stern, Ramos & Light, 1965). A better name for
it is 'sacro-coccygeal agenesis' and it was first described by Hohl
(1852). The syndrome has varying forms, but at its most serious
involves absence or hypoplasia of the femur on one or both sides,
as well as absence of some or all of the sacral vertebrae. Isolated
cases, all in IDMs, have been described by Assemany (1972),
Fields, Schwarz, Dickens & Tunnessen (1968), Passarge & Lenz
(1966) and Rusnak & Driscoll (1965). Smith (1970) regards the
condition as being due to a deficiency of the caudal paraxial
mesoderm in the early embryo, and would include urinogenital
and hindgut defects in this syndrome. There is no clear evidence
in humans to support Smith's view, however, and in rats an
apparently similar condition that I have seen arises much later in
development, by a delay in ossification of skeletal elements that
are present as mesenchyme or cartilage.

Apart from this one special type of abnormality, and the low
lecithin/sphingomyelin ratio in the lungs mentioned above, there
is no other special characteristic of IDMs at birth except that they
are often obese and overweight: in fact the delivery of an overweight
baby may be the first indication to the clinician that a woman is
diabetic or prediabetic (Pedersen, 1967).

In order to find out the possible reasons for the ill-effects that
diabetic mothers evidently have on their offspring developing *in
utero*, it has been necessary to look at diabetic mammals other than
man. Most of the experimental work on the effects of maternal
diabetes has been carried out on monkeys, mice and rats.

Experimental work on animals
Monkeys

Some outstanding work on monkeys has been carried out by Chez and his associates in the United States. Mintz, Chez & Hutchinson (1972) induced diabetes in females of the macaque (*Macaca mulatta*) in the first trimester of pregnancy by injecting them with the drug streptozotocin which attacks insulin-secreting cells of the pancreas (Rerup, 1970). Perhaps because the diabetes was induced at a stage when organogenesis in the embryo was already complete, these workers did not see any malformations in the fetuses. They found, however, that some stillbirths resulted and that both fetuses and placentae were heavier than normal. Another feature of the fetuses borne by diabetic monkeys was a high level of insulin in the blood: 2.6 ng/ml compared with the normal 0.7 ng/ml. This high basal level was also more readily elevated in response to increased amino acids and glucose, than was the basal level in fetuses from non-diabetic monkeys. High insulin levels are found in fetuses of other diabetic animals, and also in IDMs, as a response to the high level of glucose in the maternal blood. The possible effects of these high fetal insulin levels and of the elevated glucose levels on developmental processes will be discussed later in connection with my own work on rats.

Chez and his collaborators (Chez, 1979) have recently been able to keep monkey fetuses alive for short periods outside the uterus, in a chamber which allows for monitoring and alteration of the supply of metabolites. Their findings, because they are on primates, will have more obvious clinical relevance than work on other mammals. Primates are not convenient, however, for surveys of the incidence of fetal abnormalities, since their long gestation period and small numbers of offspring make the collection of data very slow. For this purpose, rats and mice are far more suitable.

Rodents

In rats, until the recent report of a spontaneously diabetic Wistar strain in Canada (Nakhooda *et al.*, 1978), no hereditarily diabetic strains have been available. But in mice, several diabetic strains are known: for instance, the yellow/obese mouse carrying the dominant A^y; the recessive *ob/ob* and *db/db* strains, and the KK and NZO strains in which diabetes is polygenically controlled (Chick & Like, 1970; Wyse & Dulin, 1970; Dulin & Wyse, 1970; Herberg & Coleman, 1977). All of these are very difficult to breed from, however, as the oestrous cycle of diabetic females is irregular and homozygous diabetics are usually infertile. Robertson (1942) and Pedersen (1974) showed that yellow/obese mice could conceive, but that the embryos, if homozygous, became abnormal at morula stages and failed to implant, the trophoblast and inner cell mass dying at the blastocyst stage. These strains do not, therefore, lend themselves to studies on late embryonic development and organogenesis.

It has now become common practice, in both rats and mice, to induce diabetes in normal animals with either the drug alloxan or, better, streptozotocin, which has fewer side effects (Rudas, 1969). Both drugs attack the pancreatic beta-cells preferentially. An animal injected intravenously with one of these drugs will develop symptoms of diabetes (i.e. thirst, excessive urination, loss in weight and blood glucose levels above 200 mg/100 ml) after 48 hours. If the pancreas is examined histologically, the islets appear shrunken and the beta-cells empty of cytoplasm. The possibility that the embryos developing in a pregnant female could also be affected by the drugs or their breakdown products has also to be borne in mind although, as we shall see later, this risk is negligible if the female is injected immediately after mating.

The best-known study on fetal abnormalities associated with maternal diabetes in mice is that of Watanabe & Ingalls (1963). They administered high doses (100 mg/kg body weight) to females on the 9th, 10th, 11th, 12th, 13th or 14th day after mating. As a result they obtained many cases of cleft palate in the offspring, a few of ectromelia and a few deformities of the ribs. The skeletal defects were more frequent if the alloxan had been injected on the 14th day of pregnancy, whereas the cleft palate cases were more frequent in animals injected on days 9–11. Horii, Watanabe & Ingalls (1966) went on to show that giving 0.4 units of insulin per day to the alloxan-treated females, thus reducing their blood glucose levels to normal (100 mg/100 ml or less) virtually eliminated the fetal abnormalities.

Like many other workers who have used alloxan or streptozotocin to induce diabetes during pregnancy, Horii et al. did not exclude the possibility that the drug itself could have affected the embryos. They claimed, however, that alloxan acted on the pancreatic beta-cells within 5–10 minutes and was then destroyed very rapidly in the body. The fact that giving insulin to the treated females prevented fetal abnormalities suggests fairly strongly, anyway, that the cause of the abnormalities had been the maternal diabetes and not the alloxan.

Rats have become more widely used than mice for studies on diabetes, perhaps because their larger size makes it easier to take blood samples from them for glucose assays. Most of the work on rats in the 1940s and 1950s was based on alloxan-induced diabetes. The chief effects observed (Davis, Fugo & Lawrence, 1947; Sinden & Longwell, 1949; Ferret, Linden & Morgans, 1950) were irregularities of the oestrous cycle in the females, deaths of fetuses if the animals did become pregnant, and also some loss of offspring before weaning. All of these effects could be prevented by giving insulin to the alloxan-treated females to cure their diabetic symptoms. None of these workers reported abnormalities in rat fetuses as a result of maternal diabetes, but it is possible that some of the intra-uterine deaths may have been due to severe malformations. In rats, embryos are resorbed rapidly after death, until eventually nothing is visible except the implantation site and a small mass of necrotic cells – so it is possible for some abnormal embryos to escape observation altogether.

In more detailed work on rats treated with alloxan, Lazarow and co-workers (Lazarow, Kim & Wells, 1960; Kim, Runge, Wells & Lazarow, 1960*a*, *b*) paid attention to body size and to the development of the pancreas in fetuses of diabetic mothers. They used low doses of the drug to produce only mild diabetes which allowed pregnancy to proceed successfully. They showed that the growth rate of fetuses in diabetic rats was retarded and that at 21 days their weight was from 13 to 40 % less than that of controls. Gestation was often prolonged by several hours, however, so that the newborn of diabetic rats could be heavier than those of controls. There was also hyperplasia of the fetal pancreatic islets, which they presumed was a response to the maternal hypergly-caemia. Later work has confirmed that fetal rat pancreases, both *in vivo* and *in vitro*, respond to increased glucose in the surround-ings by hyperplasia of the islets and increased secretion of insulin (Van Assche, 1975; de Gasparo, Pictet, Rall & Rutter, 1975).

When streptozotocin came into use as a diabetogenic drug in the

1960s it proved more satisfactory than alloxan for studies on pregnant rats because it could produce a mild, chronic diabetes with long survival of the animals. Golob and co-workers (Golob & Becker, 1969; Golob, Rishi & Becker, 1969; Golob, Rishi, Becker & Moore, 1970) gave streptozotocin to female Sprague–Dawley rats on day 5 of pregnancy, then examined fetuses on days 18, 19 and 21, as well as offspring 4 days *post partum*. They did not note any abnormalities except for smaller size of fetuses as compared with controls. There were, however, many fetal losses, and only 38.8 % of the streptozotocin-treated females had living fetuses, as compared with 70.3 % of controls. Sybulski & Maughan (1971) carried out a similar study on Wistar rats, injecting females with streptozotocin on day 6, 7, 8 or 9 of pregnancy. They also found no abnormalities in the resulting fetuses except that they were smaller than those of controls: their blood glucose levels at term were also high (265 mg/100 ml). Their placentae were also larger than normal, and the glycogen contents of both placenta and fetal liver were higher in diabetic animals than in controls. In view of this enlargement of the placentae, the small size of the fetuses is puzzling.

In more recent studies of streptozotocin-diabetic pregnant rats Van Assche (1975) confirmed the reduction in fetal size and also studied the fetal pancreases in more detail. The pancreatic islets were found to occupy a greater percentage of the pancreatic tissue area in fetuses of diabetics than in controls, but the beta-granules were depleted, indicating that more insulin than normal had been released into the blood-stream in these fetuses.

In view of the clinical findings, as well as those of Horii *et al.* (1966) on mice, it is surprising and in a sense disappointing that none of these investigators on rats was able to show an increase in the incidence of congenital malformations as a result of maternal diabetes. The most likely explanation seems to be that, just as the increased incidence in humans was at first masked by a high rate of perinatal deaths, so in the rat, deaths of conceptuses and their rapid resorption has prevented the discovery of developmental abnormalities in earlier work. When I started my own investigations of the effects of maternal diabetes on embryonic development in rats, therefore, I decided first to examine embryos immediately after the period of organogenesis in which malformations might occur, and if possible before resorption could set in.

Effects of maternal diabetes on embryos of Wistar rats

This work has been described fully elsewhere (Deuchar, 1977, 1979a, b) so I will merely summarize it briefly. Female Wistar rats were made diabetic by a single intravenous injection of 45 mg/kg alloxan or streptozotocin after they had been mated to normal males. Embryos were then examined at 11–13 days' gestation, i.e. immediately after the completion of early organogenesis. It was found that already at this stage many embryos had been resorbed in the diabetic animals. Among the survivors, however, small numbers of abnormalities of the central nervous system and the heart could be observed histologically. The central nervous system (CNS) abnormalities were failures of either the anterior or the posterior neuropore to close: conditions which could lead later to either anencephaly or spina bifida. In the heart, there were distorted or abnormally small atria. These abnormalities were not exclusive to embryos of diabetic rats: they also occurred among control embryos, but at a significantly lower frequency (Table 1).

Having seen evidence that maternal diabetes could cause increased numbers of embryonic malformations in Wistar rats, it seemed worth examining some at later fetal stages to see if abnormalities could also be detected then. A summary of the findings is given in Table 2. Perhaps the most striking fact is that in these no abnormalities of the central nervous system or the heart were seen. So evidently the cases seen at 11–13 days do not survive: they are presumably resorbed before 20 days. The number of resorptions was again relatively high in diabetic animals. There were some striking abnormalities among the surviving fetuses. One was the rare but extreme anomaly shown in Fig. 1 (a), in which all the viscera were externalized. This appeared to have resulted from deficient growth of the abdominal wall. Another even rarer but also striking anomaly was extrusion of the tonge: this was found to be associated with a shorter than normal lower jaw. The most consistent and the commonest abnormality among these 20-day fetuses of diabetic rats was revealed in X-rays of the skeleton, which showed deficient ossification of the sacral vertebrae, compared with normal fetuses (cf. Fig. 1a and b). These sacral defects are particularlyinteresting, as they resemble one of the features of the 'caudal regression syndrome' in human IDMs, discussed earlier. It is possible that

Table 1. *Incidence of abnormalities in 11–13-day Wistar rat embryos*

	Total live embryos	Resorptions	CNS abnormal	Heart abnormal
(a) Alloxan				
Diabetics	120	34	7	2
Controls	136	10	3	0
(b) Streptozotocin				
Diabetics	190	38	19	10
Controls	176	8	5	3

they may result from hyperinsulinaemia in the fetuses in response to the mother's hyperglycaemia, for it is well known that adding insulin to bone cultures *in vitro* deranges the process of ossification (Chen, 1954; Hay, 1958; Reynolds, 1972), and that high doses of insulin given to pregnant mammals (Lichtenstein, Guest & Warkany, 1951; Smithberg & Runner, 1963) can cause skeletal defects in their fetuses. Examining the pancreases of these fetuses from Wistar rats, we found that there was hyperplasia of the islets in those from diabetic mothers and that the beta-cells appeared to have discharged their insulin (Deuchar, 1979a): this confirms the findings of Van Assche (1975) and of Kim et al. (1960a) mentioned above.

The sacral ossification defect is seen in fetuses of Wistar rats made diabetic with either alloxan or streptozotocin and it is prevented, as are the other anomalies, by administering sufficient insulin to the diabetic females to reduce their blood glucose levels to normal (Table 2). These two findings support the view that it is the maternal diabetes, rather than any side-effects of the drugs themselves on the embryos, which causes these developmental

Fig. 1. (*a*) Fetus with viscera completely extruded through ventral abdominal wall. (*b*) X-ray of lumbosacral region of fetus with incompletely ossified sacral vertebrae. Note that vertebrae scarcely show at all in the region arrowed. (*c*) X-ray of lumbosacral region in normal 20-day fetus. Sacral vertebrae show clearly in arrowed region. (*d*) Sprague–Dawley fetus with viscera only slightly extruded and enclosed in remains of body stalk (arrow). (*e*) Sprague–Dawley fetus with anencephaly. Arrow indicates exposed, degenerate, brain tissue. All figures are to same magnification. Scale is indicated on (*a*).

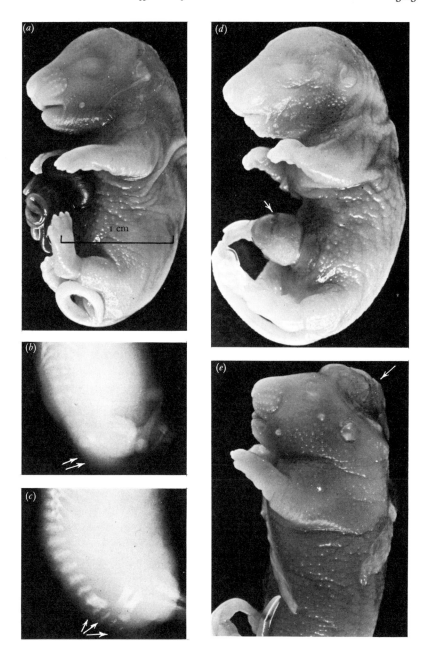

Table 2. *Incidence of abnormalities in 20-day fetuses*
of Wistar rats

	Total live fetuses	Resorp-tions	Small size	Viscera everted	Sacral defects	Jaw defects
(a) *Alloxan*						
Diabetics	333	4	12	0	27	0
Controls	141	0	2	0	0	0
(b) *Streptozotocin*						
Diabetics	450	70	35	11	74	2
Controls	339	1	1	0	0	0
Insulin-treated diabetics	180	0	0	0	0	0

abnormalities. Besides, the findings of Horii *et al.* (1966) on the rapid action and elimination of alloxan in mice, as well as work by Karunanayake, Hearse & Mellows (1976) with radioactively labelled streptozotocin in rats, which also showed a very rapid breakdown and elimination of the drug, make it most unlikely that either of these drugs could have time to affect developing embryos if, as in my experiments, the females are injected on the morning after mating. Nevertheless, I felt that some more convincing evidence was needed to eliminate the possibility that the abnormalities seen in these Wistar rat embryos and fetuses were caused by the diabetogenic drugs. I therefore carried out two types of test: first, examining embryos only a few days after injection of the female rats with alloxan or streptozotocin, and secondly, testing the effects of both these drugs on embryos grown *in vitro* during their early organogenesis stages (Deuchar, 1979*a*).

For the first test female rats which had been given the usual dose of either alloxan or streptozotocin on day 0 were sacrificed on day 4 and embryos flushed out of the uteri were examined microscopically. Embryos from control animals which had been injected with the solvents but not the drugs themselves were also examined. The results (for details see Deuchar, 1979*a*) were as follows: (1) there was no loss of embryos as a result of the drug treatments, as judged from the agreement between counts of ruptured follicles and of embryos retrieved; (2) morphological abnormalities were not significantly more frequent in embryos of drug-treated animals than in controls. It is not known yet whether there was any difference in the potentiality for further development in embryos of diabetic rats as compared with controls: this could perhaps be tested by cultures *in vitro* or by transfer to another female.

For the second type of test rat embryos of 9 and 10 days' gestation were grown *in vitro* for 24 hours, using the watch-glass culture method of New (1966). Either alloxan or streptozotocin was added to the rat serum used as culture medium, at a concentration equivalent to that in the blood of injected females on day 0. Control embryos were cultured in serum to which the solvents of the two drugs had been added. At the end of the culture period the embryos were examined externally, their dimensions were measured and they were also examined histologically. In alloxan-treated embryos there was no decrease in viability, no decrease in size, and no sign of abnormalities, as compared with controls. Embryos cultured in the presence of streptozotocin, however, did suffer some decrease in viability and in size, and also showed retarded development of the liver (Deuchar, 1978). They did not, though, show any abnormalities of the central nervous system or the heart, such as had been seen in early embryos that had developed *in vivo* in females made diabetic with streptozotocin.

These findings therefore show that alloxan, in the doses given to produce diabetes in my experiments, presents no threat to organogenesis stages in rat embryos, but that streptozotocin, if administered at these stages, could cause some abnormalities. In my own work, streptozotocin was never given later than day 0 of pregnancy. It seems fair to conclude, therefore, that the abnormalities we have seen in embryos and fetuses of Wistar rats were indeed caused by the maternal diabetes and not by the diabetogenic drugs.

The problem of the maternal effects in diabetes

The question to which scientists and clinicians alike most urgently want an answer is that of what factors in a diabetic mother's metabolism lead to the increased risk of her offspring dying or becoming abnormal before birth. Extensive data have been collected (Freinkel, 1979) on the concentrations of metabolites and

Table 3. *Cultures of 10-day embryos from diabetic and normal rats in either diabetic or normal serum*

	Total embryos	No. with cell death	Vescicle diameter	Body length	Tail length
(a) Embryos from diabetics					
Diabetic serum	52	8	3.2 ± 0.07	2.3 ± 0.09	2.1 ± 0.07
Normal serum	60	20	3.0 ± 0.08	2.1 ± 0.06	1.9 ± 0.05
(b) Embryos from controls					
Diabetic serum	30	1	3.3 ± 0.08	2.5 ± 0.09	2.2 ± 0.06
Normal serum	59	19	3.0 ± 0.07	2.2 ± 0.06	1.8 ± 0.07

hormones in the blood of pregnant diabetic women: in fact clinical work along these lines is well in advance of animal studies, perhaps because it is easier to catheterize and to obtain frequent blood samples from a conscious and unperturbed human than it is from a conscious and unperturbed rat! However, in the rat there are alternative, experimental approaches by which one may attempt to discover what serum-borne factors in diabetic females may influence the early development of embryos. My first, simple approach has been to grow 10–11-day embryos in watch-glass cultures, using serum from diabetic rats as culture medium, and to see if these embryos develop less well than in normal serum.

Contrary to my expectation, results so far have indicated that diabetic rats' serum may have *beneficial* effects on this stage of embryonic development *in vitro*. Embryos from either normal or diabetic rats, grown by New's method (1966), become larger in diabetic serum than in normal serum, and also less frequently show necrotic cells in histological section (Table 3). The apparently beneficial effect of the diabetic serum may be due to its high glucose content: in a number of measurements of glucose uptake by cultured embryos we have found that in 24 hours they can take up two to three times as much from diabetic serum as from normal serum.

The serum used for cultures was taken from rats bled at 10 days' gestation. So there are evidently no factors in the serum of diabetic rats at 10 days that derange organogenesis in the embryo. It remains possible, however, that the 10-day embryo in a diabetic mother has already acquired some ill-effect from the maternal

Table 4. *Cultures of 10-day embryos from diabetic and normal rats in normal serum*

(a) *External morphology*

	Total embryos	Good	Fair	Poor	Dead
Embryos from diabetics	60	33	17	5	5
Embryos from controls	59	45	3	4	7

(b) *Histology*

	Total embryos	Normal	Abnormal CNS/heart	Retarded	Cell death
Embryos from diabetics	52	21	4	10	20
Embryos from controls	50	28	2	3	19

blood at an earlier stage, which will increase the risks of abnormal organogenesis later. To test the possibility that 10-day embryos from diabetic rats are already potentially abnormal, I have cultured some of these in normal serum and compared their development with that of 10-day embryos from control rats. Embryos of similar size and stage of rotation were selected for comparison, and after 24 hours in culture their external appearance, size and histology were observed (Deuchar, 1979b). Table 4 summarizes the findings. It is clear from the table that embryos from diabetic rats do not grow so well in culture as those from normal rats. Fewer of them could be classed as 'Good' (i.e. large, well advanced and with a rapid circulation of red blood), and more were only 'Fair' (i.e. smaller and with little or no blood circulation), than among the controls. Histologically, fewer of the embryos from diabetic rats were entirely normal than were the control embryos and more were retarded. There was no significant difference in size between embryos from diabetic and control rats after culture, however.

From these preliminary results the only conclusion one can draw at present, pending further data, is that some deleterious factor or factors from diabetic female rats affects their embryos during the period between 4 and 10 days' gestation (since the 4-day embryos appeared normal, and at 10 days the female's serum has no ill-effects). Blood samples are now being collected from normal and from diabetic rats over the period from day 3 to day 10 of pregnancy and analysed for their content of metabolites and hormones, in the hopes of pinpointing one or more differing factors in diabetic females' blood that may be responsible for ill-effects on embryos. From clinical evidence it seems that one main difference between the sera of diabetic and of normal women during pregnancy is that the diabetic is subject to steeper fluctuations in the concentrations of glucose, insulin, amino acids and ketones. But unfortunately there are no clinical data covering the period immediately after implantation or the period of early organogenesis, because pregnancy is so rarely confirmed before 8 weeks after fertilization. So there is still a very broad spectrum of possibilities to explore, in trying to find out by what mechanism the diabetic mother causes abnormalities in her offspring.

Genetic variations and maternal diabetes

In a symposium which has been concerned with effects on embryonic development passed via the maternal gamete as well as via the uterine environment, some consideration should be given to the possibility that hereditary factors are also involved in the causation of abnormalities in embryos of diabetic mothers. In man the tendency to develop diabetes runs in families and is thought to be governed by multiple recessive factors (Creutzfeldt, Köbberling & Neel, 1976). In mice, as we saw earlier, there are diabetic strains in which the trait may be controlled by single pairs of alleles, or by polygenic effects (Coleman & Hummel, 1967; Dulin & Wyse, 1970). There are also, in rats, obese strains in which diabetes is easy to induce (Barry & Bray, 1969; Zucker & Zucker, 1961). But I do not know of any data on genetic trends in the incidence of abnormalities among embryos of either human or animal diabetic mothers. Nevertheless, a striking feature of our own findings was that the gross abnormalities such as eversion of viscera were shown by only one or two individuals in a litter. Since all members of each litter were exposed to the same maternal environment, this variation in response is difficult to explain except in terms of some genetic variation among the members of

Table 5. *Incidence of abnormalities in 20-day fetuses of Sprague–Dawley rats*

	Total fetuses	Resorptions	Small size	Viscera everted	Anencephaly	Head and jaw defects	Sacral defects
Streptozotocin diabetics	374	13	284	3	1	7	225
Controls	249	0	1	0	0	0	7

each litter. By a long inbreeding programme one might perhaps be able to select for higher incidence of these abnormalities.

Not having the time or space for such an inbreeding test on Wistar rats, I have instead used a stock of Sprague–Dawley rats to compare with the Wistars and have looked at the incidences of abnormalities in them. Table 5 summarizes the data so far. As with the Wistars, fetuses are examined at 20 days, after females have been injected with streptozotocin (or with its solvent, for controls) on day 0 of pregnancy. The two most characteristic types of abnormality seen in Wistars – namely extrusion of the viscera and deficient ossification of the sacral vertebrae – occur in these fetuses too. The incidence of sacral defects is much higher than in Wistars, however, while the extrusion of the viscera was in two cases much less extreme than in Wistars (cf. Fig. 1*a* and *d*). In addition, there was one fetus with anencephaly (Fig. 1*e*), whereas this had never been seen at fetal stages in Wistars. So evidently the different genetic background in the Sprague–Dawley strain can cause differences in the frequency and expression of certain abnormal developmental tendencies that were seen in Wistars as a result of maternal diabetes. It was again found that only one or two individuals in any one litter showed the extreme anomalies, but on the other hand there were some litters in which all the fetuses were abnormally small. Other features in which these Sprague–Dawley embryos may resemble or differ from Wistar embryos of diabetic mothers are still being examined.

Conclusions

It is clear that we are still very far from knowing what characteristics of a diabetic mother's metabolism, or what factors that may
be transmitted from her to her developing offspring, cause the
higher incidence of congenital abnormalities that is characteristic
of this condition. Clinical data do not cover early enough periods
of pregnancy to have any relevance to organogenesis, and data on
animals are complicated by the need to induce diabetes with drugs
which may themselves have deleterious effects on embryos.
Although the findings I have presented here make it unlikely that
the particular abnormalities seen in Wistar rat embryos and
fetuses were due to alloxan or streptozotocin, one cannot absolutely
rule out the possibility that breakdown products of these drugs,
circulating in the female rats for several days after the injection,
might have had side-effects on the embryos too. Finally, one has
to admit that all the evidence that maternal diabetes causes fetal
malformations, both in humans and in other mammals, is still no
more than coincidental, and can become incontrovertible only
when a mechanism for this maternal effect is discovered. Current
clinical opinion is that fluctuations in maternal nutrients and in
the insulin content of the maternal serum may be crucial factors
affecting embryonic development. This opinion could be put to
experimental test, perhaps, with rat embryos in a longer-term
culture system such as that devised by New (1967) and used with
such success recently by Cockroft (1973) for embryos up to $14\frac{1}{2}$
days' gestation age. Our own results, however, suggest that some
deleterious maternal effects reach the embryo earlier than this,
before 10 days' gestation: so it may be more useful to use cultures
of pre-implantation and 7–10-day stages. The desired fluctuations
in nutrients and insulin could be introduced via the culture
medium.

Besides such direct maternal effects, possible later contributions
from the fetus to its own ill-being must be borne in mind: for
instance, the possibility that fetal hyperinsulinism in response to
the mother's hyperglycaemia may affect ossification processes in
the fetal skeleton. Finally, as we have seen, genetic factors appear
to play some part in determining the degree and incidence of the
malformations. One useful experimental approach, to distinguish
between early and late maternal effects, and between genetic and
environmental factors, might be to follow the development of

embryos transferred at the blastocyst stage between diabetic and normal rats, of the same strain and of different strains.

I wish to thank the British Diabetic Association for supporting this work, and Mrs Carol Jeynes for excellent technical assistance.

References

ASSEMANY, S. R. (1972). Syndrome of phocomelic diabetic embryopathy (caudal dysplasia). *American Journal of Diseases of Children*, **123**, 489–91.

BARRY, W. S. & BRAY, G. A. (1969). Plasma triglycerides in genetically obese rats. *Metabolism*, **18**, 833–9.

BEARD, R. W. & HOET, J. J. (1979). Introduction. In *Pregnancy Metabolism, Diabetes and the Fetus, Ciba Foundation Symposium 63*. Amsterdam & New York: Elsevier/Excerpta Medica/North-Holland (in press).

BENNETT, P. H., WEBNER, C. & MILLER, M. (1979). Congenital anomalies in the prediabetic and diabetic pregnancy. In *Pregnancy Metabolism, Diabetes and the Fetus, Ciba Foundation Symposium 63*. Amsterdam & New York: Elsevier/ Excerpta Medica/North-Holland.

BERGSTEIN, N. A. M. (1979). The influence of preconceptional glucose values on the outcome of pregnancy. In *Pregnancy Metabolism, Diabetes and the Fetus, Ciba Foundation Symposium 63*. Amsterdam & New York: Elsevier/Excerpta Medica/North-Holland.

CHEN, J. M. (1954). The effect of insulin on embryonic limb-bones cultivated *in vitro*. *Journal of Physiology*, **125**, 148–62.

CHEZ, R. A. (1979). Fetal nutrition and macrosomia: experimental contribution. In *2nd Aberdeen International Colloquium on Carbohydrate Metabolism in Pregnancy and the Newborn*, ed. H. W. Sutherland & J. M. Stowers. Berlin: Springer-Verlag (in press).

CHICK, W. L. & LIKE, A. A. (1970). Studies on the diabetic mutant mouse. IV. DBM, a modified diabetic mutant produced by out-crossing of the original strain. *Diabetologia*, **6**, 252–62.

COCKROFT, D. L. (1973). Development in culture of rat foetuses explanted at 12.5 and 13.5 days of gestation. *Journal of Embryology and Experimental Morphology*, **29**, 473–83.

COLEMAN, D. L. & HUMMEL, K. P. (1967). Studies with the mutation, diabetes, in the mouse. *Diabetologia*, **3**, 238–48.

CREUTZFELDT, W., KÖBBERLING, J. & NEEL, J. (eds.) (1976). *The Genetics of Diabetes Mellitus*. Berlin: Springer-Verlag (in press).

DAVIS, M. E., FUGO, N. V. & LAWRENCE, K. G. (1947). Effect of alloxan diabetes on reproduction in the rat. *Proceedings of the Society for Experimental Biology and Medicine*, **66**, 638–41.

DEUCHAR, E. M. (1977). Embryonic malformations in rats, resulting from maternal diabetes: a preliminary study. *Journal of Embryology and Experimental Morphology*, **41**, 93–9.

DEUCHAR, E. M. (1978). Effects of streptozotocin on early rat embryos grown in culture. *Experientia*, **34**, 84–5.

DEUCHAR, E. M. (1979*a*). Experimental evidence relating fetal anomalies to diabetes. In *2nd Aberdeen International Colloquium on Carbohydrate Metabolism in Pregnancy and the Newborn*, ed. H. W. Sutherland & J. M. Stowers. Berlin: Springer-Verlag (in press).

DEUCHAR, E. M. (1979*b*). Culture *in vitro* as a means of analysing the effects of maternal diabetes on embryonic development in rats. In *Pregnancy Metabolism, Diabetes and the Fetus, Ciba Foundation Symposium 63*. Amsterdam & New York: Elsevier/Excerpta Medica/North-Holland (in press).

DULIN, W. E. & WYSE, B. M. (1970). Diabetes in the KK mouse. *Diabetologia*, **6**, 317-23.

FERRET, P., LINDEN, O. & MORGANS, M. E. (1950). Pregnancy in insulin-treated alloxan diabetic rats. *Journal of Endocrinology*, **7**, 100-2.

FIELDS, G. A., SCHWARZ, R. H., DICKENS, H. O. & TUNNESSEN, W. (1968). Sacral agenesis in the infant of a gestational diabetic. *Obstetrics and Gynecology*, **32**, 778-81.

FREINKEL, N. (1979). The critical implications of maternal metabolism for fetal development. In *Pregnancy Metabolism, Diabetes and the Fetus, Ciba Foundation Symposium 63*. Amsterdam & New York: Elsevier/Excerpta Medica/North-Holland (in press).

GABBE, S. G. (1977). Congenital malformations in infants of diabetic mothers. *Obstetrical and Gynecological Survey*, **32**, 125-33.

GASPARO, M. DE, PICTET, R. L., RALL, L. B. & RUTTER, W. J. (1975). Control of insulin secretion in the developing pancreatic rudiment. *Developmental Biology*, **47**, 106-22.

GOLOB, E. K. & BECKER, K. L. (1969). Streptozotocin diabetes in pregnant rats. *Federation Proceedings*, **28**, 708 (abstr.).

GOLOB, E. K., RISHI, S. & BECKER, K. L. (1969). Streptozotocin diabetes in pregnant rats. *Clinical Research*, **17**, 44 (abstr.).

GOLOB, E. K., RISHI, S., BECKER, K. L. & MOORE, C. (1970). Streptozotocin diabetes in pregnant and non-pregnant rats. *Metabolism*, **19**, 1014-19.

HAY, M. F. (1958). The effect of growth hormone and insulin on limb-bone rudiments of the embryonic chick cultivated *in vitro*. *Journal of Physiology*, **144**, 490-504.

HERBERG, L. & COLEMAN, D. (1977). Laboratory animals exhibiting obesity and diabetes. *Metabolism*, **26**, 59-99.

HOHL, A. F. (1852). *Zur Pathologie des Beckens*, vol. 1, *Das schrägovale Becken*, p. 61. Leipzig: Wilhelm Engelmann.

HORII, K., WATANABE, G. & INGALLS, T. H. (1966). Experimental diabetes in pregnant mice. Prevention of congenital malformations in offspring by insulin. *Diabetes*, **15**, 194-204.

KARUNANAYAKE, E. H., HEARSE, D. J. & MELLOWS, G. (1976). Streptozotocin: its excretion and metabolism in the rat. *Diabetologia*, **12**, 483-8.

KIM, J. N., RUNGE, W., WELLS, L. J. & LAZAROW, A. (1960*a*). Pancreatic islets and blood sugars in prenatal and postnatal offspring from diabetic rats: beta granulation and glycogen infiltration. *Anatomical Record*, **138**, 239-49.

KIM, J. N., RUNGE, W., WELLS, L. J. & LAZAROW, A. (1960*b*). Effects of experimental diabetes in the offspring of the rat. Fetal growth, birth weight, gestation period and fetal mortality. *Diabetes*, **9**, 396-404.

LAZAROW, A., KIM, J. N. & WELLS, L. J. (1960). Birth weight and fetal mortality in pregnant subdiabetic rats. *Diabetes*, **9**, 114–17.

LICHTENSTEIN, H., GUEST, G. M. & WARKANY, J. (1951). Abnormalities in the offspring of white rats given protamine zinc insulin during pregnancy. *Proceedings of the Society for Experimental Biology and Medicine*, **78**, 398–405.

MALINS, J. M. (1978). Congential malformations and fetal mortality in diabetic pregnancy. *Journal of the Royal Society of Medicine*, **71**, 205–7.

MINTZ, D. H., CHEZ, R. A. & HUTCHINSON, D. L. (1972). Subhuman primate pregnancy complicated by streptozotocin-induced diabetes mellitus. *Journal of Clinical Investigation*, **51**, 837–43.

NAEVE, R. R. L. (1979). The outcome of diabetic pregnancies: a prospective study. In *Pregnancy Metabolism, Diabetes and the Fetus, Ciba Foundation Symposium 63*. Amsterdam & New York: Elsevier/Excerpta Medica/North-Holland (in press).

NAKHOODA, A. F., LIKE, A. A., CHAPPEL, C. I., WEI, C.-N. & MARLISS, E. B. (1978). The spontaneously diabetic Wistar rat (the 'BB' rat); studies prior to and during development of the overt syndrome. *Diabetologia*, **14**, 199–208.

NEW, D. A. T. (1966). Development of rat embryos cultured in blood sera. *Journal of Reproduction and Fertility*, **12**, 509–24.

NEW, D. A. T. (1967). Development of explanted rat embryos in circulating medium. *Journal of Embryology and Experimental Morphology*, **17**, 513–25.

PASSARGE, E. & LENZ, W. (1966). Syndrome of causal regression in infants of diabetic mothers: observations of further cases. *Pediatrics*, **37**, 672–5.

PEDERSEN, J. (1967). *The Pregnant Diabetic and her Newborn*. Copenhagen: Munksgaard.

PEDERSEN, J., PEDERSEN, L. M. & ANDERSEN, B. (1974). Assessors of fetal perinatal mortality in diabetic pregnancy. Analysis of 1,332 pregnancies in the Copenhagen series, 1946–72. *Diabetes*, **23**, 302–5.

PEDERSEN, L. M., TYGSTRUP, I. & PEDERSEN, J. (1964). Congenital malformations in newborn infants of diabetic women. Correlation with maternal diabetic vascular complications. *Lancet*, **i**, 1124–6.

PEDERSEN, L. M., TYGSTRUP, I., VILLUMSEN, A. L. & PEDERSEN, J. (1971). Congenital malformations in the offspring of diabetic women. *Diabetologia*, **7**, 404 (abstr.).

PEDERSEN, R. A. (1974). Development of lethal yellow (A^yA^y) mouse embryos in vitro. *Journal of Experimental Zoology*, **188**, 307–20.

RERUP, C. S. (1970). Drugs producing diabetes through damage of the insulin-secreting cells. *Pharmacological Reviews*, **22**, 485–520.

REYNOLDS, J. J. (1972). Skeletal tissue in culture. In *The Biochemistry and Physiology of Bone*, 2nd edn, vol. 1, ed. G. H. Bourne, pp. 69–126. New York & London: Academic Press.

ROBERTSON, G. G. (1942). An analysis of the development of homozygous yellow mouse embryos. *Journal of Experimental Zoology*, **89**, 197–231.

RUDAS, B. (1969). Über das Verhalten von Ratten mit chronischen Strepto-zotocin-Diabetes. *Klinische Wochenschrift*, **47**, 1120–1.

RUSNAK, S. L. & DRISCOLL, S. G. (1965). Congenital spinal anomalies in infants of diabetic mothers. *Pediatrics*, **35**, 989–95.

SINDEN, J. A. & LONGWELL, B. B. (1949). Effect of alloxan on fertility and gestation in the rat. *Proceedings of the Society for Experimental Biology and Medicine*, **70**, 607–10.

SMITH, D. W. (1970). *Recognizable Patterns of Malformation*. Philadelphia: W. B. Saunders.

SMITHBERG, M. & RUNNER, M. N. (1963). Teratogenic effects of hypoglycaemic treatments in inbred strains of mice. *American Journal of Anatomy*, **113**, 479–89.

STERN, L., RAMOS, A. & LIGHT, I. (1965). Congenital malformations and diabetes. *Lancet*, **i**, 1393–4.

SYBULSKI, S. & MAUGHAN, G. B. (1971). Use of streptozotocin as diabetic agent in pregnant rats. *Endocrinology*, **89**, 1537–40.

VAN ASSCHE, F. A. (1975). The fetal endocrine pancreas. In *1st Aberdeen International Colloquium on Carbohydrate Metabolism in Pregnancy and the Newborn*, ed. H. W. Sutherland & J. M. Stowers, pp. 68–82. London & Edinburgh: Churchill-Livingstone.

WATANABE, G. & INGALLS, T. H. (1963). Congenital malformations in the offspring of alloxan-diabetic mice. *Diabetes*, **12**, 66–72.

WYSE, B. M. & DULIN, W. E. (1970). The influence of age and dietary conditions on diabetes in the Db mouse. *Diabetologia*, **6**, 268–73.

ZUCKER, L. M. & ZUCKER, T. (1961). Fatty, a new mutation in the rat. *Journal of Heredity*, **52**, 275–8.

The influence of changes in the utero-placental circulation on the growth and development of the fetus

COLIN T. JONES AND JEFFREY S. ROBINSON

University of Oxford, Nuffield Institute for Medical Research and
Nuffield Department of Obstetrics and Gynaecology, Headley Way,
Headington, Oxford, UK

The detailed mechanisms giving rise to the timed sequence of events of growth and differentiation of the embryonic and fetal organs are in general poorly understood. Much work in recent years has centred on the role of hormones in these processes. There are now many examples of how local hormones (e.g. mesenchymal factor and the pancreas: Rutter *et al.*, 1978) or circulating hormones such as the corticosteroids (e.g. in the liver: Jost & Picon, 1970; in the lung: DeLemos *et al.*, 1970; Kotas, Mims & Hart, 1974; in the intestine: Hardy, Daniels, Malinowska & Nathanielsz, 1973; or in the placenta: Steele, Flint & Turnbull, 1976) influence organ development. These studies and others, for instance those showing the importance of thyroxine in amphibian metamorphosis (Tata, 1970), indicate that much of growth and differentiation may be under humoral control. It is also clear that other mechanisms are involved, such as those requiring cell contact in kidney development (Lehtonen, Wartiovaara, Nordling & Saxén, 1975). In mammalian development the probable involvement of large numbers of 'growth factors' and circulating hormones, and the likelihood that modification of the proportion of several of these directs specific developmental changes, contribute to the difficulty of investigating the complex problem of the growth and differentiation of the fetus.

A method of investigation that differs from the ones most commonly used is to outline the manner in which organ maturation and fetal endocrine state is modified in conditions where growth rate is substantially reduced. Conditions of restricted intra-uterine growth occur naturally (Gruenwald, 1963; Naeye, 1965; McLaren, 1965; Widdowson, 1971). They can also be induced experimentally

395

by constriction of a part of the uterine circulation (Wigglesworth, 1964; Hohenauer & Oh, 1969; Roux, Tordet-Caridroit & Chanez, 1970) or by reduction of placental mass (Alexander, 1964; Myers *et al.*, 1971) or placental exchange area (Creasy *et al.*, 1972). This paper will discuss the effect of growth restriction on the timing of maturational changes occurring in the latter half of gestation. It will be shown that nutrient supply may influence hormone secretion and thereby modify the course of development.

General effects of growth restriction on organ growth

Restriction of intra-uterine growth by interference with the utero-placental circulation has generally similar effects in different species. Although body weight is reduced by 30–75 %, the effects on organ growth are not symmetrical. In almost all cases there is

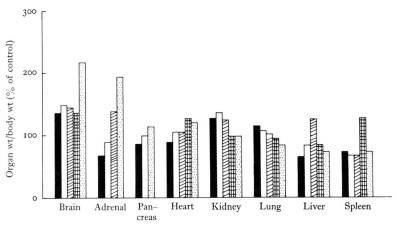

Fig. 1. The changes in the organ/body weight ratio during growth-retardation. ■ Naturally occurring intra-uterine retardation in man: from Gruenwald (1963) and Naeye (1965). □ Intra-uterine growth-retardation in the monkey caused by ligation of interplacental vessels: data from Myers *et al.* (1971). ▨ Intra-uterine growth-retardation in the sheep caused by removing endometrial caruncles prior to conception (Robinson, Kingston & Thorburn, unpublished observations: mean body weight 61 % of normal). ▦ Intra-uterine growth-retardation in rats caused by uterine artery ligation on about day 17 of pregnancy: data from Roux *et al.* (1970), Hohenauer & Oh (1969) and Rosso & Winick (1974). ▥ Intra-uterine growth retardation in guinea pigs caused by uterine artery ligation at day 30 of pregnancy; fetuses were delivered at 60–63 days. The mean body weight was 37 % of normal.

a reduction in brain weight, but this is generally small and thus the brain/body weight ratio increases (Fig. 1). Heart, kidney and lung are usually reduced in proportion to body weight, although for the severely growth-retarded guinea pig fetus the lung weight falls to a greater extent than body weight (Fig. 1). Characteristically the fall in liver and spleen weight is usually greater than the fall in body weight. The interesting contrasts are in the growth-retarded fetal sheep (Fig. 1) and in the growth-retarded fetal rabbit, where the liver/body weight ratio increases (Dietzmann & Lessel, 1976). These two situations are probably related to the need to increase hepatic haematopoietic cell content (see below). The changes in the weights of the endocrine glands are complex, with a spectrum of responses from the adrenal gland to the pancreas (Fig. 1). The large reduction in body weight must be associated with proportionately greater fall in the mass of skeletal muscle. Summarizing the general picture in this condition the growth of the brain, and to lesser extent that of the heart, kidneys and lungs, although less than normal, is maintained at the expense of the growth of the liver, spleen and skeletal muscle. The mechanisms by which these specific changes are brought about are far from clear.

The causes of the growth restriction

In the situations where placental mass is reduced it is assumed that the placenta and hence placental transport of nutrients is restricting the growth of the fetus. In the naturally occurring conditions and those caused by uterine artery constriction, it is not clear whether the placenta or its maternal blood supply is the limiting site. The placental/body weight ratio may increase or decrease, but this is not a good indicator of placental function as infarctions are frequently seen in the placenta of the growth-retarded fetus. Moreover, vascular problems in the placenta do not necessarily cause intra-uterine growth-retardation (Fox, 1975).

Fetal and placental growth later in gestation correlate well with placental blood flow and thus the rate of nutrient supply to the fetus (Duncan & Lewis, 1969; Moll & Künzel, 1973; Bruce & Abdul-Karim, 1973; Wooton, McFadyen & Cooper, 1977). Hypoxaemia and polycythaemia, together with hypoglycaemia, are frequently observed in intra-uterine growth-retardation (Table 1), indicating

Table 1. *The concentrations of some metabolic substrates in the plasma of growth-retarded fetuses*

	Guinea pig[a]		Rat[a]		Sheep[b]	
	Control	Retarded	Control	Retarded	Control	Retarded
P_{O_2} (mm Hg)	26	25	—	—	23	15
Packed cell volume (%)	45	55	—	—	30	37
Glucose (μmol/ml)	5.7	3.5	1.6	1.2[c]	1.2	0.65
Amino acid (μmol/ml)[e]	0.2	0.5	0.5	0.8[d]	7.6	7.0
Free fatty acid (μmol/ml)	0.6	0.3	0.6	0.46[c]	—	—

[a] Growth-retardation was induced by the ligation of one uterine artery.
[b] Growth-retardation was induced by removal of endometrial caruncles prior to conception (Alexander, 1964).
[c] Data taken from Roux *et al.* (1970).
[d] Data taken from Manniello, Schulman & Farrell (1977).
[e] The values for the guinea pig and rat are alanine, for the sheep they are for total α-amino nitrogen.

impaired transport of nutrients to the fetus. For the growth-retarded rat and guinea pig this view is further supported by the low free fatty acid concentrations (Table 1). In the mature fetal guinea pig there is a relatively high maternal to fetal transport of free fatty acids (Jones, 1976a). Elevations of amino acids in the plasma of growth-retarded fetuses (Table 1) are not consistent with this view but, as will be shown later, this is more likely the consequence of reduced fetal consumption. It should be noted that we cannot, at present, say whether the contribution of the placenta to restricted fetal growth is simply one of impaired transport. Preliminary evidence with radioactive isotopes suggests that there is a reduction of placental transport of glucose and amino acids after uterine artery ligation in the rat (Nitzan, Orloff & Schulman, 1977).

In contrast to the conditions discussed above, protein and calorie malnutrition sufficient to cause reduced fetal growth is associated with a general symmetrical reduction in organ weight

(Zamenhof, Van Martens & Margolis, 1968; Winick, Brasel & Rosso, 1972; Srivastava, Vu & Goswami, 1974). One interpretation of this is that modification of the vascular supply to the placenta or of the placental mass influences fetal growth in a manner not explained simply by a modification of nutrient supply. However, the signals favouring the growth of some organs and not others are still largely unknown. Are they of placental origin? It is clear that the situation is not that simple as altered nutrient supply does affect hormone secretion and, as will be discussed below, this is probably one of the major causes of modified organ maturation.

The effects of growth restriction on the endocrine state of the fetus

In man, small-for-dates newborns have poor pancreatic and β-cell growth (Kyle, 1963; Van Assche, DePrins, Aerts & Verjans, 1977). The functional significance of this is suggested by the low insulin concentration in the plasma of growth-retarded fetal sheep, rats and guinea pigs (Table 2). This is likely to be directly related to nutrient supply since in the growth-retarded sheep and guinea pig there is still a close relationship between plasma insulin and glucose concentrations. This supports the view (Bassett & Jones, 1976) that insulin, by responding to changes in glucose supply and by controlling the umbilical arterio-venous glucose difference, is a major 'growth' hormone for the fetus. In contrast to insulin, plasma glucagon concentrations are higher in growth-retarded fetal rats and particularly fetal guinea pigs (Table 2). Thus growth-retardation causes a large fall in the plasma insulin/glucagon ratio (Table 2).

As suggested by the weight changes of the adrenals of the growth-retarded fetuses (Fig. 1) the adrenal hormones' responses to growth retardation fit no consistent pattern. In small-for-dates newborn infants there is evidence for a reduction (Kenney & Preeyasombat, 1967) or no change (Reynolds & Mirkin, 1973) in the ability to secrete corticosteroids. The data from the guinea pig and sheep suggest, respectively, reduced and increased cortisol secretion (Fig. 2). However, since maternal cortisol is likely to contribute to fetal plasma cortisol in both these species and this may be influenced by alterations of the placental circulation, the

Table 2. *The effect of restricted fetal growth rate on plasma insulin
and glucagon concentrations and on insulin/glucagon ratios*

	% of control body wt	Insulin (ng/ml) Control	Insulin (ng/ml) Retarded	Glucagon (ng/ml) Control	Glucagon (ng/ml) Retarded	Insulin/glucagon Control	Insulin/glucagon Retarded
Rat[a]	66	7.5	3.5	0.49	0.63	15.3	5.5
Sheep[b]	61	1.0	0.18	—	—	—	—
Guinea pig[c]	40	7.0	2.5	0.025	0.12	280	20

[a] Growth-retardation caused by uterine artery ligation at day 17 of pregnancy, fetuses studied at last day of gestation. Data calculated from Girard et al. (1976).
[b] Growth-retardation caused by reducing the number of endometrial caruncles prior to conception (Alexander, 1964).
[c] Growth-retardation caused by uterine artery ligation at day 30 of pregnancy; fetuses studied at days 60–63 of gestation. Insulin was measured using guinea pig insulin as the standard.

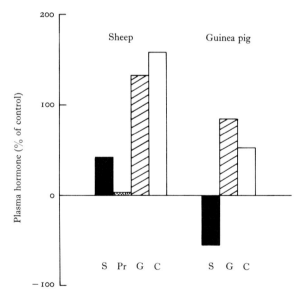

Fig. 2. Plasma hormone concentrations in growth-retarded fetuses in relation to normal values. S, sulphation promoting activity assayed with pig costal cartilage; Pr, prolactin; G, growth hormone; C, cortisol. Other details are as for Fig. 1.

changes in fetal adrenal function are not clear. It is notable that in the growth-retarded fetal guinea pig androstenedione concentrations are almost double their normal values, in contrast to the effects on plasma cortisol.

There is evidence of delayed bone development in small-for-dates newborn infants (Scott & Usher, 1964) and similar observations have been made for the growth-retarded sheep and guinea pig where delayed ossification has been observed. These data are consistent with the reduced activity of sulphation promoting activity (measured with pig cartilage bioassay) in the plasma of growth-retarded fetal sheep and guinea pigs (Fig. 2); in the latter there is inhibition of sulphate incorporation. This reduction of sulphation promoting activity occurs despite the maintenance of normal growth hormone concentrations.

One of the most striking, but as yet unexplained endocrine effects of growth-retardation in the fetus is the substantial reduction (to $< 5\%$ of control) in the plasma prolactin concentration (Fig. 2). This and the possibility that, at least in the rat, prolactin is somatotrophic, suggests that it may be an important 'growth' hormone. However, when fetal growth is restricted by maternal under-nutrition, the plasma prolactin concentrations in the foetus are higher than normal (Koritnik, Lehman & Dunn, 1976).

As might be expected from the rise in the packed cell volume in blood, there is evidence for man that growth retardation causes an elevation of plasma erythropoietin concentrations (Finne, 1966).

In summary, the endocrine features of growth-retardation are a fall in insulin and in the insulin/glucagon ratio, a fall in sulphation promoting activity but little change in growth hormone concentrations. The plasma concentrations of prolactin and thyroid hormones probably also fall, whereas that of erythropoietin rises.

Growth-retardation and organ composition and maturation

It is clear that the combined actions of nutrient supply and hormone concentrations regulate the proliferation of animals cells (Leffert, 1974). In the liver for instance, insulin, glucagon, corticosteroids, parathyroid and thyroid hormones interact with

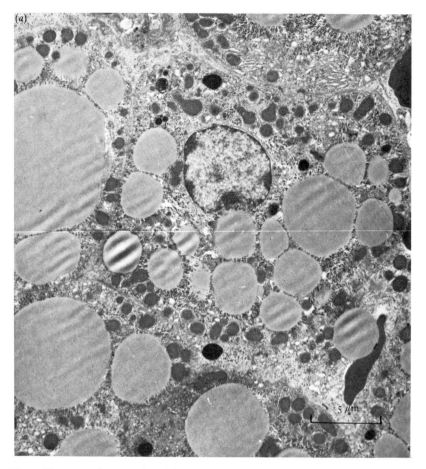

Fig. 3. Electron micrographs of the liver of (a) a normal 62-day guinea-pig fetus, and (b) a growth-retarded 61-day fetus.

various 'growth' factors and with amino acids to regulate growth (Leffert & Koch, 1977). Of these, insulin probably has the greatest effect on hepatic cell proliferation; without it the liver atrophies (Starzl *et al.*, 1973, 1975; Butcher & Swaffield, 1975; Leffert & Koch, 1977). Glucagon may be required for this insulin effect (Armato, Draghi & Andreis, 1978).

Thus, in view of the large effects of growth-retardation on fetal endocrine state, significant effects on organ growth and differentiation may be predicted. In addition to being very small, the liver of the growth-retarded fetal guinea pig has about 50% more haematopoietic cells than normal. It contains more glycogen and

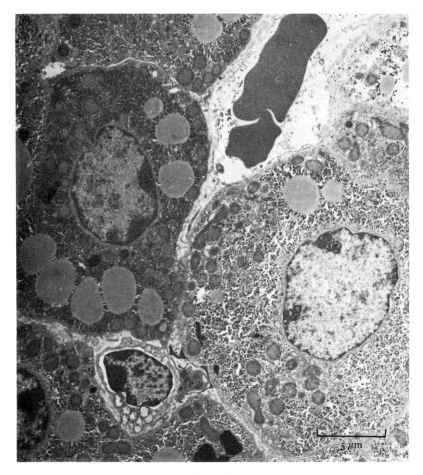

Fig. 3(*b*).

less triacylglycerol, DNA and protein than do normal-sized fetuses. The hepatocytes contain less cytosol and fewer mitochondria than usual (Fig. 3). The enzyme activities in these livers are substantially different from those in normal fetal livers (Table 3). The changes are associated with increased diversion of glucose to glycogen and reduced glucose synthesis, glycolysis, amino acid metabolism and urea synthesis. With the exception of the fall in the activity of phosphofructokinase, these enzyme changes are consistent with delays in the expression of the development profiles of enzyme appearance (Faulkner & Jones, 1975*a*; Jones, 1976*b*; Jones & Ashton, 1976; Faulkner & Jones, 1976). They are

Table 3. *The development of enzyme activities in the liver and muscle of normal and of growth-retarded fetal guinea pigs*[a]

	Enzyme activity (units/g wet wt)			
	Liver		Muscle	
	Control	Retarded	Control	Retarded
Hexokinase	0.5	0.94	0.28	0.38
Phosphofructokinase	4.2	1.6	4.8	2.4
Phosphoenolpyruvate carboxykinase	6.4	1.1	—	—
Alanine aminotransferase	3.4	0.4	2.1	0.8
Glutamate dehydrogenase	51	14	2.1	2.1
Carbamyl phosphate synthetase	2.6	0.5	—	—
Ca^{2+}-ATPase	—	—	42	22

[a] Growth-retardation was caused by ligating one uterine artery at day 30 of pregnancy; the fetuses were delivered at 60–63 days. Controls were obtained from the unoperated uterine horn; retarded fetuses were 26–40 % of normal size. Enzymes were extracted and assayed as previously described (Faulkner & Jones, 1975a, 1976; Jones & Ashton, 1976) The alanine aminotransferase activity is for the cytosolic enzyme.

not caused by the changes in haematopoietic cell content, which contributes little to total hepatic enzyme activity (Faulkner & Jones, 1979). A further illustration of this delay in the timing of development is seen in the liver of the fetal guinea pig, which has three pyruvate kinase isoenzymes PK1, PK2 and PK4 (Faulkner & Jones, 1975b). Between 50 and 60 days PK2 normally disappears and is not present in post-natal life. Intra-uterine growth-retardation delays the loss of PK2 (Fig. 4).

Skeletal muscle, like the liver, is severely affected by growth-retardation. Despite the hypoglycaemia glycogen concentrations are maintained but there are fewer mitochondria and myofibrils and the mitochondria are structurally different from those in the normal muscle. Enzyme activity is affected as in the liver and a 50 % reduction in Ca^{2+}-ATPase activity mirrors the reduced myofibrillar density (Table 3).

There are changes in other organs but these are generally less

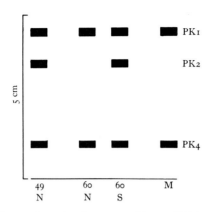

Fig. 4. Starch-gel electrophoresis of pyruvate kinase (Pk) isoenzymes of the fetal guinea pig liver. The preparation of the extract and the electrophoresis was carried out as previously described (Faulkner & Jones, 1975*b*). N, normal-size fetus of stated gestational age; S, growth-retarded fetus, M, maternal liver. Other details as for Fig. 1.

pronounced. In small fetal guinea pigs there is a delay in the onset and course of myelination in the brain. Enzyme development in the heart is generally normal although there is a delay in the appearance of alanine aminotransferase and an increase in the triacylglycerol content. Other organs have not been investigated. Thus in the growth-retarded fetus the maintenance of growth of the brain, heart and kidneys may, in part, be explained by the fall in plasma insulin concentration favouring the supply of glucose to those tissues such as the brain that are relatively insulin insensitive. (The fetal heart for much of gestation is in this category.) If the low insulin also slows the proliferation of certain tissues then the delays in maturation may be explained.

This raises the question of whether in normal development, the timing of maturational changes is flexible and partly tied to nutrient supply and hence endocrine state. There is some evidence for this from studies on the guinea pig, where changes in the activities of some of the hepatic enzymes of carbohydrate and amino acid metabolism (Fig. 5) are slower in naturally occurring small fetuses. It differs from the extreme growth-retardation situation in that the activity finally reaches the normal value. Thus, in long-gestation species such small differences in the timing of development are probably unimportant.

In conclusion, growth-retardation caused primarily by restrict-

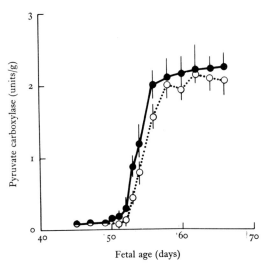

Fig. 5. Changes in pyruvate carboxylase activity in the livers of normal (●) and small (○) fetal guinea pigs. Fetuses were taken from normal pregnancies and those that were naturally small were 50–70 % of the normal weight. Enzyme activity was extracted and assayed as previously described (Jones & Ashton, 1976).

ing the utero-placental circulation, exerts profound and selective changes on fetal organ development. These suggest that the timing of some developmental changes is under nutrient, and thence endocrine, control, and that poor nutrient supply delays the maturation of organs such as the liver and skeletal muscle. The specific effects of these nutrient and endocrine changes on genetic expression in developing tissues remain to be established.

References

ALEXANDER, G. (1964). Studies on the placenta of the sheep (*Ovis aries* L.). Effect of surgical reduction in the number of caruncles. *Journal of Reproduction and Fertility*, **7**, 307–22.

ARMATO, U., DRAGHI, E. & ANDREIS, P. G. (1978). Effect of glucagon and insulin on the growth of neonatal rat hepatocytes in primary tissue culture. *Endocrinology*, **102**, 1155–66.

BASSETT, J. M. & JONES, C. T. (1976). Fetal glucose metabolism. In *Fetal Physiology and Medicine*, ed. R. W. Beard & P. W. Nathanielsz, pp. 158–72. Philadelphia: W. B. Saunders.

BRUCE, N. W. & ABDUL-KARIM, R. W. (1973). Relationships between fetal weight, placental weight and maternal placental circulation in the rabbit at different stages of gestation. *Journal of Reproduction and Fertility*, **32**, 15–24.

BUTCHER, N. L. R. & SWAFFIELD, M. N. (1975). Synergistic action of glucagon and insulin in regulation of hepatic regeneration. *Advances in Enzyme Regulation*, **13**, 281–93.

CREASY, R. K., BARRETT, C. T., DE SWIET, M., KAHANPÄÄ, K. V. & RUDOLPH, A. M. (1972). Experimental intrauterine growth retardation in the sheep. *American Journal of Obstetrics and Gynecology*, **112**, 566–73.

DELEMOS, R., SHERMETA, D., KNELSON, J., KOTAS, R. & AVERY, M. (1970). Acceleration of appearance of pulmonary surfactant in fetal lamb by administration of corticosteroids. *American Review of Respiratory Disorders*, **102**, 459–61.

DIETZMANN, K. & LESSEL, W. (1976). An experimental model for inducing fetal hypotrophy in the rabbit. *Experimental Pathology, Jena*, **12**, 309–14.

DUNCAN, S. L. B. & LEWIS, B. V. (1969). Maternal placental and myometrial blood flow in the pregnant rabbit. *Journal of Physiology*, **202**, 471–81.

FAULKNER, A. & JONES, C. T. (1975a). Changes in the activities of some glycolytic enzymes during the development of the guinea pig. *International Journal of Biochemistry*, **6**, 789–92.

FAULKNER, A. & JONES, C. T. (1975b). Pyruvate kinase isoenzymes in tissues of the developing guinea pig. *Archives of Biochemistry and Biophysics*, **170**, 228–41.

FAULKNER, A. & JONES, C. T. (1976). Hexokinase isoenzymes in tissues of the adult and developing guinea pig. *Archives of Biochemistry and Biophysics*, **175**, 477–86.

FAULKNER, A. & JONES, C. T. (1979). The distribution of enzyme and isoenzyme activities between parenchymal and haematopoietic cells in the liver of the foetal guinea pig. *Biochemical Journal*, **178**, 89–95.

FINNE, P. H. (1966). Erythropoietin levels in cord blood as an indicator of intrauterine hypoxia. *Acta Paediatrica Scandanavica*, **55**, 478–88.

FOX, H. (1975). Morphological pathology of the placenta. In *The Placenta and its Maternal Supply Line. Effects of Insufficiency on the Fetus*, ed. P. Gruenwald, pp. 197–220. Lancaster: Medical and Technical Publishing Co.

GIRARD, J. R., CHANEZ, C., KERVAN, A., TORDET-CARIDROIT, C. & ASSAN, R. (1976). Studies on experimental hypotrophy in the rat. III. Plasma insulin and glucagon. *Biology of the Neonate*, **29**, 262–6.

GRUENWALD, P. (1963). Chronic fetal distress and placental insufficiency. *Biologica Neonatorum*, **5**, 215–65.

HARDY, R. N., DANIELS, V. G., MALINOWSKA, K. W. & NATHANIELSZ, P. W. (1973). The adrenal cortex and intestinal absorption of macromolecules by the new-born animal. In *Foetal and Neonatal Physiology*, ed. R. S. Comline, K. W. Cross, G. S. Dawes & P. W. Nathanielsz, pp. 546–50. Cambridge University Press.

HOHENAUER, L. & OH, W. (1969). Body composition in experimental intrauterine growth retardation. *Journal of Nutrition*, **99**, 23–6.

JONES, C. T. (1976a). Lipid metabolism and mobilization in the guinea pig during pregnancy. *Biochemical Journal*, **156**, 357–65.

JONES, C. T. (1976b). Fetal metabolism and fetal growth. *Journal of Reproduction and Fertility*, **47**, 189–201.

JONES, C. T. & ASHTON, I. K. (1976). The appearance, properties and functions of gluconeogenic enzymes in the liver and kidney of the guinea pig during fetal and early neonatal development. *Archives of Biochemistry and Biophysics,* **174,** 506–22.

JOST, A. & PICON, L. (1970). Hormonal control of fetal development and metabolism. *Advances in Metabolic Disorders,* **4,** 123–84.

KENNY, F. M. & PREEYASOMBAT, C. (1967). Cortisol production rate. VI. Hypoglycaemia in the neonate and postnatal period, and in association with dwarfism. *Pediatrics,* **70,** 65–75.

KORITNIK, D. R., LEHMAN, T. F. & DUNN, T. G. (1976). Role of maternal dietary energy intake on ovine fetal prolactin and growth hormone levels during late gestation. *Society for the Study of Reproduction 9th Annual Meeting,* abstr. 149.

KOTAS, R. V., MIMS, L. C. & HART, L. K. (1974). Reversible inhibition of lung cell number after glucocorticoid injection into fetal rabbits to enhance surfactant appearance. *Pediatrics,* **53,** 358–61.

KYLE, F. C. (1963). Diabetes in pregnancy. *Annals of Internal Medicine,* **59,** *Supplement 3,* 1–26.

LEFFERT, H. L. (1974). Growth control of differentiated fetal rat hepatocytes in primary monolayer culture. V. Occurrence in dialyzed fetal bovine serum of macromolecules having positive and negative growth regulatory functions. *Journal of Cell Biology,* **62,** 792–801.

LEFFERT, H. L. & KOCH, K. S. (1977). Control of animal cell proliferation. In *Growth, Nutrition and Metabolism of Cells in Culture,* vol. 3, ed. G. H. Rothblat & V. J. Cristofals, pp. 225–94. New York & London: Academic Press.

LEHTONEN, E., WARTIOVAARA, J., NORDLING, S. & SAXÉN, L. (1975). Demonstration of cytoplasmic processes in Millipore filters permitting kidney tubule induction. *Journal of Embryology and Experimental Morphology,* **33,** 187–203.

MANNIELLO, R. L., SCHULMAN, J. D. & FARRELL, P. M. (1977). Amino acid metabolism in dysmature newborn rats – possible explanation for the anti-hypoglycemic effect of prenatal glucocorticoids. *Pediatric Research,* **11,** 1165–6.

McLAREN, A. (1965). Genetic and environmental effects on fetal and placental growth in mice. *Journal of Reproduction and Fertility,* **9,** 79–98.

MOLL, W. & KÜNZEL, W. (1973). The blood pressure in arteries entering the placentae of guinea pigs, rats, rabbits and sheep. *Pflügers Archiv,* **338,** 125–31.

MYERS, R. E., HILL, D. E., HOLT, A. B., SCOTT, R. E., MELLITS, E. D. & CHEEK, D. B. (1971). Fetal growth retardation produced by experimental placental insufficiency in the rhesus monkey. I. Body weight, organ size. *Biology of the Neonate,* **18,** 379–94.

NAEYE, R. L. (1965). Malnutrition. Probable cause of fetal growth retardation. *Archives of Pathology,* **79,** 284–91.

NITZAN, M., ORLOFF, S. & SCHULMAN, J. D. (1977). Materno-fetal transfer and uptake of α-deoxyglucose (DG) and α-amino-butyric acid (AIB) in intrauterine growth retardation (IUGR) associated with restricted uterine blood supply. *Pediatric Research,* **11,** 410.

REYNOLDS, J. W. & MIRKIN, B. L. (1973). Urinary steroid levels in newborn infants with intrauterine growth retardation. *Journal of Clinical Endocrinology and Metabolism,* **36,** 576–81.

Rosso, P. & Winick, M. (1974). Intrauterine growth retardation. A new systematic approach based on the clinical and biochemical characteristics of this condition. *Journal of Perinatal Medicine*, **2**, 147–60.

Roux, J. M., Tordet-Caridroit, C. & Chanez, C. (1970). Studies on experimental hypotrophy in the rat. I. Chemical composition of the total body and some organs in the rat foetus. *Biology of the Neonate*, **15**, 342–7.

Rutter, W. J., Pictet, R. L., Harding, J. D., Chirgwin, J. M., MacDonald, R. J. & Przybyla, A. E. (1978). An analysis of pancreatic development: role of mesenchymal factor and other extracellular factors. In *Molecular Control of Proliferation and Differentiation*, ed. J. Papaconstantinou & W. J. Rutter, pp. 205–27. New York & London: Academic Press.

Scott, K. E. & Usher, R. (1964). Epiphyseal development in foetal malnutrition syndrome. *New England Journal of Medicine*, **270**, 822–4.

Srivastava, U., Vu, M.-L. & Goswami, T. (1974). Maternal dietary deficiency and cellular development of progeny in the rat. *Journal of Nutrition*, **104**, 512–20.

Starzl, T. E., Francavilla, A., Halgrimson, C. G., Francavilla, F. R., Porter, K. A., Brown, T. & Putman, C. W. (1973). The origin, hormonal nature and action of portal venous hepatotrophic substances. *Surgery, Gynecology and Obstetrics*, **137**, 179–99.

Starzl, T. E., Porter, K. A., Kashiwagi, N., Lee, I. Y., Russell, W. J. I. & Putman, C. W. (1975). The effect of diabetes mellitus on portal blood hepatotrophic factors in dogs. *Surgery, Gynecology and Obstetrics*, **140**, 549–62.

Steele, P. A., Flint, A. P. F. & Turnbull, A. C. (1976). Activity of steroid C17–20 lyase in the ovine placenta: effect of exposure to fetal glucocorticoid. *Journal of Endocrinology*, **69**, 239–46.

Tata, J. R. (1970). Regulation of protein synthesis by growth and developmental hormones. In *Biochemical Actions of Hormones*, vol. 1, ed. G. Litwack, pp. 89–134. New York & London: Academic Press.

Van Assche, F. A., De Prins, F., Aerts, L. & Verjans, M. (1977). The endocrine pancreas in small-for-dates infants. *British Journal of Obstetrics and Gynaecology*, **84**, 751–3.

Widdowson, E. M. (1971). Intrauterine growth retardation in the pig. I. Organ size and cellular development at birth and after growth to maturity. *Biology of the Neonate*, **19**, 329–40.

Wigglesworth, J. S. (1964). Experimental growth retardation in the foetal rat. *Journal of Pathology and Bacteriology*, **88**, 1–13.

Winick, M., Brasel, J. A. & Rosso, P. (1972). Nutrition and cell growth. In *Current Concepts in Nutrition*, vol. 1, *Nutrition and Development*, ed. M. Winick, p. 49. New York: Wiley.

Wooton, R., McFadyen, I. R. & Cooper, J. E. (1977). Measurement of placental blood flow in the pig and its relation to placental and fetal weight. *Biology of the Neonate*, **31**, 333–9.

Zamenhof, S., Van Martens, F. & Margolis, F. L. (1968). DNA (cell number) and protein in the neonatal brain: alteration of maternal dietary protein restriction. *Science*, **160**, 322–3.

Index

Page numbers in italic type indicate references to tables or figures.

Tridacna spp., 26
Trifolium spp., *34*
triploid sex chromosome abnormalities,
297, 298–9, 300, *301*, 303
trisomies, chromosome abnormalities,
297, 300, *301*, 302, 303
Triticum spp., *35*
Triturus alpestris, 93, 94
Triturus cristatus, 57, *58*, *60*
Triturus spp., *48*; transcription during
oogenesis of, 49, 51, 52, 53, 54, 55–61
tRNA (transfer RNA), 72, 92; and
transcription in amphibian oogenesis,
49, 56, 59, 95
trophoblast barrier, and maternal immune
attack on, 324–6
tryptan blue, and teratogenic response of
embryo to, 354, 359
Tubifex spp., 11–12, 138; pole plasms of,
134, 139
Tubiflorae, *34*
Turner's syndrome, 299

Urechis caupo, 159
uridylic acid, in RNA, 57, 58–9
urodele amphibians, 168
Urticales, *34*
utero-placental circulation, and effects
on growth and development of the
fetus, 395–406; and effects on fetal
endocrine system, 399–401; growth
restriction, causes of, 397–9; and effects
on organ growth, 396–7, 421–6
UV (ultraviolet) irradiation, effects of,
169, 211, *212*; on *Drosophila* pole
plasm, 205–8, 209–10; on germinal
plasm in anuran amphibians, 170–4,
175–6, 176–88

v (vasodilation) mutant gene of the
axolotl, *244*, 249–54, 264
vegetal body, of oocytes, *130*, 131, 132,
136
vesicular aggregates, in mosaic eggs,
128–32
vinyl chloride, 294
Viola spp., *34*
Violaceae, *34*
viruses, 7–8, 26
vitamin A: pathogenesis of effects of,
364–5; and teratogenic response to,
355–6, 358–9
vitamin deficiency, and teratogenicity of,
353–4
vitamin E (a-tocopherol) deficiency, and
teratogenic response to, 358

vitellogenin, and synthesis of, 93, 95,
111–23; induction of, 114–18;
oestrogen receptor from *Xenopus
liver*, 118–22; oocytes, entry into,
112–14, 160; structure and properties
of, 112

'wild-type' allele, *see* allele, 'wild type'
Wistar rats, effects of maternal diabetes
on embryonic development of, 380,
381–5, 389, 390

X chromosome activity, in mammalian
embryos, 307–8
Xenopus borealis, 91–2, 101
Xenopus laevis, 47, *82*, 156, 261; cell
proliferation in, 66, *67*, 68; chromatin
assembly in early development of, 70–1;
DNA synthesis in eggs, *69*; gene
transcription in oocytes, 73–8; germinal
plasm of, 168, 169, 176, *177*; rDNA
amplification in oogenesis, 90–2; r
protein and r RNA synthesis in
oogenesis of, 86, *98*, *101*; UV
irradiation, effects on development,
173, *178*; vitellogenin synthesis, 111–23
Xenopus laevis–Xenopus borealis
interspecific hybrids: 5S rRNA
synthesis in, 100–4; maternal effect on
development of sterile ovaries, 104–5;
rRNA synthesis in, 99–100
Xenopus mulleri, *see Xenopus borealis*
Xenopus spp., 54, 87, *189*, 203, *243*;
chromosomal replication in, 65–78;
genome size (C-value), 49, *51*;
mitochondria, in oocytes of, 148, 150,
152, 153–8, 159, 162, 183; protein
synthesis in eggs, 72–3; ribosome
synthesis, 89, 90, 95; UV irradiation
and effects on (germinal plasm), 175,
176, 178, 179, 180, 181, 185 (embryos)
186–7, 208
X-irradiation, 28; and effects on
mammalian gamete development,
294–6, 303; teratogenic effect of, 352,
354, 357; *see also* UV irradiation

yeast: and ribopsomal protein synthesis
in, 85–6; and RNA gene structure, 87
yolk, 9, 13–15, 27, 161; platelets, 160–1,
162, *172*, 251, *252*; proteins, 113–14,
160–1, 242, *243*

Zamia spp., 31
Zea spp., *35*